Applied Mathematical Sciences

Volume 77

Applied Mathematical Sciences

1. *John:* Partial Differential Equations, 4th ed.
2. *Sirovich:* Techniques of Asymptotic Analysis.
3. *Hale:* Theory of Functional Differential Equations, 2nd ed.
4. *Percus:* Combinatorial Methods.
5. *von Mises/Friedrichs:* Fluid Dynamics.
6. *Freiberger/Grenander:* A Short Course in Computational Probability and Statistics.
7. *Pipkin:* Lectures on Viscoelasticity Theory.
9. *Friedrichs:* Spectral Theory of Operators in Hilbert Space.
11. *Wolovich:* Linear Multivariable Systems.
12. *Berkovitz:* Optimal Control Theory.
13. *Bluman/Cole:* Similarity Methods for Differential Equations.
14. *Yoshizawa:* Stability Theory and the Existence of Periodic Solution and Almost Periodic Solutions.
15. *Braun:* Differential Equations and Their Applications, 3rd ed.
16. *Lefschetz:* Applications of Algebraic Topology.
17. *Collatz/Wetterling:* Optimization Problems.
18. *Grenander:* Pattern Synthesis: Lectures in Pattern Theory, Vol I.
20. *Driver:* Ordinary and Delay Differential Equations.
21. *Courant/Friedrichs:* Supersonic Flow and Shock Waves.
22. *Rouche/Habets/Laloy:* Stability Theory by Liapunov's Direct Method.
23. *Lamperti:* Stochastic Processes: A Survey of the Mathematical Theory.
24. *Grenander:* Pattern Analysis: Lectures in Pattern Theory, Vol. II.
25. *Davies:* Integral Transforms and Their Applications, 2nd ed.
26. *Kushner/Clark:* Stochastic Approximation Methods for Constrained and Unconstrained Systems
27. *de Boor:* A Practical Guide to Splines.
28. *Keilson:* Markov Chain Models—Rarity and Exponentiality.
29. *de Veubeke:* A Course in Elasticity.
30. *Sniatycki:* Geometric Quantization and Quantum Mechanics.
31. *Reid:* Sturmian Theory for Ordinary Differential Equations.
32. *Meis/Markowitz:* Numerical Solution of Partial Differential Equations.
33. *Grenander:* Regular Structures: Lectures in Pattern Theory, Vol. III.
34. *Kevorkian/Cole:* Perturbation methods in Applied Mathematics.
35. *Carr:* Applications of Centre Manifold Theory.
36. *Bengtsson/Ghil/Källén:* Dynamic Meteorology: Data Assimilation Methods.
37. *Saperstone:* Semidynamical Systems in Infinite Dimensional Spaces.
38. *Lichtenberg/Lieberman:* Regular and Stochastic Motion.
39. *Piccini/Stampacchia/Vidossich:* Ordinary Differential Equations in R^n.
40. *Naylor/Sell:* Linear Operator Theory in Engineering and Science.
41. *Sparrow:* The Lorenz Equations: Bifurcations, Chaos, and Strange Attractors.
42. *Guckenheimer/Holmes:* Nonlinear Oscillations, Dynamical Systems and Bifurcations of Vector Fields.
43. *Ockendon/Tayler:* Inviscid Fluid Flows.
44. *Pazy:* Semigroups of Linear Operators and Applications to Partial Differential Equations.
45. *Glashoff/Gustafson:* Linear Optimization and Approximation: An Introduction to the Theoretical Analysis and Numerical Treatment of Semi-Infinite Programs.
46. *Wilcox:* Scattering Theory for Diffraction Gratings.
47. *Hale et al.:* An Introduction to Infinite Dimensional Dynamical Systems—Geometric Theory.
48. *Murray:* Asymptotic Analysis.
49. *Ladyzhenskaya:* The Boundary-Value Problems of Mathematical Physics.
50. *Wilcox:* Sound Propagation in Stratified Fluids.
51. *Golubitsky/Schaeffer:* Bifurcation and Groups in Bifurcation Theory, Vol. I.
52. *Chipot:* Variational Inequalities and Flow in Porous Media.
53. *Majda:* Compressible Fluid Flow and Systems of Conservation Laws in Several Space Variables.
54. *Wasow:* Linear Turning Point Theory.

(continued following index)

David Aldous

Probability Approximations via the Poisson Clumping Heuristic

With 25 Illustrations

Springer-Verlag
New York Berlin Heidelberg
London Paris Tokyo

David Aldous
Department of Statistics
University of California–Berkeley
Berkeley, CA 94720
USA

Mathematics Subject Classification (1980): 60C05, 60D05, 60F05, 60J20, 60K99

Library of Congress Cataloging-in-Publication Data
Aldous, D. J. (David J.)
Probability approximations via the Poisson clumping heuristic.
(Applied mathematical sciences ; v. 77)
Bibliography: p.
1. Combinatorial probabilities. 2. Stochastic geometry. 3. Markov processes. I. Title. II. Title: Poisson clumping heuristic. III. Series: Applied mathematical sciences (Springer-Verlag New York Inc.) ; v. 77.
QA1.A647 vol. 77 [QA273.45] 510 s [519.2] 88-29453

Printed on acid-free paper

Camera-ready copy prepared using LaTEX.
Printed and bound by R.R. Donnelley and Sons, Harrisonburg, Virginia.
Printed in the United States of America.

9 8 7 6 5 4 3 2 1

ISBN 0-387-96899-7 Springer-Verlag New York Berlin Heidelberg
ISBN 3-540-96899-7 Springer-Verlag Berlin Heidelberg New York

Preface

If you place a large number of points randomly in the unit square, what is the distribution of the radius of the largest circle containing no points? Of the smallest circle containing 4 points? Why do Brownian sample paths have local maxima but not points of increase, and how nearly do they have points of increase? Given two long strings of letters drawn i.i.d. from a finite alphabet, how long is the longest consecutive (resp. non-consecutive) substring appearing in both strings? If an imaginary particle performs a simple random walk on the vertices of a high-dimensional cube, how long does it take to visit every vertex? If a particle moves under the influence of a potential field and random perturbations of velocity, how long does it take to escape from a deep potential well? If cars on a freeway move with constant speed (random from car to car), what is the longest stretch of empty road you will see during a long journey? If you take a large i.i.d. sample from a 2-dimensional rotationally-invariant distribution, what is the maximum over all half-spaces of the deviation between the empirical and true distributions?

These questions cover a wide cross-section of theoretical and applied probability. The common theme is that they all deal with maxima or minima, in some sense. The purpose of this book is to explain a simple idea which enables one to write down, with little effort, approximate solutions to such questions. Let us try to say this idea in one paragraph.

(a) Problems about random extrema can often be translated into problems about sparse random sets in $d \geq 1$ dimensions.

(b) Sparse random sets often resemble i.i.d. random clumps thrown down randomly (i.e., centered at points of a Poisson process).

(c) The problem of interest reduces to estimating mean clump size.

(d) This mean clump size can be estimated by approximating the underlying random process locally by a simpler, known process for which explicit calculations are possible.

(Part (b) explains the name *Poisson clumping heuristic*).

This idea is known, though rarely explicitly stated, in several specific settings, but its power and range seems not to be appreciated. I assert that this idea provides the correct way to look at extrema and rare events in a wide range of probabilistic settings: to demonstrate this assertion, this book treats over 100 examples. Our arguments are informal, and we are not going to prove anything. This is a rather eccentric format for a mathematics book — some reasons for this format are indicated later.

The opening list of problems was intended to persuade every probabilist to read the book! I hope it will appeal to graduate students as well as experts, to the applied workers as well as theoreticians. Much of it should be comprehensible to the reader with a knowledge of stochastic processes at the non-measure-theoretic level (Markov chains, Poisson process, renewal theory, introduction to Brownian motion), as provided by the books of Karlin and Taylor (1975; 1982) and Ross (1983). Different chapters are somewhat independent, and the level of sophistication varies.

Although the book ranges over many fields of probability theory, in each field we focus narrowly on examples where the heuristic is applicable, so this work does not constitute a complete account of any field. I have tried to maintain an honest "lecture notes" style through the main part of each chapter. At the end of each chapter is a "Commentary" giving references to background material and rigorous results. In giving references I try to give a book or survey article on the field in question, supplemented by recent research papers where appropriate: I do not attempt to attribute results to their discoverers. Almost all the examples are natural (rather than invented to show off the technique), though I haven't always given a thorough explanation of how they arise.

The arguments in examples are sometimes deliberately concise. Most results depend on one key calculation, and it is easier to see this in a half-page argument than in a three-page argument. In rigorous treatments it is often necessary to spend much effort in showing that certain effects are ultimately negligible; we simply omit this effort, relying on intuition to see what the dominant effect is. No doubt one or two of our heuristic conclusions are wrong: if heuristic arguments *always* gave the right answer, then there wouldn't be any point in *ever* doing rigorous arguments, would there? Various problems which seem interesting and unsolved are noted as "thesis projects", though actually some are too easy, and others too hard, for an average Ph.D. thesis.

The most-studied field of application of the heuristic is to extremes of 1-parameter stationary processes. The standard reference work on this field, Leadbetter et al. (1983), gives theoretical results covering perhaps 10 of our examples. One could write ten similar books, each covering 10 examples from another field. But I don't have the energy or inclination to do so, which is one reason why this book gives only heuristics. Another reason is that connections between examples in different fields are much clearer in the heuristic treatment than in a complete technical treatment, and I hope this book will make these connections more visible.

At the risk of boring some readers and annoying others, here is a paragraph on the philosophy of approximations, heuristic and limit theorems. The proper business of probabilists is calculating probabilities. Often exact calculations are tedious or impossible, so we resort to approximations. A limit theorem is an assertion of the form: "the error in a certain approximation tends to 0 as (say) $N \to \infty$". Call such limit theorem *naive* if there is no explicit error bound in terms of N and the parameters of the underlying process. Such theorems are so prevalent in theoretical and applied probability that people seldom stop to ask their purpose. Given a serious applied problem involving specific parameters, the natural first steps are to seek rough analytic approximations and to run computer simulations; the next step is to do careful numerical analysis. It is hard to give any argument for the relevance of a proof of a naive limit theorem, except as a vague reassurance that your approximation is sensible, and a good heuristic argument seems equally reassuring. For the theoretician, the defense of naive limit theorems is "I want to prove something, and that's all I can prove". There are fields which are sufficiently hard that this is a reasonable attitude (some of the areas in Chapters G, I, J for example), but in most of the fields in this book the theoretical tools for proving naive limit theorems have been sufficiently developed that such theorems are no longer of serious theoretical research interest (although a few books consolidating the techniques would be useful).

Most of our approximations in particular examples correspond to known naive limit theorems, mentioned in the Commentaries. I deliberately deemphasize this aspect, since as argued above I regard the naive limit theory as irrelevant for applications and mostly trite as theory. On the other hand, explicit error bounds are plainly relevant for applications and interesting as theory (because they are difficult, for a start!). In most of our examples, explicit error bounds are not know: I regard this as an important area for future research. Stein's method is a powerful modern tool for getting explicit bounds in "combinatorial" type examples, whose potential is not widely realized. Hopefully other tools will become available in the future.

Acknowledgements: As someone unable to recollect what I had for dinner last night, I am even more unable to recollect the many people who (consciously or unconsciously) provided sources of the examples; but I thank them. Course based on partial early drafts of the book were given in Berkeley in 1984 and Cornell in 1986, and I thank the audiences for their feedback. In particular, I thank Persi Diaconis, Rick Durrett, Harry Kesten, V. Anantharam and Jim Pitman for helpful comments. I also thank Pilar Fresnedo for drawing the diagrams, and Ed Sznyter for a great job converting my haphazard two-fingered typing into this elegant LaTeX book.

Contents

A The Heuristic 1

A1 The M/M/1 queue. 1
A2 Mosaic processes on $\boldsymbol{R}^2$. 2
A3 Mosaic processes on other spaces. 5
A4 The heuristic. 5
A5 Estimating clump sizes. 7
A6 The harmonic mean formula. 8
A7 Conditioning on semi-local maxima. 10
A8 The renewal-sojourn method. 11
A9 The ergodic-exit method. 12
A10 Limit assertions. 14
A11–A21 Commentary 15

B Markov Chain Hitting Times 23

B1 Introduction. 23
B2 The heuristic for Markov hitting times. 24
B3 Example: Basic single server queue. 25
B4 Example: Birth-and-death processes. 25
B5 Example: Patterns in coin-tossing. 26
B6 Example: Card-shuffling. 27
B7 Example: Random walk on $\boldsymbol{Z}^d$ mod N. 28
B8 Example: Random trapping on $\boldsymbol{Z}^d$. 28
B9 Example: Two M/M/1 queues in series. 28
B10 Example: Large density of heads in coin-tossing. 29
B11 Counter-example. 30
B12 Hitting small subsets. 30
B13 Example: Patterns in coin-tossing, continued. 31
B14 Example: Runs in biased die-throwing. 31
B15 Example: Random walk on $\boldsymbol{Z}^d$ mod N, continued. 32
B16 Hitting sizable subsets. 32
B17 The ergodic-exit form of the heuristic for Markov hitting times. 32
B18 Example: A simple reliability model. 33
B19 Example: Timesharing computer. 33

B20 Example: Two M/M/1 queues in series. 35
B21 Another queueing example. 37
B22 Example: Random regular graphs. 38
B23–B32 Commentary 38

C Extremes of Stationary Processes **44**
C1 Classical i.i.d. extreme value theory. 44
C2 Examples of maxima of i.i.d. sequences. 45
C3 The point process formulation. 47
C4 The heuristic for dependent extrema. 48
C5 Autoregressive and moving average sequences. 48
C6 Example: Exponential tails. 49
C7 Approximate independence of tail values. 50
C8 Example: Superexponential tails. 50
C9 Example: Polynomial tails. 50
C10 The heuristic for dependent extrema (continued). 51
C11 Additive Markov processes on $[0, \infty)$. 52
C12 Continuous time processes: the smooth case. 54
C13 Example: System response to external shocks. 55
C14 Example: Uniform distribution of the Poisson process. . . . 56
C15 Drift-jump processes. 57
C16 Example: Positive, additive processes. 58
C17 Example: Signed additive processes. 58
C18 Positive, general drift processes. 58
C19 Autoregressive symmetric stable process. 59
C20 Example: The I5 problem. 60
C21 Approximations for the normal distribution. 62
C22 Gaussian processes. 63
C23 The heuristic for smooth Gaussian processes. 64
C24 Monotonicity convention. 64
C25 High-level behavior of smooth Gaussian processes. 65
C26 Conditioning on semi-local maxima. 66
C27 Variations on a theme. 67
C28 Example: Smooth $\mathcal{X}^2$ processes. 68
C29–C40 Commentary 69

D Extremes of Locally Brownian Processes **72**
D1 Brownian motion. 72
D2 The heuristic for locally Brownian processes. 73
D3 One-dimensional diffusions. 74
D4 First hitting times for positive-recurrent diffusions. 76
D5 Example: Gamma diffusion. 77
D6 Example: Reflecting Brownian motion. 78
D7 Example: Diffusions under a potential. 78
D8 Example: State-dependent M/M/1 queue. 79

D9 Example: The Ornstein-Uhlenbeck process. 80
D10 Gaussian processes. 81
D11 Example: System response to external shocks. 82
D12 Example: Maximum of self-normalized Brownian bridge. . . 82
D13 Boundary-crossing. 83
D14 Example: Boundary-crossing for reflecting Brownian motion. 84
D15 Example: Brownian LIL. 85
D16 Maxima and boundary-crossing for general Gaussian processes. 86
D17 Example: Maximum of Brownian bridge. 86
D18 Maxima of non-stationary Gaussian processes. 87
D19 Example: Maximum of Brownian Bridge with drift. 88
D20 Example: Brownian motion and quadratic boundary. 88
D21 Example: Ornstein-Uhlenbeck quadratic boundary. 89
D22 Semi-local maxima for the Ornstein-Uhlenbeck process. . . 90
D23 Example: A storage/queuing process. 91
D24 Approximation by unstable Ornstein-Uhlenbeck process. . . 93
D25 Example: Escape from a potential well. 93
D26 Example: Diffusion in random environment. 94
D27 Interpolating between Gaussian processes. 95
D28 Example: Smoothed Ornstein-Uhlenbeck. 95
D29 Boundary-crossing revisited. 97
D30 Tangent approximation for Brownian boundary-crossing. . . 99
D31–D42 Commentary 100

E Simple Combinatorics 106
E1 Introduction. 106
E2 Poissonization. 107
E3 Example: The birthday problem. 108
E4 Example: K-matches. 109
E5 Example: Unequal probabilities. 109
E6 Example: Marsaglia random number test. 110
E7 Several types of coincidence. 110
E8 Example: Similar bridge hands. 111
E9 Example: Matching K-sets. 112
E10 Example: Nearby pairs. 112
E11 Example: Basic coupon-collectors problem. 113
E12 Example: Time until most boxes have at least one ball. . . . 113
E13 Example: Time until all boxes have at least $(K+1)$ balls. . 114
E14 Example: Unequal probabilities. 114
E15 Abstract versions of CCP. 114
E16 Example: Counting regular graphs. 115
E17–E22 Commentary 116

F Combinatorics for Processes **118**
F1 Birthday problem for Markov chains. 118
F2 Example: Simple random walk on $\boldsymbol{Z}^K$. 119
F3 Example: Random walks with large step. 119
F4 Example: Simple random walk on the K-cube. 120
F5 Example: Another card shuffle. 120
F6 Matching problems. 120
F7 Matching blocks. 122
F8 Example: Matching blocks: the i.i.d. case. 122
F9 Example: Matching blocks: the Markov case. 123
F10 Birthday problem for blocks. 124
F11 Covering problems. 124
F12 Covering problems for random walks. 125
F13 Example: Random walk on $\boldsymbol{Z}^d$ modulo N. 126
F14 Example: Simple random walk on the K-cube. 126
F15 Covering problem for i.i.d. blocks. 127
F16 Example: Dispersal of many walks. 127
F17 Example: M/M/∞ combinatorics. 128
F18–F21 Commentary 129

G Exponential Combinatorial Extrema **131**
G1 Introduction. 131
G2 Example: Cliques in random graphs. 132
G3 Example: Covering problem on the K-cube. 133
G4 Example: Optimum partitioning of numbers. 134
G5 Exponential sums. 135
G6 Example: First-passage percolation on the binary tree. . . . 136
G7 Example: Percolation on the K-cube. 138
G8 Example: Bayesian binary strings. 139
G9 Example: Common cycle partitions in random permutations. 140
G10 Conditioning on maxima. 141
G11 Example: Common subsequences in fair coin-tossing. 141
G12 Example: Anticliques in sparse random graphs. 142
G13 The harmonic mean formula. 143
G14 Example: Partitioning sparse random graphs. 143
G15 Tree-indexed processes. 145
G16 Example: An additive process. 145
G17 Example: An extremal process. 146
G18–G22 Commentary 147

H Stochastic Geometry **149**
H1 Example: Holes and clusters in random scatter. 149
H2 The Poisson line process. 151
H3 A clump size calculation. 152
H4 Example: Empty squares in random scatter. 154

H5 Example: Empty rectangles in random scatter. 156
H6 Example: Overlapping random squares. 157
H7 Example: Covering K times. 159
H8 Example: Several types of particle. 159
H9 Example: Non-uniform distributions. 160
H10 Example: Monochrome squares on colored lattice. 161
H11 Example: Caps and great circles. 161
H12 Example: Covering the line with intervals of random length. 162
H13 Example: Clusters in 1-dimensional Poisson processes. . . . 164
H14–H21 Commentary 165

I Multi-Dimensional Diffusions **167**
I1 Background. 167
I2 The heuristic. 169
I3 Potential function. 169
I4 Reversible diffusions. 169
I5 Ornstein-Uhlenbeck processes. 170
I6 Brownian motion on surface of sphere. 170
I7 Local approximations. 170
I8 Example: Hitting times to small balls. 171
I9 Example: Near misses of moving particles. 171
I10 Example: A simple aggregation-disaggregation model. . . . 173
I11 Example: Extremes for diffusions controlled by potentials. . 173
I12 Example: Escape from potential wells. 176
I13 Physical diffusions: Kramers' equation. 177
I14 Example: Extreme values. 178
I15 Example: Escape from potential well. 180
I16 Example: Lower boundaries for transient Brownian motion. 182
I17 Example: Brownian motion on surface of sphere. 183
I18 Rice's formula for conditionally locally Brownian processes. 185
I19 Example: Rough $\mathcal{X}^2$ processes. 186
I20–I29 Commentary 186

J Random Fields **190**
J1 Spatial processes. 190
J2 In analysis of 1-parameter processes. 190
J3 Gaussian fields and white noise. 190
J4 Analogues of the Kolmogorov-Smirnov test. 191
J5 The heuristic. 192
J6 Discrete processes. 192
J7 Example: Smooth Gaussian fields. 193
J8 Example: 2-dimensional shot noise. 195
J9 Uncorrelated orthogonal increments Gaussian processes. . . 195
J10 Example: Product Ornstein-Uhlenbeck processes. 196
J11 An artificial example. 197

J12 Maxima of μ-Brownian sheets. 198
J13 1-parameter Brownian bridge. 198
J14 Example: Stationary × Brownian bridge processes. 200
J15 Example: Range of Brownian bridge. 201
J16 Example: Multidimensional Kolmogorov-Smirnov. 202
J17 Example: Rectangle-indexed sheets. 204
J18 Isotropic Gaussian processes. 205
J19 Slepian's inequality. 206
J20 Bounds from the harmonic mean. 207
J21 Example: Hemispherical caps. 208
J22 Example: Half-plane indexed sheets. 209
J23 The power formula. 212
J24 Self-normalized Gaussian fields. 212
J25 Example: Self-normalized Brownian motion increments. . . 212
J26 Example: Self-normalized Brownian bridge increments. . . . 213
J27 Example: Upturns in Brownian bridge with drift. 214
J28 Example: 2-parameter LIL. 215
J29–J37 Commentary 216

K Brownian Motion: Local Distributions **220**
K1 Modulus of continuity. 220
K2 Example: The Chung-Erdos-Sirao test. 221
K3 Example: The asymptotic distribution of W. 221
K4 Example: Spikes in Brownian motion. 224
K5 Example: Small increments. 225
K6 Example: Integral tests for small increments. 228
K7 Example: Local maxima and points of increase. 230
K8 Example: Self-intersections of d-dimensional Brownian motion. 233
K9–K17 Commentary 234

L Miscellaneous Examples **237**
L1 Example: Meetings of empirical distribution functions. . . . 237
L2 Example: Maximal k-spacing. 238
L3 Example: Increasing runs in i.i.d. sequences. 239
L4 Example: Growing arcs on a circle. 239
L5 Example: The LIL for symmetric stable processes. 241
L6 Example: Min-max of process. 242
L7 2-dimensional random walk 243
L8 Example: Random walk on $\boldsymbol{Z}^2$ modulo N. 244
L9 Example: Covering problems for 2-dimensional walks. . . . 244

M The Eigenvalue Method **246**
M1 Introduction. 246
M2 The asymptotic geometric clump principle. 247
M3 Example: Runs in subsets of Markov chains. 248
M4 Example: Coincident Markov chains. 248
M5 Example: Alternating runs in i.i.d. sequences. 248
M6 Example: Longest busy period in M/G/1 queue. 250
M7 Example: Longest interior sojourn of a diffusion. 250
M8 Example: Boundary crossing for diffusions. 251

Postscript **252**

Bibliography **253**

Index **267**

A The Heuristic

This chapter amplifies the one paragraph description of the heuristic given in the introduction. We develop a language, slightly abstract and more than slightly vague, in which the examples in subsequent chapters are discussed. The "Commentary" at the end of the chapter gives some perspective, relating the heuristic to more standard topics and techniques in probability theory. We illustrate the heuristic method with one simple example, the M/M/1 queue. To avoid interrupting the flow later, let us first develop notation for this process.

A1 The M/M/1 queue. This is the continuous-time Markov chain X_t on states $\{0,1,2,\dots\}$ with transition rates

$$\begin{array}{llll} i \to i+1 & & \text{(arrivals)} & \text{rate } \alpha \\ i \to i-1, & (i \geq 1) & \text{(departures)} & \text{rate } \beta > \alpha. \end{array}$$

The stationary distribution π has

$$\begin{aligned} \pi(i) &= \left(1-\frac{\alpha}{\beta}\right)\left(\frac{\alpha}{\beta}\right)^i \\ \pi[i,\infty) &= \left(\frac{\alpha}{\beta}\right)^i. \end{aligned}$$

For b large (relative to the stationary distribution — that is, for $\pi[b,\infty)$ small) consider the first hitting time

$$T_b \equiv \min\{\, t : X_t = b \,\}$$

(assume $X_0 < b$). Let

$$\lambda_b = \beta^{-1}(\beta-\alpha)^2\left(\frac{\alpha}{\beta}\right)^b.$$

There is a rigorous limit theorem for T_b as $b \to \infty$:

$$\sup_t |\boldsymbol{P}(T_b > t \mid X_0 = i) - \exp(-\lambda_b t)| \to 0 \quad \text{as} \quad b \to \infty, \qquad i \text{ fixed.} \tag{A1a}$$

The event $\{T_b > t\}$ is the event $\{\sup_{0<s\leq t} X_s < b\}$, and so we can re-write (A1a), after exploiting some monotonicity, as

$$\sup_b \left| \boldsymbol{P}\big(\max_{0\leq s\leq t} X_s < b \mid X_0 = i\big) - \exp(-t\lambda_b) \right| \to 0 \quad \text{as } t \to \infty; \qquad i \text{ fixed.} \tag{A1b}$$

We express conclusions of heuristic analyses in a form like

$$T_b \overset{\mathcal{D}}{\approx} \text{exponential}(\lambda_b). \tag{A1c}$$

We use $\approx$ to mean "is approximately equal to" in a vague sense; to any heuristic conclusion like (A1c) there corresponds a formal limit assertion like (A1a,A1b).

One component of the heuristic is the local approximation of a process. Let Z_t be a continuous-time simple random walk on $\{\ldots, -1, 0, 1, \ldots\}$ with transition rates α upwards and β downwards. Obviously, for b large we have:

given $X_0 \approx b$, the process $X_t - b$ evolves locally like Z_t.

Here "locally" means "for a short time", and "short" is relative to the hitting time of T_b we are studying. In most examples the approximating process Z will depend on the level b about which we are approximating: in this special example, it doesn't.

A2 Mosaic processes on $\boldsymbol{R}^2$. A *mosaic process* or *Boolean model* formalizes the idea of "throwing sets down at random". The recent book of Hall (1988) studies mosaic processes in their own right: we use them as approximations.

Let $\mathcal{C}$ be a random subset of $\boldsymbol{R}^2$. The simplest random subset is that obtained by picking from a list $B_1, \ldots, B_k$ of subsets with probabilities $p_1, \ldots, p_k$. The reader should have no difficulty in imagining random sets with continuous distribution. Think of the possible values B of $\mathcal{C}$ as smallish sets located near the origin 0. The values B need not be connected, or geometrically nice, sets. We ignore the measure-theoretic proprieties involved in a rigorous definition of "random set", except to comment that requiring the B's to be closed sets is more than enough to make rigorous definitions possible. Write $C =$ area$(\mathcal{C})$. So C is a random variable; $C(\omega)$ is the area of the set $\mathcal{C}(\omega)$. Assume

$$\boldsymbol{P}(C > 0) = 1; \qquad EC < \infty.$$

Given a set B and $x \in \boldsymbol{R}^2$, write $x+B$ for the translated set $\{\, x+y : y \in B \,\}$.

Now let $\lambda > 0$ be constant. Define the *mosaic process* $\mathcal{S}$ as follows.

1. Set down points y according to a Poisson point process of rate λ per unit area.

2. for each y pick a random set $\mathcal{C}_y$ distributed as $\mathcal{C}$, independent for different y.

3. Set $\mathcal{S} = \bigcup(y + \mathcal{C}_y)$

Call the y's *centers* and the $y + \mathcal{C}_y$ *clumps.* In words, $\mathcal{S}$ is the union of i.i.d.-shaped random clumps with Poisson random centers. Call λ the *clump rate.*

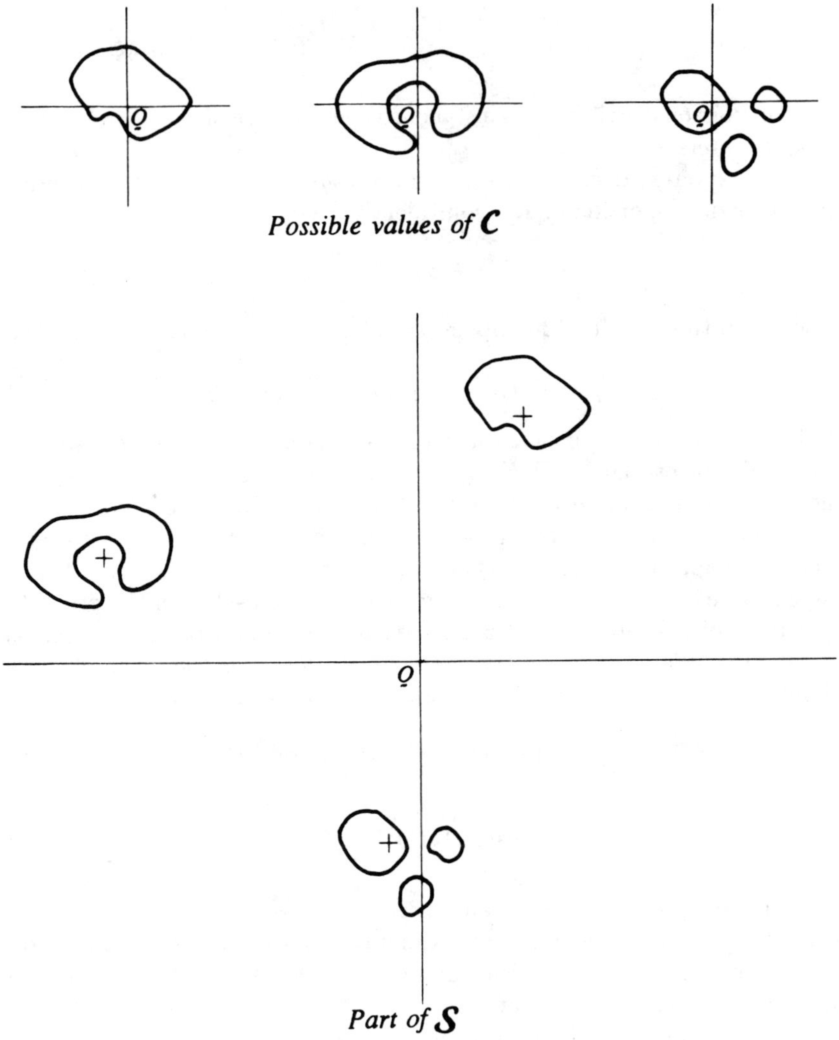

FIGURE A2a.

We need a few simple properties of mosaics.

Lemma A2.1 *For $x \in \boldsymbol{R}^2$ let N_x be the number of clumps of $\mathcal{S}$ which contain x. Then N_x has Poisson distribution with mean λEC.*

Indeed, conditioning on possible centers y gives

$$EN_x = \int \boldsymbol{P}(x \in y + \mathcal{C}_y)\lambda\, dy \;=\; \lambda \int \boldsymbol{P}(x - y \in \mathcal{C})\, dy$$
$$= \lambda \int \boldsymbol{P}(z \in \mathcal{C})\, dz = \lambda EC,$$

and the Poisson distribution is a standard consequence of the Poisson structure of centers.

We are interested in mosaics which are *sparse,* in the sense of covering only a small proportion of $\boldsymbol{R}^2$; equivalently, that

$$p \equiv \boldsymbol{P}(x \in \mathcal{S}), \qquad x \text{ fixed}$$

is small. In this case, (A2.1) implies $p = \boldsymbol{P}(N_x \geq 1) = 1 - \exp(-\lambda EC)$ and so

$$p \approx \lambda EC, \qquad \text{with error } O(p^2). \tag{A2a}$$

Although the clumps in a mosaic may overlap, in a sparse mosaic the proportion of overlap is $\boldsymbol{P}(N_x \geq 2 \mid N_x \geq 1) = O(p)$. So for a sparse mosaic we may ignore overlap, to first order approximation.

For our purpose only two properties of sparse mosaics are important. One is the approximation (A2a) above; the other is as follows. Let A be a large square or disc in $\boldsymbol{R}^2$, or more generally a fixed set with the property that most of the interior of A is not near its boundary. Let $\mathcal{S}$ be a sparse mosaic. Consider $\mathcal{S} \cap A$. We can approximate $\mathcal{S} \cap A$ as the union of those clumps of $\mathcal{S}$ whose centers lie in A, and this approximation gives

$$\boldsymbol{P}(\mathcal{S} \cap A \text{ empty}) \approx \exp(-\lambda \text{ area}(A)) \tag{A2b}$$

$$\text{area}(\mathcal{S} \cap A) \overset{\mathcal{D}}{\approx} \sum_{i=1}^{M} C_i; \tag{A2c}$$

where C_i are i.i.d. copies of C, and M has Poisson distribution with mean λ area(A). The error in these approximations arises from boundary effects — clumps which are partly inside and partly outside A — and, in the case of (A2c), from ignoring overlap.

The approximation in (A2c) involves a compound Poisson distribution; see Section A19 for sophisticated notation for such distributions.

So far we have discussed stationary mosaics. Everything extends to the non-stationary setting, where we have a rate function $\lambda(x)$ controlling the Poisson distribution of centers, and a random set distribution $\mathcal{C}_y$ with area

$\mathcal{C}_y$ from which clumps with centers y are picked. In this setting, (A2a) and (A2b) become

$$p(x) \equiv \boldsymbol{P}(x \in \mathcal{S}) \approx \lambda(x) E C_x \tag{A2d}$$

$$\boldsymbol{P}(\mathcal{S} \cap A \text{ empty}) \approx \exp(-\int_A \lambda(x)\,dx). \tag{A2e}$$

There is an implicit smoothness condition required for these approximations: $\lambda(x)$ and $\mathcal{C}_x$ should not vary much as x varies over a typical clump B.

A3 Mosaic processes on other spaces. We discussed mosaics on $\boldsymbol{R}^2$ for definiteness, and to draw pictures. The concepts extend to $\boldsymbol{R}^d, d \geq 1$, without essential change: just replace "area" by "length", "volume", etc. We can also define mosaics on the integer lattices $\boldsymbol{Z}^d$. Here the Poisson process of centers y becomes a Bernoulli process — each y is chosen with chance λ — and "area" becomes "cardinality".

Abstractly, to define a stationary mosaic on a space I all we need is a group of transformations acting transitively on I; to define a non-stationary mosaic requires no structure at all.

Many of our examples involve the simplest settings of $\boldsymbol{R}^1$ or $\boldsymbol{Z}^1$. But the d-dimensional examples tend to be more interesting, in that exact calculations are harder so that heuristic approximations are more worthwhile.

A4 The heuristic. Distributional questions concerning extrema or rare events associated with random processes may be reformulated as questions about sparse random sets; the heuristic consists of approximating these random sets by mosaics.

As a concrete class of examples, let $(X_t : t \in \boldsymbol{R}^2)$ be stationary real-valued, and suppose that we are interested in the distribution of $M_T = \sup_{t \in [0,T]^2} X_t$ for large T. Then we can define the random set $\mathcal{S}_b = \{t : X_t \geq b\}$, and we have

$$\boldsymbol{P}(M_T < b) = \boldsymbol{P}(\mathcal{S}_b \cap [0,T]^2 \text{ empty}). \tag{A4a}$$

For b large, $\mathcal{S}_b$ is a sparse stationary random set. Suppose $\mathcal{S}_b$ can be approximated by a sparse mosaic with some clump rate λ_b and some clump distribution $\mathcal{C}_b$. Then by (A2b)

$$\boldsymbol{P}(\mathcal{S}_b \cap [0,T]^2 \text{ empty}) \approx \exp(-\lambda_b T^2). \tag{A4b}$$

Assume we know the marginal distribution of X_t, and hence know

$$p_b \equiv \boldsymbol{P}(X_t \geq b) \equiv \boldsymbol{P}(x \in \mathcal{S}_b).$$

Then the approximation (A2a)

$$p_b \approx \lambda_b E C_b$$

combines with (A4a) and (A4b) to give

$$\boldsymbol{P}(M_T < b) \approx \exp\left(\frac{-p_b T^2}{EC_b}\right). \tag{A4c}$$

This approximation involves one "unknown", EC_b, which is the mean clump area for $\mathcal{S}_b$ considered as a mosaic process. Techniques for estimating EC_b are discussed later — these ultimately must involve the particular structure of (X_t).

There is a lot going on here! A theoretician would like a definition of what it means to say that a sequence $\mathcal{S}_b$ of random sets is asymptotically like a sequence $\widehat{\mathcal{S}}_b$ of mosaics (such a definition is given at Section A11, but not used otherwise). Second, one would like general theorems to say that the random sets occurring in our examples do indeed have this asymptotic behavior. This is analogous to wanting general central limit theorems for dependent processes: the best one can expect is a variety of theorems for different contexts (the analogy is explored further in Section A12). It turns out that our "sparse mosaic limit" behavior for rare events is as ubiquitous as the Normal limit for sums; essentially, it requires only some condition of "no long-range dependence".

To use the heuristic to *obtain* an approximation (and not worry about trying to justify it as a limit theorem), the only issue, as (A4c) indicates, is to estimate the mean clump size. Techniques for doing so are discussed below. Understanding that we are deriving approximations, it does no harm to treat (A2a) as an identity

$$p \equiv \boldsymbol{P}(x \in \mathcal{S}) = \lambda EC \tag{A4d}$$

which we call the *fundamental identity.* To reiterate the heuristic: approximating a given $\mathcal{S}$ as a sparse mosaic gives, by (A2b),

$$\boldsymbol{P}(\mathcal{S} \cap A \text{ empty}) \approx \exp(-\lambda \operatorname{area}(A)) \tag{A4e}$$

where λ is calculated from p and EC using (A4d). In practice, it is helpful that we need only the mean of C and not its entire distribution. If we can estimate the entire distribution of C, then (A2c) gives the extra approximation

$$\operatorname{area}(\mathcal{S} \cap A) \overset{\mathcal{D}}{\approx} \sum_{i=1}^{M} C_i; \qquad \begin{array}{c} (C_i) \text{ i.i.d. copies of } C \\ M \overset{\mathcal{D}}{=} \text{Poisson}(\lambda \operatorname{area}(A)) \end{array}. \tag{A4f}$$

In the non-stationary case, we use (A2d, A2e) instead:

$$\boldsymbol{P}(\mathcal{S} \cap A \text{ empty}) \approx \exp\left(-\int_A \lambda_x \, dx\right); \qquad \lambda_x = \frac{p(x)}{EC_x}. \tag{A4g}$$

A5 Estimating clump sizes. The next four sections give four general methods of estimating mean clump size. To qualify as a "general" method, there must be three completely different examples for which the given method is the best (other methods failing this test are mentioned at Section A20). Of course, any methods of calculating the same thing must be "equivalent" in some abstract sense, but the reader should be convinced that the methods are conceptually different. The first two methods apply in the general d-dimensional setting, whereas the last two are only applicable in 1 dimension (and are usually preferable there).

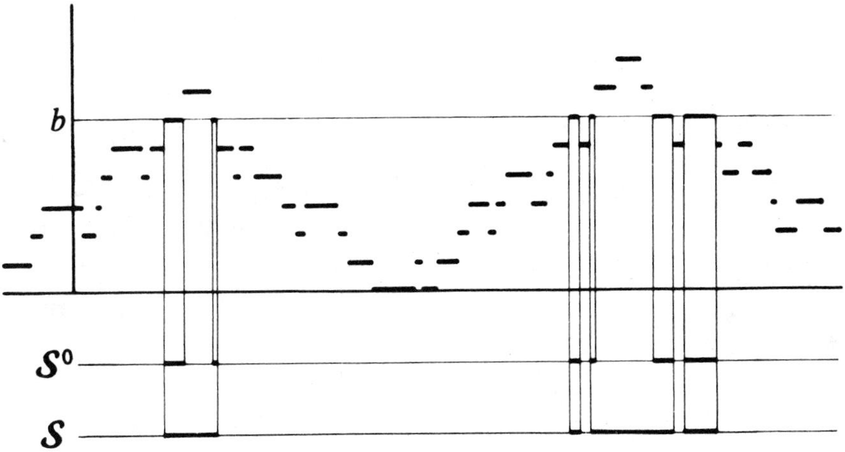

FIGURE A5a.

We illustrate with the M/M/1 queue described at Section A1. There are two random sets we can consider:

$$\begin{aligned} \mathcal{S}_b &= \{t : X_t \geq b\} \\ \mathcal{S}_b^0 &= \{t : X_t = b\}. \end{aligned}$$

For b large, we can approximate each of these as a sparse mosaic process on $\boldsymbol{R}^1$. We shall argue that the clump sizes are

$$EC^0 \approx \frac{1}{\beta - \alpha} \qquad \text{(A5a)}$$

$$C \approx \beta(\beta - \alpha)^{-2}. \qquad \text{(A5b)}$$

Then in each case we can use the fundamental identity

$$\begin{aligned} \lambda_b EC^0 &= \pi(b) &= \left(1 - \frac{\alpha}{\beta}\right)\left(\frac{\alpha}{\beta}\right)^b \\ \lambda_b EC &= \pi[b, \infty) &= \left(\frac{\alpha}{\beta}\right)^b \end{aligned}$$

to calculate the clump rate

$$\lambda_b = \beta^{-1}(\beta-\alpha)^2\left(\frac{\alpha}{\beta}\right)^b.$$

The heuristic (A4e) says

$$\boldsymbol{P}(\mathcal{S}^0 \cap [0,t] \text{ empty}) \approx \boldsymbol{P}(\mathcal{S} \cap [0,t] \text{ empty}) \approx \exp(-\lambda_b t).$$

In terms of first hitting times T_b or maxima, this is

$$\begin{aligned} \boldsymbol{P}(T_b > t) &\approx \exp(-\lambda_b t). \\ \boldsymbol{P}(\max_{0\le s\le t} X_s < b) &\approx \exp(-\lambda_b t). \end{aligned}$$

As mentioned in Section A1, these heuristic approximations correspond to rigorous limit theorems. Deriving these approximations from the heuristic involves only knowledge of the stationary distribution and one of the estimates (A5a),(A5b) of mean clump size. In some of the implementations of the heuristic the estimation of clump size is explicit, while in others it is implicit.

A6 The harmonic mean formula. Let $\mathcal{S}$ be a sparse mosaic process on $\boldsymbol{R}^2$. Let $\widetilde{\mathcal{S}}$ denote $\mathcal{S}$ conditioned on $0 \in \mathcal{S}$. $\widetilde{\mathcal{S}}$ is still a union of clumps, and by definition $0 \in \widetilde{\mathcal{S}}$, so let $\widetilde{\mathcal{C}}$ be the clump containing 0. It is important to understand that $\widetilde{\mathcal{C}}$ is different from $\mathcal{C}$, in two ways. First, suppose $\mathcal{C}$ is a non-random set, say the unit disc centered at 0. Then $\mathcal{S}$ consists of randomly-positioned unit discs, one of which may cover 0, and so $\widetilde{\mathcal{C}}$ will be a unit disc covering 0 but randomly-centered. More importantly, consider the case where $\mathcal{C}$ takes two values, a large or a small disc centered at 0, with equal probability. Then $\mathcal{S}$ consists of randomly-positioned large and small discs; there are equal numbers of discs, so the large discs cover more area, so 0 is more likely to be covered by a large disc, so $\widetilde{\mathcal{C}}$ will be a randomly-centered large or small disc, but more likely large than small.

As an aside, the definition of $\widetilde{\mathcal{C}}$ above is imprecise because 0 may be covered by more than one clump. A slick trick yields a precise definition. Imagine the clumps of a mosaic process labeled by i.i.d. variables L_y. Condition on $0 \in \mathcal{S}$ and 0 being in a clump of label l; define $\widetilde{\mathcal{C}}$ to be the clump labeled l. With this definition, results (A6a,A6b) below are precise.

Let $\widetilde{C} =$ area($\widetilde{\mathcal{C}}$). Write f, $\widetilde{f}$ for the densities of C, $\widetilde{C}$. Then

$$\widetilde{f}(a) = \frac{af(a)}{EC}, \qquad a > 0. \tag{A6a}$$

To see why, note that the proportion of the plane covered by clumps of area $\in (a, a+da)$ is $\lambda a f(a)\,da$, while the proportion of the plane covered by $\mathcal{S}$ is λEC. The ratio of these proportions is the probability that a covered point is covered by a clump of area $\in (a, a+da)$, this probability is $f(a)\,da$.

Two important observations.

1. The relationship (A6a) is precisely the renewal theory relationship between "lifetime of a component" and "lifetime of the component in use at time t", in the stationary regime (and holds for the same reason).

2. The result (A6a) does not depend on the dimensionality ; we have been using $\boldsymbol{R}^2$ in our exposition but (A6a) does not change for $\boldsymbol{R}^n$ or $\boldsymbol{Z}^n$.

Using (A6a),

$$\int_0^\infty a^{-1}\widetilde{f}(a)\,da = \int_0^\infty \frac{f(a)}{EC}\,da = \frac{1}{EC}.$$

This gives the *harmonic mean formula*

$$EC = \frac{1}{E(1/\widetilde{C})} = \text{harmonic mean}(\widetilde{C}), \tag{A6b}$$

which is the main point of this discussion.

To see the heuristic use of this formula, consider $\mathcal{S}_b^0 = \{\, t : X_t = b \,\}$ for the stationary M/M/1 queue. For large b, approximate $\mathcal{S}_b^0$ as a mosaic process on $\boldsymbol{R}^1$. Approximating $X_t - b$ by the random walk Z_t gives

$$\widetilde{\mathcal{C}} = \{\, t : Z_t = 0, -\infty < t < \infty \,\}$$

where $(Z_t, t \geq 0)$ and $(Z_{-t}, t \geq 0)$ are independent random walks with $Z_0 = 0$. In such a random walk the sojourn time of Z at 0 has exponential (rate $\beta - \alpha$) distribution. So

$$\widetilde{\mathcal{C}} = \Gamma_1 + \Gamma_2; \qquad \Gamma_i \text{ independent exponential}(\beta - \alpha)$$

and we can compute

$$\text{harmonic mean}(\widetilde{\mathcal{C}}) = \frac{1}{(\beta - \alpha)}.$$

Thus the harmonic mean formula gives the mean clump size (EC) estimate (A5a). There is a similar argument for (A5b): with $\mathcal{S} = \{\, t : X_t \geq b \,\}$ we get $\widetilde{\mathcal{C}} = \{\, t : Z_t \geq 0 \,\}$ where

$$\boldsymbol{P}(Z_0 = i) = \frac{\pi(b+i)}{\pi[b,\infty)} = \left(1 - \frac{\alpha}{\beta}\right)\left(\frac{\alpha}{\beta}\right)^i, \qquad i \geq 0; \tag{A6c}$$

given $Z_0 = i$, the processes $(\, Z_t; t \geq 0\,)$ and $(\, Z_{-t}; t \geq 0\,)$ are independent random walks (A6d)

We omit the messy calculation of harmonic mean $(\widetilde{C})$, but it does work out as (A5b).

Some examples where we use this technique are "partitioning random graphs" (Example G14), "holes in random scatter" (Example H1), "isotropic Gaussian fields" (Section J18), "integral test for small increments of Brownian motion" (Example K6). It is most useful in hard examples, since it is always applicable; its disadvantage is that exact calculation of the harmonic mean requires knowledge of the whole distribution of C, which is often hard to find explicitly. But we can get bounds: "harmonic mean $\leq$ arithmetic mean" implies

$$EC \leq E\widetilde{C} \tag{A6e}$$

and $E\widetilde{C}$ is often easy to calculate. This gives heuristic one-sided bounds which in some settings can be made rigorous — see Section A15.

In the discrete setting we have $C \geq 1$, and so

$$\text{if } E\widetilde{C} \approx 1 \text{ then } EC \approx 1. \tag{A6f}$$

This is the case where the mosaic process looks like the Bernoulli process, and the compound Poisson approximation becomes a simple Poisson approximation. This setting is rather uninteresting from our viewpoint, though it is quite prevalent in discrete extremal asymptotics.

A7 Conditioning on semi-local maxima. This second technique is easiest to describe concretely for the M/M/1 queue X_t. Fix b_0 large, and then consider b much larger than b_0. Given $X_{t_0} \geq b$, there is a last time $V_1 < t_0$ and a first time $V_2 > t_0$ such that $X_{V_i} \leq b_0$; let $x^* = \sup_{V_1 < t < V_2} X_t$ and pick t^* from $\{t \in (V_1, V_2) : X_t = x^*\}$ according to some rule. Call (t^*, x^*) a *semi-local maximum* of X. This construction yields a process of semi-local maxima (t^*, x^*): their precise definition depends of b_0 but the asymptotic $(x^* \to \infty)$ behavior is the same for any $b_0 \to \infty$ sufficiently slowly. The same idea works for d-parameter processes.

Let $L(x)$ be the rate at which semi-local maxima of height x occur (here and below x, y, u, b are integers since X is integer-valued). Then the clump rate λ_b for $\{t : X_t \geq b\}$ satisfies

$$\lambda_b = \sum_{x \geq b} L(x) \tag{A7a}$$

since each clump contains a semi-local maximum with some height $x \geq b$. Now define a conditioned process Y_t^x, $-\infty < t < \infty$, such that for $|t|$ small

$$Y_t^x \approx X_t - x \qquad \text{conditional on } (0, x) \text{ being a semi-local maximum,} \tag{A7b}$$

and such that $Y_t^x \to -\infty$ as $|t| \to \infty$. In the particular case of the M/M/1 queue, Y^x will be the random walk conditioned not to go above height 0, and in particular Y^x will not depend on x, but in more general examples Y^x will depend on x. Define

$$m(x,u) = E(\text{sojourn time of } Y^x \text{ at } -u); \qquad u \geq 0. \tag{A7c}$$

Then

$$\pi(y) = \sum_{u \geq 0} L(y+u)m(y+u,u) \tag{A7d}$$

by an ergodic argument. Indeed, $L(y+u)m(y+u,u)$ is the sojourn rate at y arising from clumps with semi-local maximum at height $y+u$, so both sides represent the total sojourn rate at y.

Assuming that the marginal distribution π is known, if we can estimate $m(x,u)$ then we can "solve" (A7d) to estimate $L(x)$ and thence to estimate λ_b via (A7a): this is our "conditioning on semi-local maxima" technique.

The actual calculation of $m(x,u)$ in the case of the M/M/1 queue, and indeed the exact conditioning in the definition of Y as a conditioned random walk, is slightly intricate; let us just record the answer

$$m(x,y) = (\beta-\alpha)^{-1}\left(2 - \left(\frac{\alpha}{\beta}\right)^{u} - \left(\frac{\alpha}{\beta}\right)^{u+1}\right); \qquad u \geq 0. \tag{A7e}$$

Since $\pi(y) = (1-\alpha/\beta)\,(\alpha/\beta)^y$, we can solve (A7d) to get

$$L(x) = \beta^{-2}(\beta-\alpha)^3\left(\frac{\alpha}{\beta}\right)^x$$

and then (A7a) gives $\lambda_b = \beta^{-1}(\beta-\alpha)^2(\alpha/\beta)^b$.

Despite its awkwardness in this setting, the technique does have a variety of uses: see smooth Gaussian fields (Example J7), longest common subsequences (Example G11), a storage/queuing model (Example D23). A closely related technique of marked clumps is mentioned at Section A20 and used in Chapter K.

A8 The renewal-sojourn method. Let $\mathcal{S}$ be a sparse mosaic process on $\boldsymbol{R}^1$. The clumps $\mathcal{C}$ consist of nearby intervals; the length C of a clump $\mathcal{C}$ is the sum of the lengths of the component intervals. The important difference between 1 and many dimensions is that in $\boldsymbol{R}^1$ each clump has a starting point (left endpoint) and an ending point (right endpoint); write span($\mathcal{C}$) for the distance between them.

If $\mathcal{S}$ is exactly a sparse mosaic, then trivially EC is the mean length of $\mathcal{S}$ from the start of a clump to the end of the clump. How do we say this idea for a random set $\mathcal{S}$ which is approximately a mosaic? Take τ large compared to the typical span of a clump, but small compared to the typical inter-clump distance $1/\lambda$;

$$E\,\text{span}(\mathcal{C}) \ll \tau \ll \frac{1}{\lambda}. \tag{A8a}$$

Then for stationary $\mathcal{S}$,

$$C \approx (\text{length}(\mathcal{S} \cap [0,\tau] \mid 0 \in \mathcal{S}, \mathcal{S} \cap (-\tau, 0) \text{ empty}). \qquad \text{(A8b)}$$

In other words, we take the starting points of clumps to be the points x in $\mathcal{S}$ such that $(x-\tau, x)$ is not touched by $\mathcal{S}$, and take the clump to be $\mathcal{S} \cap [x, x+\tau)$. This is our *renewal-sojourn* estimate (roughly, we are thinking of clump starts as points of a renewal process).

In the M/M/1 queue example, consider $\mathcal{S}^0 = \{ t : X_t = b \}$. Then

$$\begin{aligned} EC^0 &\approx E(\text{sojourn time of } X_t, 0 \le t \le \tau \text{ at } b \\ &\qquad \mid X_0 = b, X_s \ne b \text{ for all } -\tau \le s < 0) \\ &\approx E(\text{sojourn time of } X_t, 0 \le t \le \tau \text{ at } b \mid X_0 = b) \\ &\qquad \text{by Markov property} \\ &\approx E(\text{sojourn time of } Z_t, 0 \le t < \infty \text{ at } 0 \mid Z_0 = 0) \\ &= \frac{1}{\beta - \alpha.} \end{aligned}$$

If instead we considered $\mathcal{S} = \{ t : X_t \ge b \}$ then we would get

$$\begin{aligned} EC &\approx E(\text{sojourn time of } Z_t, 0 \le t < \infty \text{ in } [0,\infty) \mid Z_0 = 0) \\ &= \beta(\beta-\alpha)^2. \end{aligned}$$

This is the most natural technique for handling hitting times for Markov processes, and is used extensively in the examples in Chapters B,D,I.

In the discrete case, mosaics on $\boldsymbol{Z}^1$, (A8b) works with "length" becoming "cardinality". Note in this case we have $C \ge 1$, since we condition to have $0 \in \mathcal{S}$.

A9 The ergodic-exit method. Again consider a sparse mosaic $\mathcal{S}$ on $\boldsymbol{R}^1$. Since each clump has one start-point and one end-point, we can identify the clump-rate λ with the rate of start-points or the rate of end-points. Thus

$$\lambda = \lim_{\delta \downarrow 0} \delta^{-1} \boldsymbol{P}(\text{some clump end-point lies in } (0,\delta)).$$

For a stationary sparse random set which we are approximating as a mosaic process, we can write

$$\begin{aligned} \lambda &= \lim_{\delta \downarrow 0} \delta^{-1} \boldsymbol{P}(\text{the clump containing 0 ends in } (0,\delta),\ 0 \in \mathcal{S}) &\text{(A9a)} \\ &= p \lim_{\delta \downarrow 0} \delta^{-1} \boldsymbol{P}(\text{clump containing 0 ends in } (0,\delta) \mid 0 \in \mathcal{S}). &\text{(A9b)} \end{aligned}$$

This is the *ergodic-exit* method. There are equivalent formulations. Let C^+ be the future length of a clump containing 0, given 0 is in $\mathcal{S}$;

$$C^+ = (\text{length}(\mathcal{S} \cap [0,\tau] \mid 0 \in \mathcal{S})$$

for τ as in Section A8. Note the conditioning is different from in Section A8; 0 is not the start of the clump. Note also the difference from the harmonic mean formula (Section A6), where we used the two-sided clump length. Writing f^+ for the density of C^+, (A9b) becomes

$$\lambda = pf^+(0). \tag{A9c}$$

In the sparse mosaic case, ignoring overlap, it is easy to see the relationship between the density f of clump length C and the density f^+ of C^+:

$$f^+(a) = \frac{\boldsymbol{P}(C > a)}{EC}. \tag{A9d}$$

This is just the renewal theory relationship between "time between renewals" and "waiting time from a fixed time until next renewal". In particular, $f^+(0) = 1/EC$ and so (A9c) is just a variant of the fundamental identity (A4d).

For the M/M/1 queue, consider $\mathcal{S}^0 = \{t : X_t = b\}$.

$$\begin{aligned}
&\boldsymbol{P}(\text{clump ends in } (0,\delta) \mid X_0 = b) \\
&\quad\approx \boldsymbol{P}(Z_t < 0 \text{ for all } t > \delta \mid Z_0 = 0) \\
&\quad\approx \delta\beta\boldsymbol{P}(Z_t \le -1 \text{ for all } t \ge \delta \mid Z_\delta = -1) \\
&\quad\approx \delta\beta\left(1 - \frac{\alpha}{\beta}\right) \\
&\quad\approx \delta(\beta - \alpha).
\end{aligned} \tag{A9e}$$

So (A9b) gives clump rate

$$\lambda_b = \pi(b)(\beta - \alpha)$$

agreeing with Section A5. If instead we consider $\mathcal{S} = \{t : X_t \ge b\}$ then, since a clump can end only by a transition downwards from b,

$$\begin{aligned}
&\boldsymbol{P}(\text{clump ends in } (0,\delta) \mid X_0 \ge b) \\
&\quad\approx \boldsymbol{P}(\text{clump ends in } (0,\delta) \mid X_0 = b)\boldsymbol{P}(X_0 = b \mid X_0 \ge b) \\
&\quad\approx \delta(\beta - \alpha)\left(1 - \frac{\alpha}{\beta}\right) \qquad \text{using (A9e).}
\end{aligned}$$

So (A9b) gives the clump rate

$$\lambda_b = \pi[b,\infty)(\beta - \alpha)\left(1 - \frac{\alpha}{\beta}\right)$$

agreeing with Section A5.

Another way to look at this method is to let ψ be the rate of component intervals of clumps:

$$\psi = \lim_{\delta \downarrow 0} \delta^{-1} \boldsymbol{P}(0 \in \mathcal{S}, \delta \notin \mathcal{S}) = \lim \delta^{-1} \boldsymbol{P}(0 \notin \mathcal{S}, \delta \in \mathcal{S}).$$

Clumps have some random number $N \geq 1$ of component intervals, so clearly

$$\lambda = \frac{\psi}{EN}. \tag{A9f}$$

In practice, it is hard to use (A9f) because EN is hard to estimate. But there is a rather trite special case, where the clumps $\mathcal{C}$ are very likely to consist of a single interval rather than several intervals. In this case $N \approx 1$ and the clump rate λ can be identified approximately as the rate ψ of start-points or end-points of these component intervals:

$$\lambda = \lim_{\delta \downarrow 0} \delta^{-1} \boldsymbol{P}(0 \in \mathcal{S}, \delta \notin \mathcal{S}) = \lim_{\delta \downarrow 0} \delta^{-1} \boldsymbol{P}(0 \notin \mathcal{S}, \delta \in \mathcal{S}). \tag{A9g}$$

Examples where we use this special case are smooth continuous-path processes (Rice's formula) (C12h), 1-dimensional coverage processes (Example H12). The general case is used in additive Markov chains (like the G/G/1 queue) (Section C11), tandem queues (Example B20).

In the discrete case, mosaics on $\boldsymbol{Z}^1$, all this is simpler: replace "δ" by "1".

$$\begin{aligned} \lambda &= \boldsymbol{P}(0 \in \mathcal{S}, \text{ clump ends at } 0) \\ &= p\boldsymbol{P}(\text{clump ends at } 0 \mid 0 \in \mathcal{S}) \\ &= pf^+(0), \end{aligned} \tag{A9h}$$

where f^+ is the probability function of

$$C^+ = \text{cardinality}(\mathcal{S} \cap [1, \tau] \mid 0 \in \mathcal{S}). \tag{A9i}$$

A10 Limit assertions. Given random variables M_K and an approximation

$$\boldsymbol{P}(M_K \leq x) \approx G(K, x) \tag{A10a}$$

for an explicit function G, the *corresponding limit assertion* is

$$\sup_x |\boldsymbol{P}(M_K \leq x) - G(K, x)| \to 0 \quad \text{as } K \to \infty, \tag{A10b}$$

where the sup is taken over integers x if M_K is integer-valued. In examples, we state the conclusions of our heuristic analysis in form (A10a): the status of the corresponding limit assertion (as a known theorem or a conjecture) is noted in the commentary at the end of the chapter.

In many cases, assertions (A10b) are equivalent to assertions of the form

$$\frac{M_k - a_K}{b_K} \xrightarrow{\mathcal{D}} M \quad \text{as } K \to \infty \tag{A10c}$$

for constants a_K, b_K and non-degenerate limit distribution M. Textbooks often treat (A10c) as a definition of "limit theorem" (for distributional limits, that is), but this is a conceptual error: it is more natural to regard limit theorems as assertions that the error in an approximation tends to zero, as in (A10b).

Sometimes we use the heuristic in settings where only a single random variable M is of interest, for instance $M = \sup_{0<t<1} X_t$ for a process defined only on $[0,1]$. In this context distributional limits do not make sense. Instead, we state heuristic conclusions in the form

$$\boldsymbol{P}(M > b) \approx G(b) \qquad \text{for } b \text{ large} \tag{A10d}$$

and the corresponding limit assertion is

$$\frac{\boldsymbol{P}(M > b)}{G(b)} \to 1 \quad \text{as } b \to \infty \tag{A10e}$$

To indicate this we call such approximations *tail approximations.* They frequently occur in the non-stationary setting. If $\lambda_b(t_0)$ is the clump rate at t_0 for clumps of $\{\, t : X_t \geq b \,\}$ then the approximation (A4g)

$$\boldsymbol{P}(M \geq b) \approx \exp\left(-\int \lambda_b(t)\, dt\right); \qquad b \text{ large}$$

becomes

$$\boldsymbol{P}(M > b) \approx \int \lambda_b(t)\, dt; \qquad b \text{ large.} \tag{A10f}$$

COMMENTARY

A11 Sparse mosaic limit property. Our heuristic is based on the notion of random sets being asymptotically like sparse mosaics. Here we state a formalization of this notion. This formalization is not used in this book, but is explored in Aldous (1988b).

Fix dimension d. Write Leb for Lebesgue measure on $\boldsymbol{R}^d$. For $x \in \boldsymbol{R}^d, F \subset \boldsymbol{R}^d$ write $F - x = \{\, y - x : y \in F \,\}$. Write $\sigma_a : \boldsymbol{R}^d \to \boldsymbol{R}^d$ for the scaling map $x \to ax$. Then σ_a acts naturally on sets, random sets, point processes, etc. Let $\mathcal{F}$ be the set of closed subsets F of $\boldsymbol{R}^d$. A topology on $\mathcal{F}$ can be obtained by identifying F with the measure μ_F with density 1_F and using the topology of vague convergence of measures.

Let λ_n, a_n be constants such that

$$\lambda_n, a_n > 0; \lambda_n a_n \to 0 \quad \text{as } n \to \infty \tag{A11a}$$

Let $\mathcal{C}$ be a random closed set, let $C = \text{Leb}(\mathcal{C})$ and suppose

$$C > 0 \quad \text{a.s.}; \qquad EC < \infty. \tag{A11b}$$

Let $\mathcal{S}_n$ be random closed sets. The notion that the $\mathcal{S}_n$ are asymptotically like the mosaics with clump rates λ_n^d and clump distributions $a_n\mathcal{C}$ is formalized as follows.

Definition A11.1 *$\mathcal{S}_n$ has the sparse mosaic $(\lambda_n, a_n, \mathcal{C})$ limit property if (A11a), (A11b) above and (A11c), (A11d) below hold.*

To state the essential conditions, consider $\mathcal{S}_n$. For each $x \in \mathbf{R}^d$ let $F_x = \sigma_{1/a_n}(\mathcal{S}_n - x)$. Informally, F_x is "the view of $\mathcal{S}_n$ from x", after rescaling. Now let ξ_n be a point process defined jointly with $\mathcal{S}_n$. Then we can define a marked point process $\{\,(\lambda_n x, F_x) : x \in \xi_n\,\}$. That is, the points x of ξ_n are rescaled by λ_n and "marked" with the set F_x. There is a natural notion of weak convergence for marked point processes, using which we can state the main condition:

> There exists ξ_n such that the marked point process $(\lambda_n x, F_x),\quad x \in \xi_n$, converges weakly to the Poisson point process of rate 1 marked with i.i.d. copies of $\mathcal{C}$. (A11c)

To state the final condition, for $x \in \mathbf{R}^d$ let $\Delta_n(x)$ be the distance from x to the nearest point of ξ_n. The condition is

$$\lambda_n \sup\{\Delta_n(x) : x \in \mathcal{S}_n, x \in \lambda_n^{-1}K\} \to 0 \quad \text{as } n \to \infty; \text{ each bounded } K. \tag{A11d}$$

Condition (A11c) is the main condition, saying that the part of $\mathcal{S}_n$ near the points ξ_n is like a mosaic process; condition (A11d) ensures that all of $\mathcal{S}_n$ is accounted for in this way.

A12 Analogies with central limit theorems.

A12.1 From independence to dependence. The first limit theorems one encounters as a student are the Poisson and Normal limits of the Binomial. For independent random variables there is a complete story: the Lindeberg-Feller CLT and the Poisson limit theorem. For dependent random variables one still expects similar results under regularity conditions: what changes in the conclusions? For the CLT, what changes is the variance; for partial sums S_n of stationary mean-zero (X_i), one expects $S_n/\sqrt{n} \xrightarrow{\mathcal{D}} Normal(0, \sigma^2)$, where σ^2 is not $\text{var}(X_i)$ but is instead $\sum_{i=-\infty}^{\infty} EX_iX_0$. For rare events, what changes is that Poisson limits become compound Poisson, since when one event A_i occurs it may become likely that nearby events A_j also occur.

A12.2 Weak convergence. An extension of CLTs is to weak convergence results, asserting that normalized partial sums $S_n^*(t)$ converges to Brownian motion $B(t)$. This has two purposes. First, such results are more informative, giving limits for functionals of the partial sum process. Second, they permit different methods of proof of the underlying CLT; for instance, one may show tightness in function space and then argue that $B(t)$ is the only possible limit.

I claim that the "sparse mosaic limit property" of Section A11 is the correct analogue for rare events. One could be more simplistic and just use the compound Poisson counting process limit of counting processes, but this ignores the spatial structure of the clumps. As with weak convergence forms of the CLT, it is natural to expect that the more abstract viewpoint of sparse mosaic limits will enable different proof techniques to be employed. The compound Poisson formulation has been studied by Berman (1982a) and subsequent papers.

A12.3 Domain of application. Extensions of the CLT occupy a much larger portion of the probability theory literature than do extensions of the Poisson limit theorem. But one can argue that the latter arise in more settings. Whenever one has a CLT for a family of random variables, one expects a compound Poisson limit theorem for their extrema. But there are many settings involving rare events where there is no natural CLT. For instance, given any sample path property that Brownian motion does not possess, one can ask how nearly the property is achieved at some point in a long interval.

Of course, not all CLTs can be fitted into the weak convergence formalism: there are "combinatorial" CLTs which, roughly speaking, involve limits as the "dimension"$\to \infty$. The same happens for rare events: Chapter G treats such combinatorial examples.

A13 Large deviations. The modern theory of large deviations (see Varadhan (1984)) can crudely be described as a theory of limits of the form

$$\lim_k k^{-1} \log \boldsymbol{P}(A_k)$$

for "exponentially rare" events A_k. The domain of usefulness of the heuristic is adjacent to the domain of usefulness of large deviation ideas: the heuristic seeks to get asymptotics of $\boldsymbol{P}(A_k)$ which are exact ($\boldsymbol{P}(A_k) \sim p_k$) or exact up to a multiplicative constant, whereas large deviation estimates may be off by a polynomially large factor. To do this, the heuristic needs more specific structure: large deviation theory applies in various more general settings.

A14 Markov hitting times. Our discussion of the heuristic so far has been "general" in the sense of not assuming any particular dependence structure in the process underlying the random set $\mathcal{S}$. Many examples involve first hitting times of Markov processes on a rare subset of state space, so let us say a few words about this particular setting.

1. We usually use the renewal-sojourn form of the heuristic; this is related to an exact result concerning recurrent potential [B61].

2. Mean hitting times for a Markov process satisfy a set of difference or differential equations; a quite different type of heuristic is to try to solve these equations approximately. A set of "singular perturbation" techniques for doing so is described in Schuss (1980). There is some overlap of examples amenable to that technique and to our heuristic, though the author feels that the heuristic (where usable) gives simpler and more direct solutions.

3. Tails of hitting time distributions are typically exponential, and the exponent has an eigenvalue interpretation: see Chapter M. In our heuristic applications the entire distribution is approximately exponential, so this eigenvalue describes the entire distribution. Techniques for determining such eigenvalues are related to the techniques of (2) and of large deviation theory.

A15 The second moment method. This is based upon

Lemma A15.1 *Let $Z \geq 0$, $EZ^2 < \infty$. Then*

$$\boldsymbol{P}(Z > 0) \geq (EZ)^2/E(Z^2).$$

This follows from the Cauchy-Schwartz inequality applied to $Z1_{(Z>0)}$. Applying this to $Z = \#\{\, i : A_i \text{ occurs}\,\}$ gives the left inequality below; the right inequality is Boole's.

Lemma A15.2 *Let $(A_i; i \in I)$ be a finite family of events. Then $\mu^2/\sigma^2 \leq \boldsymbol{P}(\bigcup A_i) \leq \mu$; where $\mu = \sum_i \boldsymbol{P}(A_i)$ and $\sigma^2 = \sum_i \sum_j \boldsymbol{P}(A_i \cap A_j)$.*

This gives bounds for maxima $\max_i X_i$ of finite families of random variables, using $A_i = \{X_i \geq b\}$, $\bigcup A_i = \{\max X_i \geq b\}$. For continuous-parameter maxima, Boole's inequality gives no information, but these "second-moment" lower bounds do.

Lemma A15.3 *Let $M = \sup_{t \in T} X_t$, and let θ be a probability measure on T. Then*

$$\boldsymbol{P}(M > b) \geq \frac{\mu_b^2}{\sigma_b^2}; \qquad \text{where} \qquad \begin{aligned} \mu_b &= \textstyle\int \boldsymbol{P}(X_t > b)\theta(dt) \\ \sigma_b^2 &= \textstyle\iint \boldsymbol{P}(X_s > b, X_t > b)\theta(ds)\theta(dt). \end{aligned} \tag{A15a}$$

This follows by applying (A15.1) to $Z = \theta\{\, t : X_t > b\,\}$.

These rigorous inequalities can be related to the heuristic as follows. In the setting of (A15.3), suppose

$$\boldsymbol{P}(M > b) \sim q(b) \quad \text{as} \quad b \to \infty$$

and let C_b be the clump size:

$$C_b = \theta\{\, t : X_t > b \,\} \qquad \text{given } M > b.$$

Then the lower bound in (A15.3) is

$$\frac{\mu_b^2}{\sigma_b^2} = \frac{(q(b)EC_b)^2}{q(b)EC_b^2} = \frac{q(b)(EC_b)^2}{EC_b^2}.$$

Thus the lower bound of (A15.3) underestimates the true value by the factor $(EC_b)^2/(EC_b^2)$. Now consider the harmonic mean formula version of the heuristic (Section A6). If there we use the "harmonic mean $\leq$ arithmetic mean" inequality, we get

$$EC = \text{harmonic mean}(\widetilde{C}) \leq E\widetilde{C} = \frac{EC^2}{EC}, \tag{A15b}$$

the last equality by (A6a). Thus if we replace the clump rate p/EC by the underestimate $p/E\widetilde{C}$, the underestimation factor is $(EC)^2/(EC^2)$; this is the same as with the second moment method.

In the context of the heuristic for stationary random sets $\mathcal{S}$ in $\boldsymbol{R}^d$, we estimate

$$E\widetilde{C} \approx \int_{x \text{ near } 0} \boldsymbol{P}(x \in \mathcal{S} \mid 0 \in \mathcal{S})\, dx. \tag{A15c}$$

This is usually easy to calculate, unlike EC itself.

A16 Continuous 1-parameter bounds. The following simple rigorous result can be regarded as Boole's inequality applied to right endpoints of component intervals.

Lemma A16.1 *Let $\mathcal{S}$ be a stationary random closed subset of $\boldsymbol{R}^1$ which consists a.s. of disjoint non-trivial intervals. Then*

$$\psi = \lim_{\delta \downarrow 0} \delta^{-1} \boldsymbol{P}(0 \in \mathcal{S}, \delta \notin \mathcal{S}) \leq \infty$$

exists, and

$$\boldsymbol{P}(\mathcal{S} \cap [0,t] \textit{ non-empty}) \leq \boldsymbol{P}(0 \in \mathcal{S}) + t\psi.$$

Applied to $\mathcal{S} = \{\, t : X_t \geq b \,\}$ for stationary (X_t) with smooth paths, the lemma gives upper bounds for $\boldsymbol{P}(\max_{s \leq t} X_s > b)$. This is essentially Rice's formula [C12].

Compare with the ergodic-exit version (Section A9) of the heuristic. If the clump rate is λ and there are a random number $N \geq 1$ of component intervals in a clump, then

$$\psi = \lambda EN.$$

In using the heuristic, we can always replace λ by its upper bound ψ to get an upper bound on the clump rate, since $EN \geq 1$; this procedure corresponds to the rigorous result (A16.1).

A17 The harmonic mean formula. Here are two version of an exact result which is plainly related to the harmonic mean version of the heuristic (Section A6).

Lemma A17.1 *Let* $(A_i : i \in I)$ *be a finite family of events. Let* $C = \sum_i 1_{A_i}$. *Then* $\boldsymbol{P}(\bigcup A_i) = \sum_i \boldsymbol{P}(A_i)E(C^{-1} \mid A_i)$.

This is obtained by writing $1_{\bigcup A_i} = \sum_i C^{-1}1_{A_i}$. Here is the continuous-parameter version, which requires some technical hypothesis we shall not specify. Let $\mathcal{S}$ be a random subset of S, let θ be a probability measure on S, let $p(x) = \boldsymbol{P}(x \in \mathcal{S})$ and let $C = \theta(\mathcal{S})$. Then

$$\boldsymbol{P}(\theta(\mathcal{S}) > 0) = \int E(C^{-1} \mid x \in \mathcal{S})p(x)\theta(dx). \qquad \text{(A17a)}$$

A18 Stein's method. A powerful general method of obtaining explicit bounds for the error in Poisson approximations has been developed, and is widely applicable in "combinatorial" type examples. See Stein (1986) for the theoretical treatment, Arratia et al. (1987) and Barbour and Holst (1987) for concise applications to examples like ours, and Barbour and Eagleson (1983) for more applications. The most interesting potential applications (from our viewpoint) require extensions of the known general results to the compound Poisson setting: developing such extensions is an important research topic.

A19 Compound Poisson distribution. There is some slick notation for compound Poisson distributions. Let ν be a positive measure on $(0, \infty)$ satisfying $\int_0^\infty \min(1, x)\nu(dx) < \infty$. Say Y has POIS(ν) distribution if

$$E\exp(-\theta Y) = \exp\left(-\int_0^\infty \left(1 - e^{-\theta x}\right)\nu(dx)\right); \qquad \theta > 0.$$

To understand this, consider some examples.

1. The familiar Poisson(mean a) distribution is POIS(ν) for $\nu = a\delta_1$. (δ_x is the probability measure degenerate at x).

2. If Z_i are independent Poisson(a_i), then the random variable $\sum x_i Z_i$ has POIS($\sum a_i \delta_{x_i}$) distribution.

3. If $X_1, X_2, \ldots$ are the times of events of a non-homogeneous Poisson process of rate $g(x)$, and if $\int_0^\infty g(x)\,dx < \infty$, then $\sum X_i$ has POIS(ν) distribution, where $d\nu/dx = g(x)$.

4. If (X_i) are i.i.d. with distribution θ, if N is independent of (X_i) with Poisson (a) distribution then $\sum_{i=1}^N X_i$ has POIS($a\theta$) distribution.

In particular, approximation (A4f) can be written as

5 area$(\mathcal{S} \cap A) \stackrel{\mathcal{D}}{\approx}$ POIS$(\lambda$ area$(A)\nu_C)$; where ν_C is the distribution of C.

In most uses of the heuristic, it is difficult enough to get a reasonable estimate of the mean of C, let alone the full distribution, so we shall say little about these compound Poisson results. Note that the first two moments of area$(\mathcal{S} \cap A)$ can be obtained directly from (A4f)

$$\begin{aligned} E\ \text{area}(\mathcal{S} \cap A) &\approx \lambda\ \text{area}(A)EC \\ \text{var}\ \text{area}(\mathcal{S} \cap A) &\approx \lambda\ \text{area}(A)EC^2. \end{aligned}$$

Similarly, in terms of the Laplace transform of C

$$\psi(\theta) \equiv E\exp(-\theta C)$$

we get the Laplace transform of our heuristic compound Poisson approximation for area$(\mathcal{S} \cap A)$

$$E\exp(-\theta(\mathcal{S} \cap A)) \approx \exp(-\lambda\ \text{area}(A)(1-\psi(\theta))).$$

A20 Other forms of the heuristic. Here are several other forms of the heuristic which we do not use often enough to classify as useful general methods.

1. In the 1-dimensional setting of Section A9, where the clumps consist of a random number of component intervals, one may pretend the random set regenerates at the start of each component interval. This leads to the *quasi-Markov estimate* (Section D42).

2. In the context of maxima of stationary processes $(X_t; t \in \boldsymbol{R}^d)$, the clump rate λ_b for $\{t : X_t \geq b\}$ has

$$\begin{aligned} \lambda_b &\approx (2T)^{-d}\boldsymbol{P}(\sup_{[-T,T]^d} X_t \geq b) \\ &\approx (2T)^{-d}\int \boldsymbol{P}(\sup_{[-T,T]^d} X_t \geq b \mid X_0 = x) f_{X_0}(x)\,dx. \end{aligned}$$

If a rescaling of X_t around high levels b approximates a limit process Z_t, then we get a result relating λ_b to

$$\lim_{T\to\infty}(2T)^{-d}\int_{-\infty}^{0}\boldsymbol{P}(\sup_{[-T,T]^d} Z_t \geq 0 \mid Z_0 = z)g(z)\,dz$$

where g is a rescaled limit of f_{X_0}. This has some theoretical appeal for proving limits exist in special cases (Section J37) but is not useful in practice, since it merely replaces one "asymptotic maximum" problem by another.

3. Let $\mathcal{S}$ be approximately a mosaic process in $\boldsymbol{R}^d$ and let $f : \boldsymbol{R}^d \to [0, \infty)$ be deterministic continuous. Write

$$q(v) = \boldsymbol{P}(t \in \mathcal{S}, f(t) \leq v).$$

Each clump $\mathcal{C}$ of $\mathcal{S}$ can be "marked" by $U_\mathcal{C} = \inf_{t \in \mathcal{C}} f(t)$. Then the clump rate λ of $\mathcal{S}$ is

$$\lambda = \int_0^\infty \lambda(u)\, du; \quad \lambda(u)\, du \text{ the rate of clumps with } U \in (u, u + du).$$

Consider $C_{u,v} =$ area$\{\, t \in \mathcal{C} : f(t) \leq v \,\}$ for a clump with $U_\mathcal{C} = u$. Clearly

$$q(v) = \int_0^v \lambda(u) EC_{u,v}\, du.$$

Thus if we can find $q(v)$ and $EC_{u,v}$ then we can solve for $\lambda(u)$ and thence obtain the clump rate λ.

Call this the *marked clump* technique. It is similar to the "conditioning on semi-local maxima" techniques: there the mark was the local extreme of the underlying random process, whereas here the mark is obtained from the geometry of the clump (one can imagine still other ways to define marks, of course). This technique is used in Chapter K to study Brownian path properties.

A21 Exponential/geometric clump sizes. In the continuous 1-dimensional setting, a simple possible distribution for clump size C is the exponential(θ) distribution:

$$f_C(c) = \theta e^{-\theta c}, c > 0; \quad EC = \frac{1}{\theta}. \tag{A21a}$$

This occurred in the M/M/1 example, and occurs in other simple examples. It is equivalent to the conditioned distribution $\widetilde{C}$ of Section A6 having the form

$$f_{\widetilde{C}}(c) = \theta^2 c e^{-\theta c}, \quad c > 0; \qquad \widetilde{C} \stackrel{\mathcal{D}}{=} C_1 + C_2 \tag{A21b}$$

for independent exponential(θ) C_i. It is also equivalent to the distribution C^+ of Section A9 being

$$C^+ \text{ is exponential}(\theta). \tag{A21c}$$

In the discrete 1-dimensional setting, there are analogous results for geometric clump sizes. Equivalent are

$$\boldsymbol{P}(C = n) = \theta(1-\theta)^{n-1}, n \geq 1 \tag{A21d}$$

$$\widetilde{C} = C_1 + C_2 - 1; \quad C_i \text{ independent with distribution (A21d)} \tag{A21e}$$

$$\boldsymbol{P}(C^+ = n) = \theta(1-\theta)^n, n \geq 0, \quad \text{for } C^+ \text{ as at (A9i).} \tag{A21f}$$

B Markov Chain Hitting Times

B1 Introduction. In the context of Markov chains, the fundamental use of the heuristic is to estimate the distribution of the first hitting time to a rarely-visited state or set of states. Such problems arise in several areas of applied probability, e.g., queueing theory and reliability, as well as pure theory. The heuristic is useful in the case where the stationary distribution is known explicitly but transient calculations are difficult.

By "chain" I mean that the state space J is discrete. There is no essential difference between the discrete-time setting, where $(X_n; n \geq 0)$ is described by a transition matrix $\boldsymbol{P}(i,j)$, and the continuous-time setting, where $(X_t; t \geq 0)$ is described by a transition rate matrix $Q(i,j), \quad j \neq i$. We shall only consider irreducible positive-recurrent chains, for which there exists a unique stationary distributions π determined by the balance equations

$$\begin{aligned} \pi(j) &= \sum_i \pi(i)\boldsymbol{P}(i,j) \quad \text{[discrete]} \\ q(j)\pi(j) &= \sum_{i \neq j} \pi(i)Q(i,j) \quad \text{[continuous]}, \quad \text{where } q(j) = \sum_{k \neq j} Q(j,k). \end{aligned} \tag{B1a}$$

For a subset of states $A \subset J$, the first hitting time T_A is

$$T_A = \min\{\, t \geq 0 : X_t \in A \,\}.$$

Positive-recurrence implies $E_i T_A < \infty$; here $E_i(\quad)$, $\boldsymbol{P}_i(\quad)$ denote expectation, probability given $X_0 = i$. The mean hitting times are determined by elementary equations. For fixed A, the means $h(i) = E_i T_A$ satisfy

$$\begin{aligned} h(i) &= 1 + \sum_j \boldsymbol{P}(i,j)h(j) \quad \text{[discrete]} \qquad \text{for } i \notin A \\ q(i)h(i) &= 1 + \sum_{j \neq i} Q(i,j)h(j) \quad \text{[continuous]} \qquad \text{for } i \notin A \\ h(i) &= 0 \qquad \text{for } i \in A. \end{aligned} \tag{B1b}$$

At first sight, one might think that equations (B1a) and (B1b) were about equally hard to solve. But in practical examples there is often special structure which enables us to find the stationary distribution π without solving equations (e.g., reversibility; double-stochasticity; certain queueing networks), whereas there are no such useful tricks for hitting times. The purpose of this chapter is to show how knowledge of the stationary distribution can be used in many cases to get simple approximations for hitting times, when $\pi(A)$ is small.

The natural form of the heuristic is the renewal-sojourn method (Section A8). Fix A and consider the random set $\mathcal{S}$ of times t such that $X_t \in A$, for the stationary process X. If $\mathcal{S}$ does look like a Poisson clump process then the clump size C is the sojourn time in A during a clump of nearby visits to A. Since $p = \boldsymbol{P}(t \in \mathcal{S}) = \pi(A)$, the fundamental identity $p = \lambda EC$ implies that the clump rate λ is

$$\lambda = \frac{\pi(A)}{EC}.$$

So the waiting time for the first clump, i.e. the first hit on A, has approximately exponential distribution with mean $1/\lambda$. To summarize:

B2 The heuristic for Markov hitting times. *If $\pi(A)$ is small then*

(i) T_A has approximately exponential distribution;

(ii) $ET_A \approx EC/\pi(A)$;

(iii) these hold for any initial distribution not close to A.

Typically, in examples we see that the "local" behavior of X around the set A can be approximated by the local behavior of some *transient* process X^* around A. If so, we can approximate the sojourn time C by the total sojourn time $(0 \le t < \infty)$ of X^* in A. If A is a singleton $\{k\}$ then we take $X_0^* = k$; for general A there is a technical issue of what distribution on A to give to X_0^*, which we defer until Section B16. The point of this procedure is:

> *when π is known, the heuristic converts the "global" problem of solving (B1b) into a "local" problem of estimating EC.*

Of course Assertion B2 is not a theorem; there are certainly examples where $\pi(A)$ is small but T_A is not approximately exponential (e.g., for simple symmetric random walk (Section B11)). But I don't know any natural example where the transient approximation heuristic is applicable but gives an erroneous conclusion. In other words, in cases where T_A is not approximately exponential there is no natural transient approximation with which to implement the heuristic. The fundamental transient process, which will frequently be used for approximations, is the simple asymmetric random walk on the integers. Here are some elementary facts about this process. Let $0 < a < b$. Consider the continuous time chain X_t with $Q(i, i+1) = a$, $Q(i, i-1) = b$, $i \in \boldsymbol{Z}$. For $A \subset \boldsymbol{Z}$ let $S(A)$ be the total amount of time spent in A.

(i) $\boldsymbol{P}_0(\text{hit 1 sometime}) = a/b$

(ii) $\boldsymbol{P}_0(\text{never return to 0}) = (b-a)/(b+a)$

(iii) $E_0 S(0) = (b-a)^{-1}$

(iv) $E_0 S[0,\infty) = b(b-a)^{-2}$.

Exactly the same results hold for the discrete time walk with $P(i,i+1) = a$, $P(i,i-1) = b$, $P(i,i) = 1-a-b$.

Let us now start the examples by repeating the argument at Section A8 for the M/M/1 queue, where we can compare the result obtained via the heuristic with the exact result.

B3 Example: Basic single server queue. Here the states are $\{0,1,2,\ldots\}$, $Q(i,i+1) = a$, $Q(i,i-1) = b$ for $i \geq 1$; the parameters a and b represent the arrival and service rates; and $a < b$ for stability. The stationary distribution is geometric: $\pi(i) = (1-a/b)(a/b)^i$. We want to estimate T_K, the time until the queue length first reaches K, where K is sufficiently large that $\pi[K,\infty) = (a/b)^K$ is small; that is, the queue length rarely exceeds K. We apply the heuristic with $A = \{K\}$. Around K, the queue process behaves exactly like the asymmetric random walk. So EC is approximately $E_K S(K)$ for the random walk, which by (B2iii) equals $(b-a)^{-1}$. So the heuristic B2 says

$$T_K \text{ has approximately exponential distribution, mean } \frac{b}{(b-a)^2}\left(\frac{b}{a}\right)^K.$$

In this case the exact mean, obtained by solving (B1b), is

$$E_i T_K = \frac{b}{(b-a)^2}\left(\left(\frac{b}{a}\right)^K - \left(\frac{b}{a}\right)^i\right) - \frac{K-i}{b-a}$$

In using the heuristic we assumed $(a/b)^K$ is small, and then we see that the heuristic solution has indeed picked up the dominant term of the exact solution.

B4 Example: Birth-and-death processes. We can use the same idea for more general birth-and-death processes, that is processes with transition rates of the form $Q(i,i+1) = a_i$, $Q(i,i-1) = b_i$, $i \geq 1$, $Q(i,j) = 0$ otherwise. Suppose we want to study T_K, where K is sufficiently large that $\pi[K,\infty)$ is small. Provided the rates a_i, b_i vary smoothly with i, we can approximate the behavior of the process around K by the "linearized" random walk which has $Q(i,i+1) = a_K$ and $Q(i,i-1) = b_K$. This random walk has $E_K S(K) = (b_K - a_K)^{-1}$. Using this as a (rough!) estimate of EC, the heuristic B2 gives

$$ET_K \approx ((b_K - a_K)\pi(K))^{-1}. \tag{B4a}$$

For concreteness, consider the infinite server queue (M/M/∞). Here $Q(i, i+1) = a$ and $Q(i, i-1) = ib$, for arbitrary $a, b > 0$. The stationary distribution π is Poisson(a/b). Estimate (B4a) gives

$$ET_k \approx (bK - a)^{-1}(b/a)^K K! e^{a/b}, \tag{B4b}$$

provided $\pi[K, \infty)$ is small (which implies $K > a/b$).

There is a general expression for $E_i T_K$ in an arbitrary birth and death process (see Karlin and Taylor (1975) p. 149) which gives a complicated exact result; as before, the dominant term for large K is just the heuristic approximation.

B5 Example: Patterns in coin-tossing. Tossing a fair coin generates a sequence of heads H and tails T. Given a pattern (i), for example

TTTT (1) TTTH (2) THTH(3),

let X_i be the number of tosses needed until pattern i first occurs. This is a first hitting-time problem for the 16-state discrete Markov chain of overlapping 4-tuples; generating function arguments give exact distributions (Section B26.3), but let us see the heuristic approximations.

Let $\mathcal{S}_i$ be the random set of n such that tosses $(n-3, n-2, n-1, n)$ form pattern i. Clearly

$$p_i \equiv \boldsymbol{P}(n \in \mathcal{S}_i) = \frac{1}{16}.$$

To determine clump size C_i, condition on pattern i occurring initially, and let C_i be the number of times pattern i appears in a position overlapping this initial pattern, including the initial pattern itself. Then

$$\begin{aligned}
\boldsymbol{P}(C_1 = n) &= \frac{1}{2^n} \quad (n = 1, 2, 3), \boldsymbol{P}(C_1 = 4) = \frac{1}{8}; & EC_1 &= \frac{15}{8} \\
\boldsymbol{P}(C_2 = 1) &= 1; & EC_2 &= 1 \\
\boldsymbol{P}(C_3 = 1) &= \frac{3}{4} \quad \boldsymbol{P}(C_3 = 2) = \frac{1}{4}; & EC_3 &= \frac{5}{4}.
\end{aligned}$$

So the clump rates λ_i can be calculated from the fundamental identity $\lambda_i = p_i / EC_i$:

$$\lambda_1 = \frac{1}{30} \quad \lambda_2 = \frac{1}{16} \quad \lambda_3 = \frac{1}{20}.$$

Bearing in mind the constraint $X_i \geq 4$, the heuristic gives

$$\boldsymbol{P}(X_i \geq 4 + m) \approx (1 - \lambda_i)^m \approx \exp(-\lambda_i m); \qquad m \geq 0. \tag{B5a}$$

The examples above are unimpressive because exact mean hitting times are calculable. We now consider some doubly-stochastic chains, for which the stationary distribution is necessarily uniform. Here the heuristic can give effortless approximations which are harder to obtain analytically.

B6 Example: Card-shuffling. Repeated shuffling of an N-card deck can be modeled by a Markov chain whose states are the $N!$ possible configurations of the deck and whose transition matrix depends on the method for doing a shuffle. Particular methods include

(a) "top to random": the top card is removed and replaced in a uniform random position.

(b) "random transposition": two cards are picked at random and interchanged.

(c) "riffle": the usual practical method, in which the deck is cut into two parts which are then interleaved. A definite model can be obtained by supposing all 2^N possible such riffles are equally likely.

Regardless of the method, we get a doubly-stochastic chain (in fact, a random walk on the permutation group), so $\pi(i) = 1/N!$ for each configuration i. Consider the number of shuffles T needed until a particular configuration i is reached. The heuristic B2 says that T has approximately exponential distribution with mean

$$ET \approx N!\,EC$$

Here C = 1 plus the mean number of returns to an initial configuration in the short term. But for any reasonable shuffling method, the chance of such returns will be small, so

$$ET \approx N!$$

More sharply, in case (a) we see

$$ET \approx N!\,(1 + 1/N)$$

taking into account the chance $1/N$ that the first shuffle puts the top card back on top. In case (b)

$$ET \approx N!\,(1 + 2/N^2)$$

taking into account the chance that the same two cards are picked on the first two shuffles.

For our next example, another well-known transient process is simple symmetric random walk on $\boldsymbol{Z}^d$, $d \geq 3$. For this walk started at 0, let

$$R_d = \text{ mean total number of visits to 0 in time } 0 \leq n < \infty. \qquad \text{(B6a)}$$

These are certain constants: $R_3 \approx 1.5$, for instance.

B7 Example: Random walk on $\boldsymbol{Z}^d$ mod N. For fixed $d \geq 3$ and large N, consider simple symmetric walk on the set of d-dimensional integers modulo N; that is, on the integer cube of side N with periodic boundary conditions. Here the state space J has $|J| = N^d$; again the chain is doubly-stochastic, so $\pi(i) = N^{-d}$ for each i. Consider the time T taken to hit a specified state i from a uniform random start. The heuristic B2 says T has approximately exponential distribution with mean $ET \approx N^d EC$. Around i, the chain behaves like the unrestricted random walk on $\boldsymbol{Z}^d$, so we estimate EC as R_d in (B6a), to obtain

$$ET \approx R_d N^d.$$

B8 Example: Random trapping on $\boldsymbol{Z}^d$. Consider the complete lattice $\boldsymbol{Z}^d$, $d \geq 3$, and let $\mathcal{R}$ be a stationary random subset of $\boldsymbol{Z}^d$ with $q = \boldsymbol{P}(i \in \mathcal{R})$ small. Let T be the time taken for simple random walk X_n started at 0 to first hit $\mathcal{R}$. In the special case where $\mathcal{R}$ is a random translate of the set $\{(j_1 N, \ldots, j_d N), j_k \in \boldsymbol{Z}\}$, a moment's thought reveals this is equivalent to the previous example, so

$$ET \approx R_d q^{-1}. \tag{B8a}$$

In fact we can apply the heuristic to more general $\mathcal{R}$ by considering the random set $\mathcal{S}$ of times n such that $X_n \in \mathcal{R}$. The reader should think through the argument: the conclusion is that (B8a) remains true provided the "trap" points of $\mathcal{R}$ are typically not close together (if they are close, the argument at Example B15 can be used).

We now turn to consideration of hitting times of a chain X_t on a set A. Before treating this systematically, here is an obvious trick. Suppose we can define a new process $Y_t = f(X_t)$ such that, for some singleton k,

$$X_t \in A \text{ iff } Y_t = k \tag{B8b}$$

Around k, the process Y_t can be approximated by a known transient Markov chain $\widehat{Y}_t$. (B8c)

Then C, the local sojourn time of X in A, equals the local sojourn time of Y at k, and can be approximated by the total sojourn time of $\widehat{Y}$ at k. Note that Y need not be exactly Markov. Here are two illustrations of this idea.

B9 Example: Two M/M/1 queues in series. Suppose each server has service rate b, and let the arrival rate be $a < b$. Write (X_t^1, X_t^2) for the queue lengths process. It is well known that the stationary distribution (X^1, X^2) has independent geometric components:

$$\pi(i,j) = (1 - a/b)^2 (a/b)^{i+j}; \qquad i, j \geq 0.$$

Suppose we are interested in the time T_k until the combined length $X_t^1+X_t^2$ first reaches k; where k is sufficiently large that the stationary probability $\pi\{(i,j) : i+j \geq k\}$ is small. We want to apply our basic heuristic B2 to $A = \{(i,j) : i+j = k\}$. Since $\pi(A) = (k+1)(1-a/b)^2(a/b)^k$, we get

$$ET_k \approx (k+1)^{-1}(1-a/b)^{-2}(b/a)^k EC.$$

Consider the combined length process $Y_t = X_t^1 + X_t^2$. I claim that around k, Y behaves like the asymmetric simple random walk with up-rate a and down-rate b, so that by (B2) $EC = (b-a)^{-1}$ and so

$$ET_k \approx (k+1)^{-1}b^{-1}(1-a/b)^{-3}(b/a)^k. \tag{B9a}$$

To justify the claim, note that when $X_t^2 > 0$ the process Y_t is behaving precisely as the specified asymmetric walk. Fix t_0 and condition on $Y_{t_0} = k$. Then $(X_{t_0}^1, X_{t_0}^2)$ is uniform on $\{(i,j) : i+j = k\}$, so X^2 is unlikely to be near 0. Moreover in the short term after t_0, X^2 behaves as the simple *symmetric* walk (up-rate = down-rate = b) and so has no tendency to decrease to 0. So in the short term it is not very likely that X^2 reaches 0, thus justifying our approximation of Y.

B10 Example: Large density of heads in coin-tossing. Fix K, L with L large and $K/L = c > 1/2$, and $K - L/2$ large compared to $L^{1/2}$. So in L tosses of a fair coin we are unlikely to get as many as K heads. Now consider tossing a fair coin repeatedly; what is the number T of tosses required until we first see a block of L successive tosses containing K heads?

Let X_n record the results of tosses $(n-L+1,\ldots,n)$ and let Y_n be the number of heads in this block. Then T is the hitting time of Y on K. I claim that, around K, Y_n behaves like the asymmetric random walk with $\boldsymbol{P}(\text{up}) = (1-c)/2$, $\boldsymbol{P}(\text{down}) = c/2$. Then by (B2) the mean local sojourn time at K is

$$EC = \left(\frac{1}{2}c - \frac{1}{2}(1-c)\right)^{-1} = (c-\frac{1}{2})^{-1}.$$

Since $\boldsymbol{P}(Y_n = K) = \binom{L}{K}/(2^L)$, the heuristic gives

$$\begin{aligned} ET &\approx \frac{EC}{\boldsymbol{P}(Y_n = K)} \\ &\approx \left(c-\frac{1}{2}\right)^{-1} 2^L \Big/ \binom{L}{K}. \end{aligned} \tag{B10a}$$

To justify the claim, fix n_0 and condition on $Y_{n_0} = K$. Then there are exactly K heads in tosses $(n-L+1,\ldots,n)$ whose positions are distributed uniformly. In the short term, sampling without replacement is like sampling with replacement, so the results of tosses $n-L+1, n-L+2, \ldots,$ are like tosses of a coin with $\boldsymbol{P}(\text{heads}) = K/L = c$. Since tosses $n+1, n+2, \ldots$ are of a fair coin, our approximation for Y is now clear: $Y_{n_0+i} - Y_{n_0+i-1} = U_i - V_i$, where

$$\begin{aligned} \boldsymbol{P}(U_i = 1) &= \frac{1}{2}, \qquad & \boldsymbol{P}(U_1 = 0) &= \frac{1}{2}, \\ \boldsymbol{P}(V_i = 1) &\approx c, \qquad & \boldsymbol{P}(V_i = 0) &\approx 1 - c, \end{aligned}$$

the (U_i) are i.i.d. and the (V_i) are approximately i.i.d.

B11 Counter-example. It is time to give an example where the heuristic does not work. Consider simple symmetric random walk on the 1-dimensional integers mod$N$ (i.e. on the discrete N-point circle). Here the limiting distribution of $E_0 T_{[\frac{1}{2}N]}$, say, as $N \to \infty$ is not exponential (think of the Brownian approximation), so the heuristic B2 gives the wrong conclusion. Of course, any attempt to use the heuristic will fail, in that the natural approximating process is simple symmetric random walk on the whole line, for which $EC = \infty$. In a sense, one can regard the heuristic as saying

$$\frac{E_0 T_{[\frac{1}{2}N]}}{N} \to \infty \quad \text{as} \quad N \to \infty$$

which is correct, though not very precise.

B12 Hitting small subsets. We now study hitting times T_A for a set A of small cardinality (as well as small $\pi(A)$). For $i, j \in A$ let $E_i C_j$ be the mean local sojourn time in state j for the chain started at i. Let $(\lambda_i; i \in A)$ be the solutions of the equations

$$\sum_{i \in A} \lambda_i E_i C_j = \pi(j); \qquad j \in A. \tag{B12a}$$

Let $\lambda = \sum_{i \in A} \lambda_i$. In this setting, the heuristic B2 becomes

$$T_A \text{ has approximately exponential distribution, rate } \lambda; \tag{B12b}$$

$$\text{the hitting place distribution } X_{T_A} \text{ satisfies } \boldsymbol{P}(X_{T_A} = i) \approx \frac{\lambda_i}{\lambda}. \tag{B12c}$$

To see this, define λ_i as the rate of clumps which begin with a hit on i. Then $\lambda_i E_i C_j$ is the long-run rate of visits to j in clumps starting at i; so the left side of (B12a) is the total long-run rate of visits to j. This identifies (λ_i) as the solution of (B12a). So λ is the rate of clumps of visits to A. Then (B12b) follows from the heuristic B2, and (B12c) follows by regarding clumps started at different states $i \in A$ as occurring independently. Incidently, equation (B12a) has a matrix solution (Section B30), though it is not particularly useful in examples.

B13 Example: Patterns in coin-tossing, continued. In the setting of Example B5 we can consider a set A of patterns. Toss a fair coin until some pattern in A first occurs. One can ask about the number T_A of tosses required, and the probabilities of each pattern being the one first seen. For instance, consider pattern 2 (TTTH) and pattern 3 (THTH) of Example B5. Apply (B12b,B12c) to the Markov chain X_n which records the results of tosses $(n-3,\ldots,n)$. Counting "clumps" to consist only of patterns overlapping the initial pattern, we see

$$E_2C_2 = 1 \quad E_2C_3 = \tfrac{1}{4} \quad E_3C_2 = 0 \quad E_3C_3 = \tfrac{5}{4}.$$

Solving (B12a) gives $\lambda_2 = 5/80$, $\lambda_3 = 3/80$, $\lambda = 8/80$. So we conclude

1. $\boldsymbol{P}$(pattern 3 occurs before pattern 2) $\approx 3/8$
2. $T_A - 4$ has approximately exponential distribution, mean 10.

More sophisticated applications of technique B12 appear in chapter F. For the moment, let us just point out two simple special cases, and illustrate them with simple examples. First, suppose A is such that $E_iC_j \approx 0$ for $i \neq j \quad (i,j \in A)$. Then the solution of (B12a) is just $\lambda_i = \pi(i)/E_iC_i$. In other words, in this setting clumps of visits to A involve just one element of A, and we take these clumps to be independent for different elements.

B14 Example: Runs in biased die-throwing. Let $(Y_n; n \geq 1)$ be i.i.d. with some discrete distribution, and let $q_u = \boldsymbol{P}(Y_n = u)$. Fix k and let $T_k = \min\{\, n : Y_n = Y_{n-1} = \ldots = Y_{n-k+1}\,\}$. To apply the heuristic, let $X_n = (Y_{n-k+1},\ldots,Y_n)$ and let A be the set of $i_u = (u,u,\ldots,u)$. Then $\pi(i_u) = q_u^k$, $E_{i_u}C_{i_u} = (1-q_{i_u})^{-1}$ and $E_{i_u}C_{i_u} \approx 0$ for $u \neq u'$. So as above, $\lambda_{i_u} \approx \pi(i_u)/E_{i_u}C_{i_u} = (1-q_u)q_u^k$, and technique B12 gives

$$ET_k \approx \left(\sum_u (1-q_u)q_u^k\right)^{-1}. \qquad \text{(B14a)}$$

As a second case of technique B12, consider the special case where $\sum_{j\in A} E_iC_j$ does not depend on $i \in A$; call it EC. In this case, one can show from technique B12 that $\lambda = \pi(A)/EC$. Of course, this is just the special case in which the mean local sojourn time in A does not depend on the initial state $i \in A$, and the conclusion $ET_A \approx EC/\pi(A)$ is just the basic heuristic B2. However, (B12c) yields some more information: we find that λ_i is proportional to $\pi(i)$, and so the hitting distribution X_{T_A} is just the relative stationary distribution

$$\boldsymbol{P}(X_{T_A} = i) \approx \frac{\pi(i)}{\pi(A)}.$$

B15 Example: Random walk on Z^d mod N, continued. In the setting of Example B7, let A be an adjacent pair $\{i_0, i_1\}$ of lattice points. For the transient random walk, the mean sojourn time E_iC in A is the same for $i = i_0$ or i_1. And

$$E_{i_0}C \approx (1+q)R_d,$$

for R_d at (B6a) and $q = \boldsymbol{P}_{i_0}(\text{the transient walk ever hits } i_1)$. But by conditioning on the first step, $\boldsymbol{P}_{i_0}(\text{the transient walk ever returns to } i_0) = q$ also. So $R_d = (1-q)^{-1}$, and we find

$$E_{i_0}C = 2R_d - 1.$$

Since $\pi(A) = 2N^{-d}$, the heuristic gives

$$ET_A \approx \frac{EC}{\pi(A)} \approx \left(R_d - \frac{1}{2}\right)N^d. \tag{B15a}$$

Note that we can use the same idea in Example B8, to handle sparse random trapping sets $\mathcal{R}$ whose points are not well-separated.

B16 Hitting sizable subsets. When the set A has large cardinality ($\pi(A)$ still small), the method of estimating ET_A via technique B12 becomes less appealing; for if one cares to solve large matrix equations, one may as well solve the exact equations (B1b). To apply the heuristic (Section B2) directly, we need to be able to calculate $E_\rho C$, where C is the local sojourn time in A, and where ρ is the hitting place distribution X_{T_A} (from X_0 stationary, say). Since ρ may be difficult to determine, we are presented with two different problems. But there is an alternative method which presents us with only one problem. Define the *stationary exit distribution* μ_A from $A \subset J$ to be

$$\mu_A(j) = \sum_{i\in A} \pi(i)\frac{Q(i,j)}{Q(A,A^C)}; \qquad j \in A^C$$

where $Q(A,A^C) = \sum_{i\in A}\sum_{j\in A^C} \pi(i)Q(i,j)$. (This is in continuous time: in discrete time, replace Q by P and the results are the same.)

Define f_A as the probability that, starting with a distribution μ_A, the chain does not re-enter A in the short term. Then we get:

B17 The ergodic-exit form of the heuristic for Markov hitting times.

$$ET_A \approx (f_A Q(A,A^C))^{-1}.$$

This is an instance of the ergodic-exit form of the heuristic in 1-dimensional time. In the notation of Section A9, C^+ is the future local sojourn time in

A, given $X_0 \in A$. So for small δ,

$$\begin{aligned} \boldsymbol{P}(C^+ \leq \delta) &\approx \boldsymbol{P}\left(\begin{matrix}\text{chain exits } A \text{ before time } \delta \text{ and} \\ \text{does not re-enter in short term}\end{matrix} \,\middle|\, X_0 \in A\right) \\ &\approx f_A \boldsymbol{P}(X_\delta \in A^C \mid X_0 \in A) \\ &\approx \frac{f_A Q(A, A^C)\delta}{\pi(A)} \end{aligned}$$

and so $f^+(0) = f_A Q(A, A^C)/\pi(A)$. Then (A9c) gives clump rate $\lambda_A = f_A Q(A, A^C)$ and hence (B17).

When π is known explicitly we can calculate $Q(A, A^C)$, and so to apply (B17) we have only the one problem of estimating f_A.

B18 Example: A simple reliability model. Consider a system with K components. Suppose components fail and are repaired, independently for different components. Suppose component i fails at exponential rate a_i and is repaired at exponential rate b_i, where $(\max a_i)/(\min b_i)$ is small. Then the process evolves as a Markov chain whose states are subsets $B \subset \{1, 2, \ldots, k\}$ representing the set of failed components. There is some set $\mathcal{F}$ of subsets B which imply system failures, and we want to estimate the time $T_{\mathcal{F}}$ until system failure. Consider the hypothetical process which does not fail in $\mathcal{F}$; this has stationary distribution

$$\pi(B) = D^{-1}\Big(\prod_{i \in B^C} b_i\Big)\Big(\prod_{i \in B} a_i\Big); \qquad D = \prod_i (a_i + b_i)$$

using independence of components. And so

$$Q(\mathcal{F}, \mathcal{F}^C) = \sum_{B \in \mathcal{F}} \sum_{\substack{i \in B \\ B \setminus \{i\} \notin \mathcal{F}}} \pi(B) b_i.$$

The assumption that a_i/b_i is small implies that, in any state, repairs are likely to be finished before further failures occur. So $f_{\mathcal{F}} \approx 1$ and (B17) says

$$ET \approx (Q(\mathcal{F}, \mathcal{F}^C))^{-1}.$$

B19 Example: Timesharing computer. Users arrive at a free terminal, and alternate between working at the terminal (not using the computer's CPU) and sending jobs to the CPU (and waiting idly until the job is completed); eventually the user departs. The CPU divides its effort equally amongst the jobs in progress.

Let Y_t be the number of jobs in progress, and X_t the number of users working at terminals, so that $X_t + Y_t$ is the total number of users. A crude model is to take (X_t, Y_t) to be a Markov chain with transition rates

$$\begin{array}{lll} (i,j) \to (i+1,j) & \text{rate } a & \\ (i,j) \to (i-1,j) & \text{rate } bi & \\ (i,j) \to (i-1,j+1) & \text{rate } ci & (i \geq 1) \\ (i,j) \to (i+1,j-1) & \text{rate } d & (j \geq 1) \end{array}$$

What does this mean? New users arrive at rate a. Jobs take mean CPU time $1/d$, though of course each user will have to wait more real time for their job to be finished, since the CPU is sharing its effort. After a job is returned, a user spends mean time $1/(b+c)$ at the terminal, and then either submits another job (chance $c/(b+c)$) or leaves (chance $b/(b+c)$).

The stationary distribution (obtained from the detailed balance equations — see Section B28) is, provided $ac < bd$,

$$\pi(i,j) = \left(1 - \frac{ac}{bd}\right)\left(\frac{ac}{bd}\right)^j e^{-a/b}\frac{(a/b)^i}{i!}.$$

That is, at stationarity (X,Y) are independent, X is Poisson and Y is geometric. Think of a/b as moderately large — roughly, a/b would be the mean number of users if the computer worked instantaneously. Think of c/d as small — roughly, c/d is the average demand for CPU time per unit time per user. Thus ac/bd is roughly the average total demand for CPU time per unit time, and the stability condition $ac/bd < 1$ becomes more natural.

Suppose there are a total of K terminals, where K is somewhat larger than d/c. We shall estimate the time T_K until the total number of users $X_t + Y_t$ first reaches K. Let $A = \{\,(i,j) : i+j \geq K\,\}$. Then

$$\begin{aligned} Q(A, A^C) &= \sum_{i=1}^{K} bi\pi(i, K-i) \\ &\approx a\left(1 - \frac{ac}{bd}\right)\left(\frac{ac}{bd}\right)^{K-1}\exp\left(\frac{d}{c} - \frac{a}{b}\right) \qquad \text{(B19a)} \end{aligned}$$

after some algebra; and the exit distribution μ_A for (X,Y) has $X \stackrel{\mathcal{D}}{\approx}$ Poisson(d/c), $Y = K-1-X$. (The approximations here arise from putting $\boldsymbol{P}(\text{Poisson}(d/c) < K) \approx 1$). Now consider the process started with distribution μ_A. X_0 has mean $i_0 \approx d/c$, and the motion parallel to the line $i+j = K$ tends to push X_t towards i_0. So $X_t + Y_t$, the motion of the process non-parallel to that line, can be approximated by the random walk with rate a upward and rate bi_0 downward; comparing with (B2i), we estimate

$$f_A \approx 1 - \frac{a}{bi_0} \approx 1 - \frac{ac}{bd}. \qquad \text{(B19b)}$$

Inserting (B19a, B19b) into (B17) gives an estimate of ET_K:

$$ET_K \approx a^{-1}\left((1-\rho)^{-2}\rho^{1-K}\exp\left((\rho - 1)\frac{d}{c}\right)\right), \qquad \rho = \frac{ac}{bd} < 1. \quad \text{(B19c)}$$

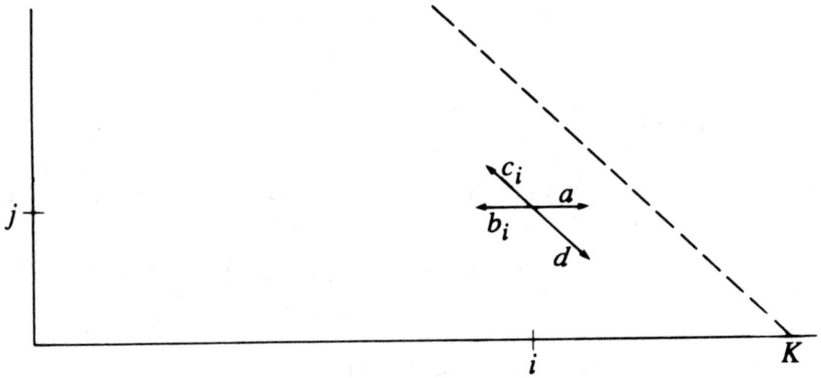

FIGURE B19a.

For our next example we need an expanded version of (B17). For $j \in A^C$ let $f_A(j)$ be the probability, starting at j, of not entering A in the short term. And let $Q(A,j) = \sum_{i \in A} \pi(i)Q(i,j)$. Then $f_A = \sum_{j \in A^C} \mu_A(j) f_A(j)$, and a little algebra shows (B17) is equivalent to

$$ET_A \approx \lambda^{-1}; \qquad \lambda = \sum_{j \in A^C} Q(A,j) f_A(j). \tag{B19d}$$

B20 Example: Two M/M/1 queues in series. Here we consider a different question than in Example B9. Let a be the arrival rate, and let b_1, b_2 be the service rates ($b_1, b_2 < a$). Fix K_1, K_2 such that $(a/b_u)^{K_u}$ is small ($u = 1, 2$) and consider

$$T = \min\{\, t : X_1(t) = K_1 \text{ or } X_2(t) = K_2 \,\}.$$

That is, imagine that queue u has capacity $K_u - 1$; then T is the first time a capacity is exceeded. Now $T = \min\{T_1, T_2\}$, where $T_u = \min\{\, t : X_u(t) = K_u \,\}$ has, by Example B3, approximately exponential (λ_u) distribution with

$$\lambda_u = \left(1 - \frac{a}{b_u}\right)^2 b_u \left(\frac{a}{b_u}\right)^{K_u}; \qquad u = 1,\, 2. \tag{B20a}$$

The heuristic implies T has approximately exponential(λ) distribution for some λ, which must satisfy

$$\max(\lambda_1, \lambda_2) < \lambda < \lambda_1 + \lambda_2, \tag{B20b}$$

the right-hand inequality indicating the positive correlation between T_1 and T_2. We need consider only the case where λ_1 and λ_2 are of similar orders of magnitude (else (B20b) says $\lambda \approx \max(\lambda_1, \lambda_2)$). We shall give an argument for the case $b_1 < b_2$ and b_1/b_2 not close to 1. In this case, the conclusion is

$$\lambda = \lambda_1 + \lambda_2 - \lambda_{12};$$
$$\lambda_{12} = \left(1 - \frac{a}{b_1}\right)\left(1 - \frac{a}{b_2}\right) b_2 \left(\frac{a}{b_1}\right)^{K_1} \left(\frac{b_1}{b_2}\right)^{K_2}. \tag{B20c}$$

We argue as follows. We want T_A for $A = \{ (i,j) : i \geq K_1 \text{ or } j \geq K_2 \}$. The exit distribution μ_A is concentrated on $B_1 \cup B_2$, for

$$\begin{aligned} B_1 &= \{ (K_1 - 1, j) : 1 \leq j < K_2 \}; \\ B_2 &= \{ (i, K_2 - 1) : 0 \leq i < K_1 \}. \end{aligned}$$

We use (B19d) to estimate λ. Consider first the contribution to λ from states in B_2. On B_2 the exit distribution is $(X_1, K_2 - 1)$ where X_1 has its stationary distribution. The chance of re-entering A across B_1 is therefore small and will be neglected. But then, considering only the possibility of re-entering A across B_2 is tantamount to considering queue 2 in isolation, and so the contribution to λ must be λ_2 to be consistent with the result for a single queue.

In considering exits onto B_1 we have to work. Write $\underset{\sim}{j}$ for the state $(K_1 - 1, j)$, $j \geq 1$. Then

$$\begin{aligned} Q(A, \underset{\sim}{j}) &= \left(1 - \frac{a}{b_1}\right)\left(\frac{a}{b_1}\right)^{K_1}\left(1 - \frac{a}{b_2}\right)\left(\frac{a}{b_2}\right)^{j-1} b_1 \\ f_A(\underset{\sim}{j}) &= \boldsymbol{P}(\Omega_1 \cap \Omega_2 \mid X_1(0) = K_1 - 1, X_2(0) = j), \end{aligned}$$

where Ω_u is the event that the chain does not enter A across B_u in the short term,

$$= \boldsymbol{P}_{\underset{\sim}{j}}(\Omega_1 \cap \Omega_2), \qquad \text{say.}$$

To calculate $\boldsymbol{P}_{\underset{\sim}{j}}(\Omega_1)$ we need only consider queue 1; approximating by simple asymmetric random walk and using (B2i),

$$\boldsymbol{P}_{\underset{\sim}{j}}(\Omega_1) \approx 1 - \frac{a}{b_1}.$$

To estimate $\boldsymbol{P}_{\underset{\sim}{j}}(\Omega_2)$, watch queue 2. Given $X_1(0) = K_1 - 1$ for large K_1, queue 2 starts out with arrival rate b_1. We want the chance, starting at j that queue 2 reaches K_2 in the short term. Approximating by the simple random walk with up-rate b_1 and down-rate b_2 gives

$$\boldsymbol{P}_{\underset{\sim}{j}}(\Omega_2) \approx 1 - \left(\frac{b_1}{b_2}\right)^{K_2 - j}.$$

The dependence between Ω_1 and Ω_2 is unclear. Noting that more arrivals make both events less likely, while faster service by server 1 makes Ω_1

more likely but Ω_2 less likely, it seems not grossly inaccurate to take them independent and put

$$f_A(\underset{\sim}{j}) \approx \boldsymbol{P}_{\underset{\sim}{j}}(\Omega_1)\boldsymbol{P}_{\underset{\sim}{j}}(\Omega_2).$$

We can now evaluate $\sum_{j \in B_1} Q(A, j) f_A(j)$, which after a little algebra becomes approximately $\lambda_1 - \lambda_{12}$. Thus (B19d) yields the estimate (B20c).

Remarks

1. If $b_1/b_2 \approx 1$ our estimate of $\boldsymbol{P}_{\underset{\sim}{j}}(\Omega_2)$ breaks down. For $b_1 = b_2$ and $K_1 = K_2$ we have $\lambda_1 = \lambda_2$ and our estimate (B20c) is $\lambda \approx \lambda_1$. Though this is probably asymptotically correct, since from properties of symmetric random walk one expects $\lambda = \lambda_1(1 + O(K^{-1/2}))$, it will be inaccurate for the practical range of K.

2. λ_{12} is the rate of "clumps" of visits to A which contain both visits to $\{X_1 \geq K_1\}$ and visits to $\{X_2 \geq K_2\}$. It is natural to try to estimate this directly; for $\lambda_{12}/\lambda_1 = \boldsymbol{P}_\rho(X_2(t) = K_2$ for some t in the short term), where ρ is the hitting distribution of (X_1, X_2) on A. The difficulty is estimating ρ.

B21 Another queueing example. We have treated queueing examples where there is a simple expression for the stationary distribution. Even where the stationary distribution is complicated, the heuristic can be used to relate hitting times to the stationary distribution, and thus reduce two problems to one. Here is a simple example.

Consider two queues, each with Poisson $(\frac{1}{2}a)$ arrivals, and one server with exponential (b) service rate; suppose the server works at one queue until it is empty and then switches to the other queue. Describe this process as $(X_1(t), X_2(t), Y(t))$, where X_i = length of queue i and $Y \in \{1, 2\}$ indicates the queue being served. Here the stationary distribution π is complicated, but we can easily use the heuristic to estimate, say, $T \equiv \min\{t : \max(X_1(t), X_2(t)) = K\}$ in terms of π. Indeed, $T = T_A$ for $A = \{\max(X_1, X_2) \geq K\}$. When the process exits A, with server serving queue 1 say, we must have $X_1 = K - 1$ and so we can use the usual random walk approximation (B2i) to get

$$f_A \equiv \boldsymbol{P}_{K-1}(X_1 \text{ does not hit } K \text{ in short term}) \approx 1 - \frac{a}{2b}.$$

So by (B17)

$$\begin{aligned} ET_A &\approx (f_A Q(A, A^C))^{-1} \\ &\approx \left(\pi(A)\left(b - \frac{a}{2}\right)\right)^{-1}. \end{aligned}$$

B22 Example: Random regular graphs. Here is another example in the spirit of (Examples B6–B8). Another well-behaved transient chain $\widehat{X}_n$ is simple symmetric random walk on the r-tree, that is on the infinite tree with r edges at each vertex, and where we take $r \geq 3$. It is easy to show that the mean number of visits to the initial vertex is

$$R_r = \frac{r-1}{r-2}. \tag{B22a}$$

Now consider the set of all graphs which are r-regular (i.e. have r edges at each vertex) and have N vertices (N large). Pick a graph at random from this set (see Bollobas (1985) for discussion of such random regular graphs), and let (X_n) be simple symmetric random walk on this graph. For distinct vertices i, j, the mean hitting time E_iT_j has (I assert)

$$E_iT_j \approx N\frac{r-1}{r-2}, \qquad \text{for most pairs } (i,j). \tag{B22b}$$

For the stationary distribution is uniform: $\pi(j) = 1/N$. And a property of random regular graphs is that, with probability $\to 1$ as $N \to \infty$, they look "locally" like the r-tree. So the chain X_n around j behaves like the transient walk $(\widehat{X}_n)$, and our heuristic says $E_iT_j \approx R_r/\pi(j)$, giving (B22b).

COMMENTARY

B23 General references. From the viewpoint of Markov chain theory as a whole, our topic of hitting times on rare sets is a very narrow topic. Thus, while there are several good introductory text on Markov chains, e.g., Isaacson and Madsen (1976), Hunter (1983), none of them really treat our topic. At a more advanced level, Keilson (1979) treats Markov chain models with applications to reliability in a way which partly overlaps with our treatment. Kelly (1979) describes many models in which the stationary distribution can be found explicitly using time-reversibility, and which are therefore amenable to study via our heuristic; this would be a good thesis topic.

B24 Limit theorems for hitting times on rare sets. It is remarkable that there are at least 4 different ways to study exponential limit theorems.

B24.1 The regeneration method. Successive excursions from a fixed state i_0 in a Markov chain are i.i.d. So the time T_A to hit a set A can be regarded as the sum of the lengths of a geometric number of excursions which do not hit A, plus a final part of an excursion which does not hit A. As $\pi(A)$ becomes small, the contribution from the geometric number of excursions becomes dominant, giving an exponential limit. This is formalized in the result below. Let T_i^+ denote first return time.

Proposition B24.1 *Let (X_n) be an irreducible positive-recurrent Markov chain with countable infinite state space J and stationary distribution π. Let (A_K) be decreasing subsets of J with $\bigcap_K A_K$ empty. Fix i_0 and let $t_K = \{\pi(i_0)\boldsymbol{P}_{i_0}(T_{A_K} < T_{i_0}^+)\}^{-1}$. Then for any fixed initial distribution,*

$$\frac{ET_{A_K}}{t_K} \to 1 \text{ as } K \to \infty;$$
$$\frac{T_{A_K}}{t_K} \xrightarrow{\mathcal{D}} \text{exponential}(1) \text{ as } K \to \infty.$$

This regeneration technique can be extended to prove similar results for Harris-recurrent Markov processes on general state spaces; the key fact is that such processes have a *distribution* which "regenerates". See Korolyuk and Sil'vestrov (1984); Cogburn (1985) for the exponential limit result; and Asmussen (1987) for the general regeneration idea.

B24.2 The small parameter method. For the second type of limit theorem, we fix the target set A and vary the process. Here is the simplest result of this type.

Proposition B24.2 *Let $\boldsymbol{P}_\epsilon$, $\epsilon > 0$ be transition matrices on a finite set J. Suppose $\boldsymbol{P}_\epsilon$ is irreducible for $\epsilon > 0$; $\boldsymbol{P}_0$ has an absorbing state i_0 and a transient class $J \setminus \{i_0\}$; and suppose $\boldsymbol{P}_\epsilon \to \boldsymbol{P}_0$ as $\epsilon \downarrow 0$. Fix $A \in J$, $i_0 \notin A$. Let T_ϵ be the first hitting time on A, starting at i_0 under $\boldsymbol{P}_\epsilon$. Let $t_\epsilon = (\boldsymbol{P}_{i_0}(T_\epsilon < T_{i_0}^+))^{-1}$. Then as $\epsilon \downarrow 0$,*

$$\frac{ET_\epsilon}{t_\epsilon} \to 1; \qquad \frac{T_\epsilon}{t_\epsilon} \xrightarrow{\mathcal{D}} \text{exponential}(1).$$

Such results have been studied in reliability theory, where one seeks limits as the ratio failure rate/repair rate tends to 0. Gertsbakh (1984) surveys such results. Extensions to the continuous time and space setting are more sophisticated and lead into large deviation theory.

B24.3 The mixing technique. Convergence to stationarity implies a "mixing" property, that events greatly separated in time should be roughly independent. One of several possible formalizations of the notion of "the time τ taken to approach stationarity" is given below. The result says that, for a set A which is sufficiently rare that its mean hitting time is large compared to τ, the hitting time distribution is approximately exponential.

Proposition B24.3 *For an irreducible continuous-time Markov chain with stationary distribution π define*

$$\tau = \min\{t : \sum_j |\boldsymbol{P}_i(X_t = j) - \pi(j)| \le e^{-1} \quad \textit{for all } i\}.$$

Then for any A,

$$\sup_{t\geq 0} |\boldsymbol{P}_\pi(T_A > t) - \exp(-t/E_\pi T_A)| \leq \psi\left(\frac{\tau}{ET_A}\right)$$

where $\psi(x) \to 0$ as $x \to 0$ is an absolute function, not depending on the chain.

See Aldous (1982; 1983b).

B24.4 The eigenvalue method. The transition matrix of a discrete-time chain killed on A has a largest eigenvalue $1 - \lambda$ for some $\lambda > 0$; in continuous time we get $-\lambda$ instead. A heuristic used in applications to the natural sciences is that, for rare sets A, the hitting distribution should be approximately exponential with rate λ. Chapter M gives more details.

B25 Remarks on formalizing the heuristic. The conclusions of our heuristic analyses of the examples could be formulated as limit assertions: as $K \to \infty$, $ET_K \sim$ some specified t_K and $T_K/t_K \xrightarrow{\mathcal{D}}$ exponential(1). In most cases, one can appeal to general theorems like those above to prove that T_K/ET_K does indeed converge to exponential. In fact, the regenerative method (section B24.1) yields this in the queueing examples (examples B3,B4,B9,B19,B20,B21); the small parameter method (section B24.2) in the reliability example (example B18); and the mixing technique (Section B24.3) in the doubly-stochastic and i.i.d. examples (Examples B5,B6,B7,B13,B22). Only for the random trapping example (Example B8) with general $\mathcal{R}$ is there any serious issue in proving asymptotic exponential distributions.

But in working the examples, our main concern was to derive a heuristic estimate t_K of ET_K. Proving $ET_K/t_K \to 1$ as $K \to \infty$ is harder. In fact, while numerous analytic methods for estimating mean hitting times have been developed in different contexts (see Kemperman (1961) for a classical treatment), these do not amount to a general theory of asymptotic mean hitting times. Proving $ET_K/t_K \to 1$ in our examples requires ad hoc techniques.

This raises the question of whether our heuristic method itself can be formalized. In the context of the mixing technique (Section B24.3) one could make a precise definition of "local sojourn time" C by cutting off at time τ; and then seek bounds on $|\pi(A)E_\pi T_A/EC - 1|$ analogous to that in (B24.3). The author has unpublished results of this type. But as yet they are not very useful in real examples since it is hard to pass from our heuristic idea of approximating by a transient process to the more formal "cut off at τ" definition.

B26 Notes on the examples.

B26.1 Card-shuffling. For specific methods of shuffling cards, it is of interest to estimate the size of parameters τ representing the number of shuffles needed to make the deck well-shuffled; see Aldous and Diaconis (1986; 1987) for surveys, in the more general context of random walks on finite groups. The specific problems of hitting times were treated probabilistically in Aldous (1983b) and analytically in Flatto et al. (1985).

B26.2 Random trapping. There is a large physics literature on this subject; Huber (1983) and den Hollander (1984) are places to start.

B26.3 Coin-tossing, etc.. There is a large literature on the first occurrence of patterns in coin-tossing and more generally in finite Markov chains; some recent papers are Li (1980), Gerber and Li (1981), Guibas and Odlyzka (1980), Blom and Thornburn (1982), Benveneto (1984), Gordon et al. (1986), Biggins and Cannings (1987). The "long runs" example (Example B14) is treated more abstractly in Anisimov and Chernyak (1982)

B26.4 Queuing examples. Anantharam (private communication) has done simulations with Example B20 and found our heuristic estimate to be quite accurate. It would be an interesting project to extend the ideas in Examples B9,B20 to more general Jackson networks, and to compare with other estimates. I do not know any survey article on rare events for queuing networks.

The heuristic can be used to approximate optimal buffer allocation: see Anantharam (1988) for related rigorous arguments.

Morrison (1986) gives a detailed treatment of a model related to our Example B19 (timesharing computer).

B27 The "recurrent potential" estimate of mean hitting times. For an irreducible aperiodic finite-state chain (X_n) with stationary distribution π, the limits

$$Z_{i,j} = \lim_{n\to\infty} \sum_{m=0}^{n} (\boldsymbol{P}_i(X_m = j) - \pi(j)) \tag{B27a}$$

exist: see e.g. Kemeny and Snell (1959), Hunter (1983). In terms of this "recurrent potential" Z there are some exact formulas for mean hitting times, for example

$$E_\pi T_j = \frac{Z_{j,j}}{\pi(j)}. \tag{B27b}$$

The exact expressions are calculable only in very special cases, but can be used as a starting point for justifying approximations. In the "mixing" setting of Section B24.3 one can argue that the sum in (B27a) up to $n = O(\tau)$ is close to its limit; so if $\tau\pi(j)$ is small,

$$Z_{j,j} \approx \sum_{m=0}^{O(\tau)} \boldsymbol{P}_j(X_m = j)$$

and the right side is essentially our mean clump size $E_j C_j$.

B28 Time-reversibility. A continuous-time Markov chain is *reversible* if the stationary distribution π satisfies the *detailed balance equations*

$$\pi(i)Q(i,j) = \pi(j)Q(j,i); \qquad \text{all } i \neq j.$$

This concept has both practical and theoretical consequences. The practical use is that for a reversible chain it is usually easy to find π explicitly; indeed, for complicated chains it is unusual to be able to get π explicitly without the presence of reversibility or some related special property. Kelly (1979) has many examples. Although reversibility is at first sight an "equilibrium" property, it has consequences for short-term behavior too. The special structure of reversible chains is discussed extensively in Keilson (1979).

B29 Jitter. At (B17) we gave one formulation of the ergodic-exit form of the heuristic for mean hitting times:

$$ET_A \approx \frac{1}{f_A Q(A, A^C)}.$$

A small variation of this method, using (A9f), gives

$$ET_A \approx \frac{EN}{Q(A, A^C)} \tag{B29a}$$

for $Q(A, A^C)$ as in (B17), and where N is the number of entries into A during a clump of visits to A. Keilson (1979) calls the $EN > 1$ phenomenon "jitter", and uses (B29a) to estimate mean hitting times in some queueing and reliability examples similar to ours. Clearly (B29a) is related to (B17) — crudely, because if after each exit from A there were chance f_A of not re-entering locally, then N would be geometric with mean $1/f_A$, although the exact connection is more subtle. But (B17) seems more widely useful than (B29a).

B30 First hitting place distribution. Equations (B12a), used for finding the hitting time on place for small A, can be "solved" as follows. Suppose we approximate X around A by a transient chain $\widehat{X}$ and estimate $E_i C_j$ as $Z_{i,j} = \widehat{E}_i(\text{total number of visits to } j)$. In vector-matrix notation, (B12a) is $\lambda Z = \pi$. But $Z = (I - R)^{-1}$, where $R_{i,j} = \widehat{\boldsymbol{P}}_i(X_S = j, S < \infty)$ for $S = \min\{n \geq 1 : \widehat{X}_n \in A\}$. And so $\lambda = \pi(I - R)$. Of course, such "solutions" are not very useful in practice, since it is not easy to calculate R.

B31 General initial distribution. Our basic heuristic is designed for stationary processes, so in estimating mean hitting times we are really estimating $E_\pi T_A$. We asserted at (B2iii) that $E_\mu T_A \approx E_\pi T_A$ for any initial distribution μ not close to A. To say this more sharply, from initial distribution μ there is some chance q_μ that the chain hits A in time $o(E_\pi T_A)$; given this does not happen, the distribution of T_A is approximately that obtained by starting with π. In other words,

$$\boldsymbol{P}_\mu(T_A > t) \approx (1-q_\mu)\exp(-t/E_\pi T_A) \quad \text{for } t \neq o(E_\pi T_A); \quad \text{(B31a)}$$
$$E_\mu T_A \approx (1-q_\mu)E_\pi T_A. \quad \text{(B31b)}$$

This can be formalized via limit theorems in the settings described in Section B24.

We can use the approximation above to refine our heuristic estimates. Consider the basic single server queue (Example B3), and consider $E_j T_K$ for $j < K$. Approximating by the random walk, $q_j \equiv \boldsymbol{P}_j(\text{hit } K \text{ in short term}) \approx (a/b)^{K-j}$. Then (B31b) and the previous estimate of $E_\pi T_K$ give

$$E_j T_K \approx b(b-a)^{-2}\left(\left(\frac{b}{a}\right)^K - \left(\frac{b}{a}\right)^j\right). \quad \text{(B31c)}$$

and our heuristic has picked up the second term of the exact expression.

B32 Compound Poisson approximation for sojourn times in 1-dimensional processes. For simple random walk Z in continuous time with up-rate a and down-rate $b > a$, the total sojourn time at 0 has an exponential distribution. So for 1-dimensional processes X whose behavior around high levels can be heuristically approximated by simple random walk, (A4f) says the sojourn time $\text{Leb}\{\, t : 0 \le t \le T, X(t) = x \,\}$ at a high level x is approximately compound Poisson where the compounded distribution is exponential. In the birth-and-death contest this is easy to formalize; see Berman (1986a). It is perhaps more natural to consider sojourn time $\{\, t : 0 \le t \le T, X(t) \ge x \,\}$ spent at or above a high level x. Here the natural compound Poisson approximation involves the sojourn time in $[0,\infty)$ for simple random walk:

$$C = \text{Leb}\{\, t \ge 0 : Z(t) \ge 0 \,\}.$$

This C is closely related to the busy period B in the M/M/1 queue: precisely, C is the sum of a geometric number of B's. From standard results about B (e.g., Asmussen (1987) III.9) one can obtain formulas for the transform and distribution of C; these are surely well-known, though I do not know an explicit reference.

C Extremes of Stationary Processes

Consider a stationary real-valued process $(X_n; n \geq 1)$ or $(X_t; t \geq 0)$. Time may be discrete or continuous; the marginal distribution may be discrete or continuous; the process may or may not be Markov. Let

$$M_n = \max_{1 \leq j \leq n} X_j; \quad M_t = \sup_{0 \leq s \leq t} X_s.$$

We shall study approximations to the distribution of M for large n, t. Note this is precisely equivalent to studying hitting times

$$T_b = \min\{\, t : X_t \geq b \,\}.$$

For $\boldsymbol{P}(M_t < b) = \boldsymbol{P}(T_b > t)$, at least under minor path-regularity assumptions in the continuous case. It turns out that the same ideas allow us to study boundary crossing problems, i.e.

$$T = \min\{\, t : X_t \geq b(t) \,\} \qquad \text{for prescribed } b(t).$$

It also turns out that many non-stationary processes can be made approximately or exactly stationary by deterministic space and time changes (e.g., Brownian motion can be transformed into the Ornstein-Uhlenbeck process), and hence we can study boundary-crossings for such processes also.

This is a large area. We divide it by deferring until Chapter D problems involving "locally Brownian" processes; i.e. diffusions, Gaussian processes similar to the Ornstein-Uhlenbeck process, and other processes where to do calculations we resort to approximating the process locally by Brownian motion with drift.

C1 Classical i.i.d. extreme value theory. Suppose $(X_i; i \geq 1)$ are i.i.d. Write

$$F(x) = \boldsymbol{P}(X_1 \leq x); \quad \overline{F}(x) = \boldsymbol{P}(X > x).$$

Of course we can write down the exact distribution of M_n:

$$\boldsymbol{P}(M_n \leq x) = F^n(x). \tag{C1a}$$

Seeking limit theorems for M_n is *prima facie* quite silly, since the purpose of limit theorems is to justify approximations, and we don't need approximations when we can write down a simple exact result. However, it turns out that the limit behavior of M_n for dependent sequences (where the exact distribution can't be written down easily) is often closely related to that for i.i.d. sequences. Thus we should say a little about the classical i.i.d. theory, even though (as will become clear) I regard it as a misleading approach to the real issues.

Classical theory seeks limit theorems of the form

$$\frac{M_n - c_n}{s_n} \xrightarrow{\mathcal{D}} \boldsymbol{\xi} \qquad \text{(non-degenerate)} \tag{C1b}$$

where c_n are centering constants and s_n are scaling constants. In freshman language, M_n will be around c_n, give or take an error of order s_n. Again, seeking limits of form (C1b) is *prima facie* rather silly: for *sums*, means and variances add, so linear renormalization is natural; for *maxima* there is no intrinsic linear structure and therefore no natural reason to consider linear rescalings, while the non-linear rescaling provided by the inverse distribution function reduces the general case to the trivial $U(0,1)$ case. It is merely fortuitous that many common distributions do admit limits of form (C1b). It turns out that only 3 essentially different limit distributions can occur: the *extreme value distributions*

$$\begin{array}{lll} \boldsymbol{\xi}_1^{(\alpha)}: \text{ support } (-\infty, 0), & \boldsymbol{P}(\boldsymbol{\xi}_1 \le x) = \exp(-(-x)^\alpha), & x < 0. \\ \boldsymbol{\xi}_2^{(\alpha)}: \text{ support } [0, \infty), & \boldsymbol{P}(\boldsymbol{\xi}_2 \le x) = \exp(-x^{-\alpha}), & x > 0. \\ \boldsymbol{\xi}_3: \text{ support } (-\infty, \infty), & \boldsymbol{P}(\boldsymbol{\xi}_3 \le x) = \exp(-e^{-x}). & \end{array} \tag{C1c}$$

where $0 < \alpha < \infty$.

The complete theory of which distributions F are "attracted" to which limit law is given in Galombos (1978); unlike central limit theory for sums, this involves only elementary but tedious real analysis. We shall merely record some examples to illustrate the qualitatively different types of behavior. Note that, since $F(x) = 1 - n\overline{F}(x)/n$, (C1a) implies

$$\boldsymbol{P}(M_n \le x) \approx \exp(-n\overline{F}(x)), \tag{C1d}$$

and then to prove (C1b) the only issue is to show

$$n\overline{F}(c_n + s_n y) \to -\log \boldsymbol{P}(\boldsymbol{\xi} \le y) \qquad \text{as } n \to \infty;\ y \text{ fixed.} \tag{C1e}$$

In concrete examples, such as the following, this is just easy calculus.

C2 Examples of maxima of i.i.d. sequences.

1. Take X_1 uniform on $[-1, 0]$. Then

$$nM_n \xrightarrow{\mathcal{D}} \boldsymbol{\xi}_1^{(1)};$$

so the centering is $c_n = 0$ and the scaling is $s_n = n^{-1}$.

2. If X is discrete and has some maximum possible value x_0, then

$$\boldsymbol{P}(M_n = x_0) \to 1.$$

3. If X_1 (discrete or continuous) has $F(x) \sim Ax^{-\alpha}$ as $x \to \infty$, then

$$\frac{M_n}{s_n} \xrightarrow{\mathcal{D}} \boldsymbol{\xi}_2^{(\alpha)}; \qquad s_n = (An)^{1/\alpha}.$$

Note here the scaling constants $s_n \to \infty$.

4. Take X_1 such that $\boldsymbol{P}(X_1 > x) \sim \exp(-x^{1/2})$. Then

$$\frac{M_n - c_n}{s_n} \xrightarrow{\mathcal{D}} \boldsymbol{\xi}_3; \qquad c_n = \log^2(n), \quad s_n = 2\log n.$$

Here the centering and scaling both $\to \infty$.

5. Take X_1 continuous and such that $\boldsymbol{P}(X_1 > x) \sim A\exp(-bx)$. Then

$$\frac{M_n - c_n}{s_n} \xrightarrow{\mathcal{D}} \boldsymbol{\xi}_3; \qquad c_n = b^{-1}\log(An), \quad s_n = b^{-1}.$$

Note here the scaling is constant.

6. Take X_1 integer-valued and such that $\boldsymbol{P}(X_1 > u) \sim A\rho^u$. Then the limit theorem for M_n is

$$\max_u |\boldsymbol{P}(M_n \le u) - \exp(-nA\rho^u)| \to 0 \quad \text{as } n \to \infty.$$

This limit theorem cannot be put into form (C1b). Let c_n be the integer closest to $\log(An)/\log(1/\rho)$; then $M_n - c_n$ is tight as $n \to \infty$, but discreteness forces "oscillatory" rather than convergent behavior.

7. Take X_1 continuous and such that $\boldsymbol{P}(X_1 > x) \sim A\exp(-q(x))$ for some polynomial q of degree ≥ 2. Then

$$\frac{M_n - c_n}{s_n} \xrightarrow{\mathcal{D}} \boldsymbol{\xi} \qquad \text{for some } c_n \to \infty,\ s_n \to 0.$$

In the particular case where X_1 has standard Normal distribution,

$$\begin{aligned} c_n &= (2\log n)^{\frac{1}{2}} - \frac{1}{2}(2\log n)^{-\frac{1}{2}}(\log(4\pi) + \log\log n) \\ s_n &= (2\log n)^{-\frac{1}{2}} \end{aligned}$$

will serve. Note here that $s_n \to 0$; contrary to what is familiar from the central limit theorem, the distribution of M_n becomes *less* spread out as n increases.

8. Take X_1 to have Poisson(θ) distribution: $\boldsymbol{P}(X_1 = u) = e^{-\theta}\theta^u/u! = p(u)$ say. Then M_n satisfies the limit theorem

$$\max_u \left|\boldsymbol{P}(M_n \le u) - \exp(n\overline{F}(u))\right| \to 0 \qquad \text{as } n \to \infty,$$

but this cannot be put into form (C1b). Let c_n be the integer $u > \theta$ for which $|\log(np(u))|$ is smallest; then $\boldsymbol{P}(M_n = c_n \text{ or } c_n + 1) \to 1$.

C3 The point process formulation. The analytic story above has a probabilistic counterpart which is more informative and which extends to dependent processes. Given a function $\phi(x) \ge 0$, we can define an associated "time-space" non-homogeneous Poisson process on $\boldsymbol{R}^2$ with intensity $\phi(x)$; that is, the chance of a point falling in $[t, t+dt] \times [x, x+dx]$ is $\phi(x)\,dt\,dx$. Let $\mathcal{N}_1$, $\mathcal{N}_2$, $\mathcal{N}_3$, be the Poisson processes associated with the following functions.

$$\begin{aligned}
\mathcal{N}_1 : \phi_1(x) &= \alpha(-x)^{\alpha-1}, & x<0 \quad (0<\alpha<\infty)\\
&= 0, & x>0 \\
\mathcal{N}_2 : \phi_2(x) &= \alpha x^{-\alpha-1}, & x>0 \quad (0<\alpha<\infty)\\
&= 0, & x<0 \\
\mathcal{N}_3 : \phi_3(c) &= e^{-x}.
\end{aligned}$$

Define the maximal processes

$$\boldsymbol{\xi}_u(t) = \max\{\, x : (s,x) \in \mathcal{N}_u \text{ for some } s \le t\,\}. \tag{C3a}$$

Then

$$\begin{aligned}
\boldsymbol{P}(\boldsymbol{\xi}_u(t) \le x) &= \boldsymbol{P}(\mathcal{N}_u \cap [0,t] \times (x,\infty) \text{ empty}) \\
&= \exp(-t\int_x^\infty \phi(y)\,dy) \\
&= \begin{cases} \exp(-t(-x)^\alpha) & \text{in case } u=1 \\ \exp(-tx^{-\alpha}) & \text{in case } u=2\,. \\ \exp(-te^{-x}) & \text{in case } u=3 \end{cases}
\end{aligned}$$

In particular, the $\boldsymbol{\xi}_u(1)$ have the extreme value distribution (C1c). Next, let $L_{c,s}\colon \boldsymbol{R} \to \boldsymbol{R}$ be the linear map

$$L_{c,s}(x) = c + sx \tag{C3b}$$

and let $L_{c,s}^{-1}$be the inverse map $x \to (x-c)/s$. These maps act on point processes in a natural way: $L_{c,s}(\mathcal{N})$ has a point at $(t, c+sx)$ whenever $\mathcal{N}$ has a point at (t,x). Similarly, let $\tau_n : [0,\infty) \to [0,\infty)$ map $t \to nt$; let τ_n^{-1} map $t \to t/n$, and let these maps act on point processes too.

Finally, note that any sequence $(X_i; i \ge 1)$ of random variables can be regarded as a point process $\mathcal{N}_X$ with points (i, X_i). We can now state the probabilistic version of (C1b).

Lemma C3.1 *Let (X_i) be i.i.d. Then*

$$(M_n - c_n)/s_n \xrightarrow{\mathcal{D}} \xi_u \text{ iff } \tau_n^{-1} L_{c_n,s_n}^{-1}(\mathcal{N}_X) \xrightarrow{\mathcal{D}} \mathcal{N}_u.$$

(There is a natural notion of convergence in distribution for point processes)

Informally, (C1b) says that $M_n \overset{\mathcal{D}}{\approx} c_n + s_n \xi_u$; whereas (C3.1) says that $N_X \overset{\mathcal{D}}{\approx} \tau_n L_{c_n,s_n}(\mathcal{N}_u)$. This is much more informative than (C1b), since for instance one can write down the distribution of positions and heights of the K highest points x of $\mathcal{N}_u$ in $[0,t] \times R$, and then (C3.1) gives the asymptotic distribution of the K largest values in $(X_1, \ldots, X_n)$ and their positions.

C4 The heuristic for dependent extrema. Consider now a stationary discrete or continuous time process (X_t). Assume an (informally stated) property of "no long-range dependence". That is, the value X_t may depend strongly on X_s for s near t, but for some large τ the value X_t does not depend much on $(X_s : |s-t| > \tau)$. This notion can be formalized in terms of mixing conditions. The theoretical literature tends to be dominated by the technical issues of formulating mixing conditions and verifying them in particular settings, thus disguising the fact that the conclusions are intuitively rather easy; establishing this last fact is our goal.

To say the heuristic, fix b such that $\overline{F}(b-) \equiv \boldsymbol{P}(X_t \geq b)$ is small and consider the random set $\mathcal{S}_b \equiv \{\, t : x_t \geq b \,\}$. Pretend $\mathcal{S}_b$ is a mosaic process with some clump rate λ_b and clump size C_b, related by the fundamental identity $\lambda_b EC_b = \boldsymbol{P}(t \in \mathcal{S}_b) = \overline{F}(b-)$. The three events "$M_t < b$", "$T_b > t$" and "$\mathcal{S}_b \cap [0,t]$ empty" are essentially the same, and the last has probability $\approx \exp(-t\lambda_b)$ by the Poisson property. Thus our heuristic approximation is

$$\boldsymbol{P}(M_t < b) = \boldsymbol{P}(T_b > t) \approx \exp(-t\lambda_b), \quad \text{where } \lambda_b = \frac{\overline{F}(b-)}{EC_b} = \frac{\boldsymbol{P}(X_t \geq b)}{EC_b}. \tag{C4a}$$

In working examples, we will merely estimate λ_b, leaving the reader to insert it into (C4a) and obtain the approximation to the distribution of M_t or T_b. It is possible to go from (C4a) to a limit assertion of the form $(M_t - c_t)/s_t \xrightarrow{\mathcal{D}} \xi_u$ but from a practical viewpoint this is rather pointless, since our aim is to get an approximation for the distribution of M_t and (C4a) does this directly.

C5 Autoregressive and moving average sequences. Let (Y_i) be i.i.d. and let (c_i) be constants. If $\sum_{i=0}^{\infty} c_i Y_i$ converges (in particular, if $EY_1 = 0$, $\text{var}(Y_1) < \infty$ and $\sum c_i^2 < \infty$), then the moving average process

$$X_n = \sum_{i=0}^{\infty} c_i Y_{n-i} \tag{C5a}$$

is a stationary process. In general (X_n) is not Markov, but the particular case

$$X_n = \sum_{i=0}^{\infty} \theta^i Y_{n-i} \qquad (|\theta| < 1) \tag{C5b}$$

is Markov; this is the autoregressive process

$$X_n = \theta X_{n-1} + Y_n. \tag{C5c}$$

It is not entirely trivial to determine the explicit distribution of X_1 from that of Y_1; we shall assume the distribution of X_1 is known. To use (C4a) to approximate the distribution of $M_n = \max(X_1, \ldots, X_n)$, the issue is to estimate EC_b. We treat several examples below.

C6 Example: Exponential tails. Suppose $\boldsymbol{P}(X_1 > x) \sim A_1 e^{-ax}$ as $x \to \infty$; under mild conditions on (c_i), this happens when $\boldsymbol{P}(Y_1 > y) \sim A_2 e^{-ay}$. We shall argue that for large b, $EC_b \approx 1$, implying that the maximum M_n behaves asymptotically as if the X's were i.i.d.; explicitly

$$\lambda_b \approx A_1 e^{-ab}; \qquad \boldsymbol{P}(M_n < b) \approx \exp(-nAe^{-ab}). \tag{C6a}$$

To show $EC_b \approx 1$, it will suffice to show

given $X_0 > b$, it is unlikely that any of $(X_1, X_2, \ldots)$ are $\geq b$ in the short term. (C6b)

Note that, writing $\widehat{X}_0$ for the distribution of $X_0 - b$ given $X_0 > b$, the exponential tail property implies that for large b,

$$\widehat{X}_0 \overset{\mathcal{D}}{\approx} \text{exponential}(a). \tag{C6c}$$

Consider first the autoregressive case (C5c). Write $(\widehat{X}_u; u \geq 0)$ for the conditional distribution of $(X_u - b)$ given $X_0 > b$. Then $(\widehat{X}_u)$ is distributed exactly like $(\theta^n b - b + X'_u)$, where (X'_u) is a copy of the autoregressive process (C5c) with $X'_u = \widehat{X}_0$. Since (X'_0) does not depend on b, while for $u \geq 1$ we have $\theta^n b - b \to -\infty$ as $b \to \infty$, it follows that for large b the process $(\widehat{X}_u; u \geq 1)$ will stay negative in the short term, establishing (C6b).

Consider now a more general finite moving average process (C5a) with $c_1 > c_2 > c_3 > \cdots > c_k > 0$. Writing $\theta_k = \max c_{i+1}/c_i$ we have

$$X_n \leq \theta_k X_{n-1} + Y_n,$$

and the argument above goes through to prove (C6b).

For more general moving averages these simple probabilistic arguments break down; instead, one resorts to analytic verification of (C7a) below.

C7 Approximate independence of tail values. The example above exhibits what turns out to be fairly common in discrete-time stationary sequences; the events $\{X_n > b\}$, $n \geq 0$, become approximately independent for large b and hence the maximum M_n is asymptotically like the maximum of i.i.d. variables with the same marginal distribution. The essential condition for this is

$$\boldsymbol{P}(X_u > b \mid X_0 > b) \to 0 \quad \text{as } b \to \infty; \qquad u \text{ fixed} \tag{C7a}$$

(more carefully, we need a "local sum" over small u of these conditional probabilities to tend to 0). Heuristically, this implies $EC_b \approx 1$ (because $E\widetilde{C}_b \approx 1$, in the notation of Section A6) and hence via (C4a)

$$\boldsymbol{P}(M_n < b) \approx \exp(-n\boldsymbol{P}(X_1 \geq b)). \tag{C7b}$$

There is no difficulty in formalizing this result under mixing hypotheses (Section C31). From our viewpoint, however, this is the "uninteresting" case where no clumping occurs; our subsequent examples focus on "interesting" cases where clumping is present. Note that this "approximate independence" property is strictly a discrete-time phenomenon, and has no parallel in continuous time (because the discrete lower bound $C \geq 1$ has no parallel in continuous time).

Returning to the discussion of autoregressive and moving average processes, in the setting of Section C5, let us mention two other cases.

C8 Example: Superexponential tails. If $\boldsymbol{P}(Y_1 > y) \to 0$ faster than exponentially (in particular, in the Gaussian case), then (C7a) holds and again the asymptotic maxima behave as if (X_n) were i.i.d. In fact, in the autoregressive case one can argue as in Example C6; here $\widehat{X}_0 \stackrel{\mathcal{D}}{\to} 0$ as $b \to \infty$.

C9 Example: Polynomial tails. Suppose $\boldsymbol{P}(Y_1 > y), \boldsymbol{P}(Y_1 < -y) \sim Ay^{-\alpha}$ as $y \to \infty$. The moving average (C5a) exists if $\sum c_i^\alpha < \infty$, and it can be shown that

$$\boldsymbol{P}(X_1 > x) \sim A\left(\sum c_i^\alpha\right) x^{-\alpha}. \tag{C9a}$$

The extremal behavior of (X_n) turns out to be very simple, but for a different reason than in the cases above. For (C9a) says

$$\boldsymbol{P}\left(\sum c_i Y_{n-i} > x\right) \approx \sum_i \boldsymbol{P}(c_i Y_{n-i} > x), \tag{C9b}$$

which leads to the important qualitative property: the sum $X_n = \sum c_i Y_{n-i}$ is large iff the maximal summand is large, and then $X_n \approx \max(c_i Y_{n-i})$.

Now write $c = \max c_i$, and fix b large. Then this qualitative property implies that each clump of times n such that $X_n > b$ is caused by a single value $Y_n > b/c$. Thus the clump rate λ_b for X is just

$$\lambda_b = \boldsymbol{P}(Y_n > b/c) \approx A(b/c)^{-\alpha}, \tag{C9c}$$

and so

$$\boldsymbol{P}(\max_{i \le n} X_i > b) \approx \boldsymbol{P}(\max_{i \le n} Y_i > b/c) \approx \exp(-n\lambda_b) \approx \exp(-nAc^\alpha b^{-\alpha}). \tag{C9d}$$

This argument does not use our clumping heuristic, but it is interesting to compare with a slightly longer argument which does. Suppose $c = c_0$ for simplicity. Condition on a clump of visits of X to $[b, \infty)$ starting at time n_0. The qualitative property and the polynomial tails imply that, conditionally,

$$X_{n_0} \approx cY_{n_0}; \quad cY_{n_0} \overset{\mathcal{D}}{\approx} bV \qquad \text{where } \boldsymbol{P}(V > v) = v^{-\alpha},\ v \ge 1. \tag{C9e}$$

The following terms X_{n_0+u} are dominated by the contribution from Y_{n_0}:

$$X_{n_0+u} \approx c_u Y_{n_0} \overset{\mathcal{D}}{\approx} \left(\frac{bc_u}{c}\right) V.$$

So the clump size C_b is approximately the number of $u \ge 0$ for which $(bc_u/c)V \ge b$, that is for which $V \ge c/c_u$. So

$$EC_b = \sum_{u \ge 0} \boldsymbol{P}(V \ge c/c_u) = c^{-\alpha} \sum_i c_i^\alpha. \tag{C9f}$$

Our heuristic (C4a) is

$$\begin{aligned} \lambda_b &= \frac{\boldsymbol{P}(X_1 \ge b)}{EC_b} \\ &\approx Ac^\alpha b^{-\alpha} \qquad \text{by (C9a) and (C9f),} \end{aligned}$$

and this recovers the same rate as the previous argument for (C9c).

C10 The heuristic for dependent extrema (continued). Returning to the discussion of Section C4, our concern is to obtain estimates of the rate λ_b of clumps of times n that $X_n \ge b$ in a discrete-time stationary process without long-range dependence. In cases where clumping does indeed occur (as opposed to (C7a) which implies it doesn't), the most useful form of the heuristic is the ergodic-exit method (Section A9), which gives

$$\lambda_b = f_b \boldsymbol{P}(X \ge b) \tag{C10a}$$

where f_b is the probability, given $X_0 \ge b$, that in the short term $X_1, X_2, \ldots$ are all $< b$. In the context of limit theorems we typically have $f_b \to f$

as $b \to \infty$. In this case, f is the *extremal index* of the process: see e.g. Leadbetter and Rootzen (1988). Its interpretation is that the maximum of the first n values of the process behaves like the maximum of fn i.i.d. variables.

The following setting provides a nice application.

C11 Additive Markov processes on $[0, \infty)$. Let (Y_n) be i.i.d. continuous random variables with $EY < 0$. Let (X_n) be a Markov process on state space $[0, \infty)$ such that

$$\boldsymbol{P}(X_{n+1} = x + Y_{n+1} \mid X_n = x) \to 1 \qquad \text{as } x \to \infty.$$

More precisely, let (X_n) have transition kernel $\boldsymbol{P}^*(x, A)$ such that

$$\sup_{A \subset [0,\infty)} |\boldsymbol{P}^*(x, A) - \boldsymbol{P}(x + Y \in A)| \to 0 \qquad \text{as } x \to \infty.$$

Call such processes *additive Markov.* In particular, this happens if

$$X_{n+1} = X_n + Y_{n+1} \text{ whenever } X_n + Y_{n+1} \geq 0.$$

Informally, X_n evolves as sums of i.i.d. variables when it is away from 0, but has some different "boundary behavior" around 0. Under weak conditions on the boundary behavior, a stationary distribution for X exists.

Here are some examples.

C11.1 Waiting times in a G/G/1 queue. Consider a G/G/1 queue with U_n = time between arrivals of $(n-1)$'st customer and n'th customer; V_n = service time of n'th customer. Let X_n = waiting time of n'th customer. Then

$$X_{n+1} = (X_n + V_n - U_{n+1})^+.$$

This is an additive process with $Y = V - U$.

C11.2 Sojourn times in a G/G/1 queue. In the setting above, let $\widehat{X}_n$ be the total (waiting + service) time customer n spends in the system. Then

$$\widehat{X}_{n+1} = (\widehat{X}_n - U_{n+1})^+ + V_{n+1}.$$

Again this is an additive process with $Y = V - U$; but the boundary behavior is different from the previous example.

C11.3 Storage/dam models. A simple model for a reservoir of capacity b assumes the inflow U_n and demand V_n in the n'th period are such that (U_n) and (V_n) are i.i.d., $EU > EV$. Let X_n be the unused capacity at the end of the n'th period ($X_n = 0$ indicates the reservoir is full, $X_n = b$ indicates it is empty). Then

$$X_{n+1} = X_n - U_{n+1} + V_{n+1}$$

except for boundary conditions at 0 and b. If we are only interested in the time T_b until the reservoir becomes empty, then the boundary conditions at b are irrelevant and can be removed, giving an additive process.

C11.4 A special construction. Suppose Y is such that there exists some $\theta > 0$ with

$$E\exp(\theta Y) = 1. \qquad \text{(C11a)}$$

Then there exists a prescription of boundary behavior which makes the stationary distribution of X exactly exponential(θ). For if Z has exponential(θ) distribution, then $Z+Y$ has density $g(x)$ such that $g(x) \le \theta e^{-\theta x}$ on $x \ge 0$. Normalize $h(x) = \theta e^{-\theta x} - g(x)$, $x \ge 0$, to make it the density of some distribution μ. Then

$$\boldsymbol{P}^*(x, A) = \boldsymbol{P}(x+Y \in A) + \mu(A) \cdot \boldsymbol{P}(x+Y < 0)$$

is the transition kernel of an additive process with exponential(θ) stationary distribution.

C11.5 Stationary distributions. For a general additive process the exact stationary distribution is complicated. But it turns out (see Section C33) that under condition (C11a) the tail of the stationary distribution X_0 is of the form

$$\boldsymbol{P}(X_0 > x) \sim D\exp(-\theta x). \qquad \text{(C11b)}$$

The constant D depends on the boundary behavior, but θ depends only on the distribution of Y.

C11.6 The heuristic analysis. We can now use the heuristic for extrema of the stationary process (X_n). Define

$$M = \max_{n \ge 0}\left(\sum_{i=1}^{n} Y_i\right) \ge 0, \qquad \text{interpreting the "}n=0\text{" sum as 0.} \quad \text{(C11c)}$$

Fix b large and condition on $X_0 \ge b$. Then $Z = X_0 - b$ has approximately exponential(θ) distribution, by (C11b). And $(X_n; n \ge 0)$ behaves locally like $(b + Z + \sum_{i=1}^n Y_i; n \ge 0)$. Thus in (C10a) we estimate

$$\begin{aligned} f_b = f &= \boldsymbol{P}(Z + \sum_{i=1}^{n} Y_i < 0 \text{ for all } n \ge 1) \\ &= \boldsymbol{P}(Z + Y + M < 0) \quad \text{by separating the } Y_1 \text{ term} \qquad \text{(C11d)} \end{aligned}$$

In (C11d) we take (Z, Y, M) independent with $Z \stackrel{\mathcal{D}}{=}$ exponential(θ), Y the additive distribution and M as at (C11c). Then (C10a) gives

$$\lambda_b = fD\exp(-\theta b), \tag{C11e}$$

and as usual we can substitute into (C4a) to get approximations for $\max(X_1, \ldots, X_n)$ or T_b.

Note that f, like θ, depends only on Y and not on the boundary behavior. Although (C11e) is less explicit than one would like, it seems the best one can do in general. One can estimate f numerically; for some boundary conditions there are analytic expressions for D, while one could always estimate D by simulation.

C12 Continuous time processes: the smooth case. We now start to study extremes of continuous-time stationary processes $(X_t; t \geq 0)$. Here the notion of "X_t i.i.d. as t varies" is not sensible. Instead, the simplest setting is where the process has smooth sample paths. So suppose

$$\text{the velocity } V_t = dX_t/dt \text{ exists and is continuous;} \tag{C12a}$$

$$(X_t, V_t) \text{ has a joint density } f(x, v). \tag{C12b}$$

Fix a level x. Every time the process hits x it has some velocity V; by (C12b) we can neglect the possibility $V = 0$, and assume that every hit on x is part of an *upcrossing* ($V > 0$) or a *downcrossing* ($V < 0$). Define

$$\rho_x = \text{rate of upcrossings of level } x \tag{C12c}$$

$$g_x(v) = \text{density of } V \text{ at upcrossings of level } x. \tag{C12d}$$

More precisely, let $V_1, V_2, \ldots$ be the velocities at successive upcrossings of x; then $g_x(v)$ is the density of the limiting empirical distribution of (V_i). This is *not* the same as the distribution of (X_t, V_t) given $X_t = x$ and $V_t > 0$. In fact the relation is given by

Lemma C12.1 *Under conditions (C12a-C12d) above,*

$$f(x, v) = \rho_x v^{-1} g_x(v). \tag{C12e}$$

In particular, (C12e) implies

$$\rho_x = \int_0^\infty v f(x, v)\, dv = E(V_t^+ \mid X_t = x) f_X(x) \tag{C12f}$$

$$g_x(v) = \rho_x^{-1} v f(x, v). \tag{C12g}$$

These are exact, not approximations; (C12f) is the classical *Rice's formula.* The nicest proof uses the ergodic argument. An upcrossing with velocity v spends time $v^{-1}\,dx$ in $[x, x+dx]$. So associated with each upcrossing is a mean time $(g_x(v)\,dv)(v^{-1}\,dx)$ for which $X \in [x, x+dx]$ and $V \in [v, v+dv]$. So the long-run proportion of time for which $X \in [x, x+dx]$ and $V \in [v, v+dv]$ is $\rho_x(g_x(v)\,dv)(v^{-1}\,dx)$. But this long-run proportion is $f(x,v)\,dx\,dv$ by ergodicity.

There is an alternative, purely "local", argument — see Section C34.

C12.1 The heuristic for smooth processes. For a process (X_t) as above, the heuristic takes the form: for b large,

$$\boldsymbol{P}(M_t \leq b) = \boldsymbol{P}(T_b \geq t) \approx \exp(-t\rho_b). \tag{C12h}$$

To see this, note that each clump of time t that $X_t \geq b$ consists of a number N_b of nearby intervals, and then (A9f) says the clump rate λ_b is related to the upcrossing rate ρ_b by

$$\lambda_b = \frac{\rho_b}{EN_b}.$$

For smooth processes one invariably finds $EN_b \approx 1$ for b large, so $\lambda_b \approx \rho_b$, and then the usual assumption of no long-range dependence leads to the exponential form of (C12h).

This heuristic use (C12h) of Rice's formula is standard in engineering applications. The simplest setting is for Gaussian processes (Section C23) and for "response" models like the following.

C13 Example: System response to external shocks. Consider a response function $h(t) \geq 0$, where $h(0) = 0$, $h(t) \to 0$ rapidly as $t \to \infty$, and h is smooth. Suppose that a "shock" at time t_0 causes a response $h(t - t_0)$ at $t \geq t_0$, so that shocks at random times τ_i cause total response $X_t = \sum_{\tau_i \leq t} h(t - \tau_i)$, and suppose we are interested in the maximum M_t. If the shocks occur as a Poisson process then X_t is stationary and

$$(X_0, V_0) \stackrel{\mathcal{D}}{=} \Big(\sum_{\tau_i > 0} h(\tau_i), \sum_{\tau_i > 0} h'(\tau_i)\Big).$$

In principle we can calculate ρ_b from (C12f) and apply (C12h).

Curiously, Rice's formula seems comparatively unknown to theoreticians, although it is often useful for obtaining bounds needed for technical purposes. For it gives a rigorous bound for a smooth continuous-time stationary process:

$$\boldsymbol{P}(\max_{s \leq t} X_s \geq b) \leq \boldsymbol{P}(X_0 \geq b) + t\rho_b. \tag{C13a}$$

Because

$$\begin{aligned} \boldsymbol{P}(\max_{s \le t} X_s \ge b) - \boldsymbol{P}(X_0 \ge b) &= \boldsymbol{P}(U \ge 1), \\ &\text{where } U \text{ is the number of upcrossings over } b \text{ during } [0,t]; \\ &\le EU \\ &= t\ \rho_b. \end{aligned}$$

Indeed, the result extends to the non-stationary case, defining $\rho_b(t)$ as at (C12f) using the density f_t of (X_t, V_t):

$$\boldsymbol{P}(\max_{s \le t} X_s \ge b) \le \boldsymbol{P}(X_0 \ge b) + \int_0^t \rho_b(s)\, ds. \qquad \text{(C13b)}$$

C14 Example: Uniform distribution of the Poisson process. A standard type of application of probability theory to pure mathematics is to prove the existence of objects with specified properties, in settings where it is hard to explicitly exhibit any such object. A classical example is Borel's normal number theorem. Here is a related example. For real $x, t > 0$ let $x \bmod t$ be the y such that $x = jt + y$ for some integer j and $0 \le y < t$. Call a sequence $x_n \to \infty$ *uniformly distributed* $\bmod t$ if as $n \to \infty$ the empirical distribution of $\{x_1 \bmod t, \ldots, x_n \bmod t\}$ converges to the uniform distribution on $[0, t)$. Call (x_n) *uniformly distributed* if it is uniformly distributed $\bmod t$ for all $t > 0$. It is not clear how to write down explicitly some uniformly distributed sequence. But consider the times (τ_i) of a Poisson process of rate 1; we shall sketch a (rigorous) proof that (τ_i) is, with probability 1, uniformly distributed.

Let $\mathcal{H}$ be the set of smooth, period 1 functions $h\colon \boldsymbol{R} \to \boldsymbol{R}$ such that $\int_0^1 h(u)\,du = 0$. Let $h_t(u) = h(u/t)$. The issue is to show that, for fixed $h \in \mathcal{H}$,

$$\sup_{1 \le t \le 2} n^{-1} \sum_{i=1}^{n} h_t(\tau_i) \to 0 \quad \text{a.s.} \quad \text{as } n \to \infty. \qquad \text{(C14a)}$$

For then the same argument, with $[1, 2]$ replaced by $[\delta, 1/\delta]$, shows

$$\boldsymbol{P}\left(n^{-1} \sum_{i=1}^{n} h_t(\tau_i) \to 0 \text{ for all } t\right) = 1;$$

extending to a countable dense subset of $\mathcal{H}$ establishes the result. To prove (C14a), first fix t. The process $\tau_i \bmod t$ is a discrete-time Markov process on $[0, t)$. Large deviation theory for such processes implies

$$\boldsymbol{P}\left(n^{-1} \sum_{i=1}^{n} h_t(\tau_i) > \epsilon\right) \le A \exp(-Bn); \qquad \text{all } n \ge 1 \qquad \text{(C14b)}$$

where the constants $A < \infty$, $B > 0$, depend on h, ϵ and t; but on $1 \leq t \leq 2$ we can take the constants uniform in t. Now fix n and consider

$$X^n(t) = n^{-1} \sum_{i=1}^{n} h_t(\tau_i); \qquad 1 \leq t \leq 2$$

as a random continuous-time process. We want to apply (C13b). Since $\tau_n / n \to 1$ a.s., we can assume $\tau_n \leq 2n$. Then

$$V^n(t) = \frac{d}{dt} X^n(t) \leq 2Cn; \qquad C = \sup h'(u) < \infty.$$

Applying (C12f) and (C13b) gives $\rho_b \leq 2Cnf_{X_t}(b)$ and

$$\boldsymbol{P}(\max_{1 \leq t \leq 2} X^n(t) \geq b) \leq \boldsymbol{P}(X^n(1) \geq b) + 2Cn \int_1^2 f_{X_t^n}(b)\, dt.$$

Integrating b over $[\epsilon, 2\epsilon)$ gives

$$\epsilon \boldsymbol{P}(\max_{1 \leq t \leq 2} X^n(t) \geq 2\epsilon) \leq \boldsymbol{P}(X^n(1) \geq \epsilon) + 2Cn \int_1^2 \boldsymbol{P}(X^n(t) \geq \epsilon)\, dt.$$

Now (C14b) implies (C14a).

C15 Drift-jump processes. Above, we studied continuous-time processes with smooth paths. In Chapter B we saw some continuous-time integer-valued Markov chains, which moved only by jumping. In this section we consider another class of continuous-time processes, which move both continuously and with jumps.

Let (ξ_t) be a compound Poisson counting process. So $\xi_t = \sum_{n \leq N_t} Y_n$, where (Y_n) are i.i.d. with some distribution Y, and N_t is a Poisson process of some rate ρ. Let $r(x)$ be a continuous function. We can define a Markov process X_t by

$$dX_t = -r(X_t)\, dt + d\xi_t. \tag{C15a}$$

In other words, given $X_t = x$ we have $X_{t+dt} = x - r(x)\, dt + Y_\eta$, where $\boldsymbol{P}(\eta = 1) = \rho\, dt$ and $\boldsymbol{P}(\eta = 0) = 1 - \rho\, dt$. Under mild conditions, a stationary distribution exists. The special case $r(x) = ax$ is the continuous-time analog of the autoregressive sequence (C5c); the special case $r(x) = a$ is the analog of the additive sequences (Section C11), at least when X_t is constrained to be non-negative.

The special cases, and to a lesser extent the general case, admit explicit but complicated expressions for the stationary distribution and mean hitting times (Section C35). Our heuristic makes clear the asymptotic relation between these quantities: here are some examples. Let

$f(x)$ be the density of the stationary distribution X_0; (C15b)

$$M_a = \sup_{t \geq 0} (\xi_t - at) \tag{C15c}$$

$$g(a, \xi) = \boldsymbol{P}(M_a = 0). \tag{C15d}$$

C16 Example: Positive, additive processes. Here we consider (C15a) with $r(x) = a$ and $Y > 0$. Consider clumps C of time spent in $[b, \infty)$, for large b. Clumps end with a continuous downcrossing of b which is not followed by any jump upcrossing of b in the short term. The rate of downcrossings is $af(b)$; the chance a downcrossing is not followed by an upcrossing is approximatedly $g(a, \xi)$; hence the primitive form (Section A9) of the ergodic-exit estimate of clump rate is

$$\lambda_b = af(b)g(a, \xi). \qquad \text{(C16a)}$$

C17 Example: Signed additive processes. If in the example above we allow Y to be negative, the argument above is inapplicable because clumps may end with a jump downcrossing of b. But let us consider the setting analogous to Section C11; suppose there exists $\theta > 0$ such that

$$E\exp(\theta(\rho Y - a)) = 1. \qquad \text{(C17a)}$$

Then as in Section C11 we expect

$$\boldsymbol{P}(X_0 > b) \sim D\exp(-\theta b) \qquad \text{as } b \to \infty. \qquad \text{(C17b)}$$

The rate of jump downcrossing of b is $\rho \boldsymbol{P}(X_0 \geq b, X_0 + Y < b)$. A downcrossing to $X_0 + Y$ causes a clump end if $X_0 + Y + M_a < b$. Thus the rate λ_b^J of clump ends caused by jump downcrossing is

$$\lambda_b^J = \rho \boldsymbol{P}(X_0 \geq b, X_0 + Y + M_a < b).$$

Writing Z for an exponential(θ) variable, and using (C17b) and its implication that $(X_0 - b \mid X_0 \geq b) \overset{\mathcal{D}}{\approx} Z$, gives

$$\lambda_b^J = \rho D\exp(-\theta b)\boldsymbol{P}(Z + Y + M_a < 0).$$

Adding the expression (C16a) for the rate λ_b^c of clump ends caused by continuous downcrossings, we find

$$\lambda_b = De^{-\theta b}\left(a\theta g(a, \xi) + \rho \boldsymbol{P}(Z + Y + M_a < 0)\right). \qquad \text{(C17c)}$$

C18 Positive, general drift processes. Consider now the case of Section C15 where $Y > 0$ and $r(x) > 0$. Around b, the process X_t can be approximated by the process with constant drift $-a = r(b)$, and then the argument for (C16a) gives

$$\lambda_b = r(b)f(b)g(r(b), \xi). \qquad \text{(C18a)}$$

Now consider the case where $r(x) \to \infty$ as $x \to \infty$. For large a, there is a natural approximation for $g(a, \xi)$ which considers only the first jump of ξ;

$$g(a, \xi) \approx \boldsymbol{P}(Y_i \leq a\tau), \qquad \text{where } \tau \overset{\mathcal{D}}{=} \text{exponential}(\rho)$$

$$\begin{aligned} &= E\exp(-\rho Y/a) \\ &\approx 1 - \frac{\rho EY}{a}. \end{aligned}$$

Thus in the case where $r(x) \to \infty$ as $x \to \infty$, (C18a) becomes: for large b,

$$\begin{aligned} \lambda_b &\approx f(b)(r(b) - \rho EY) && \text{(C18b)} \\ &\approx f(b)r(b), \quad \text{to first order.} && \text{(C18c)} \end{aligned}$$

For a simple explicit example, consider the autoregressive case $r(x) = ax$ and take Y to have exponential(β) distribution. Then the stationary density works out as

$$f(x) = \frac{\beta^{\rho/a} e^{-\beta x} x^{\rho/a - 1}}{(\rho/a - 1)!}.$$

By rescaling space and time, we can take $\rho = \beta = 1$. Then our estimate of the mean first hitting time T_b is

$$\begin{aligned} ET_b &\approx \lambda_b^{-1} \\ &\approx (f(b)(ab-1))^{-1} \qquad \text{using (C18b)} \\ &\approx (a^{-1})!\left(1 + \frac{1}{ab}\right) b^{-1/a} e^b. && \text{(C18d)} \end{aligned}$$

It turns out (Section C35) that this approximation picks up the first two terms of the exact asymptotic expansion of ET_b as $b \to \infty$.

C19 Autoregressive symmetric stable process. In Section C15 we can replace the compound Poisson process (ξ_t) by a more general "pure jump" process with stationary independent increments. A natural example is the symmetric stable process (ξ_t) of exponent $1 < \alpha < 2$:

$$E\exp(i\theta\xi_t) = \exp(-t|\theta|^\alpha).$$

This has the properties

$$\begin{aligned} \xi_t &\stackrel{\mathcal{D}}{=} t^{1/\alpha}\xi_1; \\ \boldsymbol{P}(\xi_1 > x) &\sim K_\alpha x^{-\alpha} \text{ as } x \to \infty; && \text{(C19a)} \\ K_\alpha &= (\alpha - 1)!\pi^{-1}\sin(\alpha\pi/2) \end{aligned}$$

Consider the autoregressive case $r(x) = ax$ of Section C15; that is

$$X_t = \int_0^\infty e^{-as}\, d\xi_{t-s}.$$

Then the stationary distribution X_0 is also symmetric stable:

$$X_0 = (\alpha a)^{-1/\alpha}\xi_1. \qquad \text{(C19b)}$$

The stationary autoregressive process (X_t) here is the continuous-time analog of Example C9; the arguments at (C9), modified appropriately, give

$$\lambda_b \approx \boldsymbol{P}(\xi_1 \geq b) \approx K_\alpha b^{-\alpha}. \tag{C19c}$$

Here is an amusing, less standard example.

C20 Example: The I5 problem. In driving a long way on the freeway, what is the longest stretch of open road you will see in front of you? To make a model, suppose vehicles pass a starting point at the times of a Poisson(α) process, and have i.i.d. speeds V with density $f_V(v)$. You drive at speed v_0. Let X_t be the distance between you and the nearest vehicle in front at time t; we want an approximation for $M_t = \max_{s \leq t} X_s$.

The first observation is that, at a fixed time, the positions of vehicles form a Poisson process of rate β, where

$$\beta = \alpha E(1/V). \tag{C20a}$$

The rate calculation goes as follows. Let N_L be the number of vehicles in the spatial interval $[0, L]$. A vehicle with speed v is in that interval iff it passed the start within the previous time L/v. The entry rate of vehicles with speeds $[v, v + dv]$ is $\alpha f(v)\,dv$, so

$$EN_L = \int_0^\infty \frac{L}{v}\alpha f(v)\,dv.$$

This gives formula (C20a), for the rate β. The Poisson property now implies

$$\boldsymbol{P}(X_t > x) = \exp(-\beta x). \tag{C20b}$$

Next, consider the diagram below which shows the trajectories of the other vehicles relative to you. Consider the "pass times", that is the times that a faster vehicle (shown by an upward sloping line) passes you, or a slower vehicle (downward sloping line) is passed by you. The pass times form a Poisson process of rate

$$\widehat{\alpha} = \alpha E\left|1 - \frac{v_0}{V}\right|. \tag{C20c}$$

To argue this, suppose you start at time 0. Consider a vehicle with speed $v > v_0$. It will pass you during time $[0, t]$ iff it starts during time $[0, t - tv_0/v]$. The entry rate of vehicles with speeds $[v, v + dv]$ is $\alpha f(v)\,dv$, so the mean number of vehicles which pass you during time $[0, t]$ is $\int_{v_0}^\infty t(1 - v_0/v)\alpha f(v)\,dv$. A similar argument works for slower vehicles, giving (C20c).

Fix $b \geq 0$. Let C_b have the distribution of the length of time intervals during which $X > b$ (the thick lines, in the diagram). As at Section A9

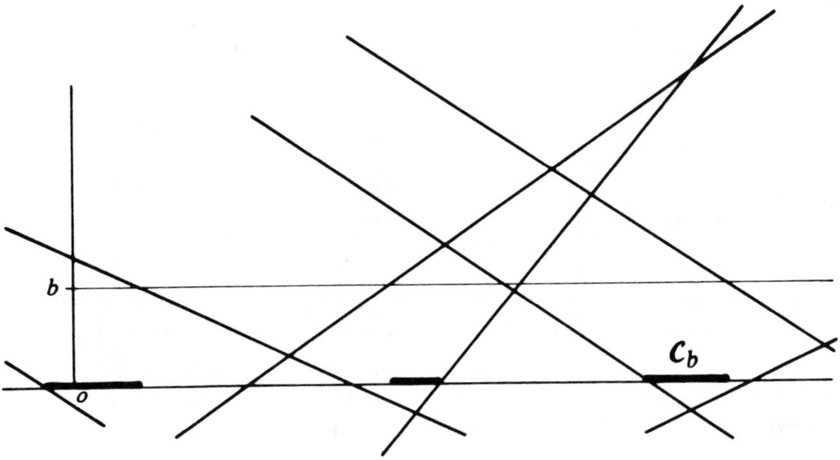

FIGURE C20a.

let C_b^+ be the distribution of the future time interval that $X > b$, given $X_0 > b$; that is,

$$C_b^+ = \min\{\, t > 0 : X_t \leq b \,\} \qquad \text{given } X_0 > b.$$

The key fact is

$$\text{the distribution } C_b^+ \text{ does not depend on } b. \tag{C20d}$$

For an interval $\{\, t : X_t > b \,\}$ ends when either a line upcrosses 0 or a line downcrosses b; and these occur as independent Poisson processes whose rate does not depend on b, by spatial homogeneity.

From (C20c) we know that C_0 and C_0^+ have exponential($\widehat{\alpha}$) distributions, so (C20d) implies C_b^+ and hence C_b have exponential($\widehat{\alpha}$) distribution, and

$$EC_b = \frac{1}{\widehat{\alpha}}. \tag{C20e}$$

So far we have exact results. To apply the heuristic, it is clear that for b large the clumps of time that $X_t > b$ tend to be single intervals, so we can identify the clump size with C_b. So applying directly our heuristic fundamental identity gives the clump rate λ_b as

$$\lambda_b = \frac{\boldsymbol{P}(X_0 > b)}{EC_b} = \widehat{\alpha}\exp(-\beta b) \tag{C20f}$$

and then $\boldsymbol{P}(M_t \leq b) \approx \exp(-t\lambda_b)$ as usual.

In the original story, it is more natural to think of driving a prescribed distance d, and to want the maximum length of open road $M_{(d)}$ during the time $t = d/v_0$ required to drive this distance. Putting together our estimates gives the approximation

$$P(M_{(d)} \le b) \approx \exp\left(-d\alpha E\left|\frac{1}{V} - \frac{1}{v_0}\right| \exp(-\alpha b E(1/V))\right). \qquad \text{(C20g)}$$

It is interesting to note that, as a function of v_0, $M_{(d)}$ is smallest when $v_0 = \text{median}(V)$.

C21 Approximations for the normal distribution. Our final topic is the application of Rice's formula to smooth Gaussian processes. It is convenient to record first some standard approximations involving the Normal distribution. Write $\phi(x) = (2\pi)^{-1/2}\exp(-x^2/2)$ for the standard Normal density; implicit in this is the integral formula

$$\int_{-\infty}^{\infty} \exp(-ax^2)\,dx = \left(\frac{\pi}{a}\right)^{\frac{1}{2}}. \qquad \text{(C21a)}$$

Write Z for a r.v. with Normal(0,1) distribution and $\overline{\Phi}(x) = P(Z \ge x)$. Then

$$\overline{\Phi}(b) \approx \frac{\phi(b)}{b} \qquad \text{for } b \text{ large} \qquad \text{(C21b)}$$

and there is an approximation for the "overshoot distribution":

$$\text{distribution}(Z - b \mid Z \ge b) \approx \text{exponential}(b). \qquad \text{(C21c)}$$

Both (C21b) and (C21c) are obtained from the identity

$$\phi(b+u) = \phi(b)e^{-bu}e^{-\frac{1}{2}u^2}$$

by dropping the last term to get the approximation

$$\phi(b+u) \approx \phi(b)e^{-bu}; \qquad b \text{ large}, u \text{ small}. \qquad \text{(C21d)}$$

This is the most useful approximation result, and worth memorizing. Next, let $f(t)$ have a unique minimum at t_0, with $f(t_0) > 0$ and $\phi(f(t_0))$ small, and let $f(t)$ and $g(t)$ be smooth functions with $g > 0$; then

$$\int_{-\infty}^{\infty} g(t)\phi(f(t))\,dt \approx g(t_0)\left(f(t_0)f''(t_0)\right)^{-\frac{1}{2}} \exp(-f^2(t_0)/2). \qquad \text{(C21e)}$$

This is obtained by writing

$$f(t_0+u) \approx f(t_0) + \frac{1}{2}f''(t_0)u^2; \quad u \text{ small}$$

$$\phi(f(t_0+u)) \approx \phi(f(t_0))\exp(-f(t_0)\frac{1}{2}f''(t_0)u^2) \qquad \text{using (C21d)},$$

and then approximating $g(t)$ as $g(t_0)$ and evaluating the integral by (C21a). Here we are using *Laplace's method:* approximating an integral by expanding the integrand about the point where its maximum is attained. We use this method at many places throughout the book.

Finally,

$$E|Z| = 2E\max(0, Z) = \left(\frac{2}{\pi}\right)^{\frac{1}{2}}. \qquad \text{(C21f)}$$

C22 Gaussian processes. A process X_t is *Gaussian* if its finite-dimensional distributions are multivariate Normal. Here we record some standard facts about such processes. If X is Gaussian then its entire distribution is determined by the *mean function*

$$m(t) = EX_t$$

and the *covariance function*

$$R(s,t) = \text{cov}(X_s, X_t).$$

Unless otherwise specified, we will take $m(t) \equiv 0$. We can then specify conditional distributions simply:

given $X_s = x$, the distribution of X_t is Normal with mean $xR(s,t)/R(s,s)$ and variance $R(t,t) - R^2(s,t)/R(s,s)$. (C22a)

For a stationary Gaussian process we have $R(s,t) = R(0, t-s)$, and so we need only specify $R(0,t) = R(t)$, say. Also, for a stationary process we can normalize so that $R(t) = \text{var}\, X_t \equiv 1$ without loss of generality. Such a normalized process will be called *smooth* if

$$R(t) \sim 1 - \frac{1}{2}\theta t^2 \quad \text{as } t \to 0; \qquad \text{some } \theta > 0. \qquad \text{(C22b)}$$

(Of course this is shorthand for $1 - R(t) \sim \frac{1}{2}\theta t^2$). Smoothness corresponds to the sample paths being differentiable functions; writing $V_t = dX_t/dt \equiv \lim \delta^{-1}(X_{t+\delta} - X_t)$ it is easy to calculate

$$EV_t = 0; \quad EV_tX_t = 0; \quad EV_t^2 = \theta$$

and hence (for each fixed t)

V_t has Normal$(0,\theta)$ distribution and is independent of X_t. (C22c)

Here we will consider only smooth Gaussian processes: others are treated in the next chapter.

C23 The heuristic for smooth Gaussian processes. Let X_t be as above, a mean-zero Gaussian stationary process with

$$R(t) \equiv EX_0X_t \sim 1 - \frac{1}{2}\theta t^2 \quad \text{as } t \to 0. \tag{C23a}$$

For $b > 0$ Rice's formula (C12f) gives an exact expression for the upcrossing rate ρ_b over b:

$$\begin{aligned} \rho_b &= E(V_t^+ \mid X_t = b) f_{X_t}(b) \\ &= \theta^{\frac{1}{2}}(2\pi)^{-\frac{1}{2}}\phi(b) \qquad \text{using (C22c).} \end{aligned} \tag{C23b}$$

As at (C12h), the heuristic idea is that for large b the clumps of $\mathcal{S} = \{t : X_t \geq b\}$ will consist of single intervals (see Section C25) and so we can identify the clump rate λ_b with the upcrossing rate (for large b):

$$\lambda_b = \theta^{\frac{1}{2}}(2\pi)^{-\frac{1}{2}}\phi(b). \tag{C23c}$$

As usual, the heuristic then gives

$$\boldsymbol{P}(\max_{0 \leq s \leq t} X_s \leq b) \approx \exp(-\lambda_b t); \qquad t \text{ large.} \tag{C23d}$$

To justify the corresponding limit assertion, we need a condition that $R(t) \to 0$ not too slowly as $t \to \infty$, to prevent long-range dependence.

The heuristic also gives an approximation for sojourn times. Let C have standard Rayleigh distribution:

$$f_C(x) = xe^{-\frac{1}{2}x^2}, \qquad x \geq 0. \tag{C23e}$$

We shall show in Section C25 that the clump lengths C_b of $\{t : X_t \geq b\}$ satisfy

$$C_b \overset{\mathcal{D}}{\approx} 2\theta^{-\frac{1}{2}}b^{-1}C. \tag{C23f}$$

Let $S(t, b)$ be the sojourn time of $(X_s : 0 \leq s \leq t)$ in $[b, \infty)$ Then as at Section A19 the heuristic gives a compound Poisson approximation:

$$\frac{1}{2}\theta^{\frac{1}{2}}bS(t,b) \overset{\mathcal{D}}{\approx} \mathrm{POIS}(t\lambda_b \text{distribution}(C)). \tag{C23g}$$

C24 Monotonicity convention. Here is a trivial technical point, worth saying once. Our heuristic for maxima $M_t = \max_{0 \leq s \leq t} X_s$ typically takes the form

$$\boldsymbol{P}(M_t \leq b) \approx \exp(-\lambda_b t), \qquad t \text{ large} \tag{C24a}$$

where there is some explicit expression for λ_b. The corresponding limit assertion, which in most cases is an established theorem in the literature, is

$$\sup_b |\boldsymbol{P}(M_t \leq b) - \exp(-\lambda_b t))| \to 0 \quad \text{as } t \to \infty. \tag{C24b}$$

There is a slight problem here, exemplified by the previous example; the formula (C23c) for λ_b has $\lambda_b \to 0$ as $b \to -\infty$, and the approximation (C24a) is wrong for large negative b. This has no real significance, since in using the heuristic we specify that b be large positive, but it does make (C24b) formally incorrect. We can make (C24b) correct by adopting the *monotonicity convention*: whenever we derive a formula for clump rates λ_b (in the context of maxima), let b_0 be the smallest real such that λ_b is decreasing for $b > b_0$, and re-define $\lambda_b = \lambda_{b_0}$ for $b < b_0$. This makes λ_b monotone and assertion (C24b) takes on a legitimate form.

C25 High-level behavior of smooth Gaussian processes. Returning to the setting of Section C23, we can calculate

$$\begin{aligned} E(V_t - V_0 \mid X_0 = b) &\sim -\theta b t && \text{as } t \to 0 \\ \text{var}(V_t - V_0 \mid X_0 = b) &= O(t^2) && \text{as } t \to 0. \end{aligned}$$

So given X_0 and V_0 with $X_0 = b$, large, we have

$$V_t = V_0 - (\theta b + O(1))t \qquad \text{as } t \to 0$$

and so, integrating,

$$X_t = b + V_0 t - (\tfrac{1}{2}\theta b + O(1))t^2 \qquad \text{as } t \to 0$$

In other words, given $X_0 = b$ large, the local motion of X_t follows a parabola

$$X_t \approx b + V_0 t - \frac{1}{2}\theta b t^2; \qquad t \text{ small.} \tag{C25a}$$

This implies the qualitative property that clumps of $\{ t : X_t \geq b \}$ are single intervals for large b; it also enables us to estimate the lengths C_b of these intervals as follows. Given $X_0 = b$ and $V_0 = v > 0$, (C25a) implies that $C_b \approx$ the solution $t > 0$ of $vt - \frac{1}{2}\theta b t^2 = 0$, that is $C_b \approx 2v/(\theta b)$. Thus

$$C_b \overset{\mathcal{D}}{\approx} \frac{2V}{\theta b}; \text{ where } V \text{ is the velocity at an upcrossing of } b. \tag{C25b}$$

Using Rice's formula (C12g), V has density

$$\begin{aligned} g(v) &= v f_{V_0}(v) \frac{\phi(b)}{\rho_b} \\ &= v f_{V_0}(v) \theta^{-\frac{1}{2}} (2\pi)^{\frac{1}{2}} && \text{by (C23b)} \\ &= v\theta^{-1} \exp\left(\frac{-v^2/2}{\theta} \right) && \text{using (C22c).} \end{aligned}$$

Thus

$$V \overset{\mathcal{D}}{=} \theta^{\frac{1}{2}} C; \qquad \text{where } C \text{ has Rayleigh distribution (C23e)} \tag{C25c}$$

and then (C25b) and (C25c) give $C_b \overset{\mathcal{D}}{\approx} 2\theta^{-1/2} b^{-1} C$, as stated at (C23f).

C26 Conditioning on semi-local maxima. At Section A7 we discussed this technique for the M/M/1 queue: here is the set-up for a stationary continuous space and time process X_t. There are semi-local maxima (t^*, x^*), where $x^* = X_{t^*}$, whose rate is described by an intensity function $L(x)$ as follows:

$$\boldsymbol{P}\begin{pmatrix}\text{some semi-local maximum} \\ (t^*, x^*) \text{ in } [t, t+dt] \times [x, x+dx]\end{pmatrix} = L(x)\, dx\, dt. \tag{C26a}$$

The heuristic idea is that at high levels the point process of semi-local maxima can be approximated by the Poisson point process of rate $L(x)$ (recall Section C3 for space-time Poisson point processes). Each clump of $\{\, t : X_t \geq b \,\}$ corresponds to one semi-local maximum of height $> b$, and so the clump rate λ_b relates to $L(x)$ via

$$\lambda_b = \int_b^\infty L(x)\, ds; \qquad L(x) = \frac{-d\lambda_x}{dx}. \tag{C26b}$$

Now suppose the process around a high-level semi-local maximum of height x can be approximated by a process Z^x:

$$\text{given } (t^*, x^*) \text{ is a semi-local maximum, } X_{t^*+t} \approx x^* - Z_t^{x^*} \text{ for } t \text{ small.} \tag{C26c}$$

Supposing $Z_t^{x^*} \to \infty$ as $|t| \to \infty$, write $m(x, y)$ for its mean sojourn density:

$$m(x, y)\, dy = E \text{ sojourn time of } (Z_t^x; -\infty < t < \infty) \text{ in } (y, y + dy). \tag{C26d}$$

Writing f for the marginal density of X_t, the obvious ergodic argument as at Section A7 gives

$$f(y) = \int_y^\infty L(x) m(x, x - y)\, dx. \tag{C26e}$$

Thus if we are able to calculate $m(x, y)$ then we can use (C26e) to solve for $L(x)$ and hence λ_x; this is our heuristic technique "conditioning on semi-local maxima".

Let us see how this applies in the smooth Gaussian case (Section C23). By (C25a) the approximating process Z^x is

$$Z_t^x = \frac{1}{2}\theta t^2. \tag{C26f}$$

It follows that $m(x, y) = (\frac{1}{2}\theta x y)^{-1/2}$. So (C26e) becomes

$$\phi(y) = \int_y^\infty L(x) \left(\frac{1}{2}\theta x(x - y)\right)^{-\frac{1}{2}} dx. \tag{C26g}$$

We anticipate a solution of the form $L(x) \sim a(x)\phi(x)$ for polynomial $a(x)$, and seek the leading term of $a(x)$. Writing $x = y + u$, putting $\phi(y+u) \approx \phi(y)e^{-yu}$ and recalling that polynomial functions of y vary slowly relative to $\phi(y)$, (C26g) reduces to

$$1 \approx a(y)\left(\frac{\theta y}{2}\right)^{-\frac{1}{2}} \int_0^\infty u^{-\frac{1}{2}} e^{-yu}\, du$$

which gives $a(y) \approx \theta^{1/2}(2\pi)^{-1/2}y$. So by (C26b),

$$\begin{aligned} \lambda_x &= \theta^{\frac{1}{2}}(2\pi)^{-\frac{1}{2}} \int_x^\infty y\phi(y)\, dy \\ &= \theta^{\frac{1}{2}}(2\pi)^{-\frac{1}{2}}\phi(x). \end{aligned}$$

This recovers the clump rate obtained earlier(C23c) using Rice's formula. We don't get any new information here; however, in the multiparameter setting, this "conditioning on semi-local maxima" argument goes over unchanged (Example J7), and is much easier than attempting to handle multiparameter analogues of upcrossings.

C27 Variations on a theme. Restating our basic approximation for stationary mean-zero Gaussian processes with $EX_0X_t \sim 1 - \frac{1}{2}\theta t^2$:

$$\boldsymbol{P}(\max_{0\le s\le t} X_s \le b) \approx \exp(-\lambda_b t); \qquad \lambda_b = \theta^{\frac{1}{2}}(2\pi)^{-\frac{1}{2}}\phi(b). \tag{C27a}$$

There are many variations on this basic result: three are treated concisely below. These type of variations are of more interest in the context of "locally Brownian" processes, and in that context are treated at greater length in the next chapter.

First, suppose we make the process non-stationary but keep the variance at 1, so that

$$EX_tX_{t+u} \sim 1 - \frac{1}{2}\theta_t u^2 \quad \text{as } u \to 0.$$

Then we have a non-stationary clump rate $\lambda_b(t)$, and (C27a) becomes

$$\boldsymbol{P}(\max_{0\le s\le t} X_s \le b) \approx \exp\left(-(2\pi)^{-\frac{1}{2}}\phi(b)\int_0^t \theta_s^{\frac{1}{2}}\, ds\right). \tag{C27b}$$

Second, suppose we are back in the stationary case, but are interested in a slowly sloping barrier $b(t)$ with $b(t)$ large. Then Rice's formula gives the rate at t for upcrossings of $X(t)$ over $b(t)$:

$$\begin{aligned} \rho_{b(t)} &= \int (v - b'(t))^+ f(b,v)\, dv \\ &= E(V_t - b'(t))^+\phi(b(t)) \\ &\approx \left((2\pi)^{-\frac{1}{2}}\theta^{\frac{1}{2}} - \frac{1}{2}b'(t)\right)\phi(b(t)) \qquad \text{if } b'(t) \text{ small.} \end{aligned}$$

Then

$$\boldsymbol{P}(X_s \text{ does not cross } b(s) \text{ during } [0,t]) \approx \exp(-\int_0^t \rho_{b(s)}\,ds). \qquad \text{(C27c)}$$

Thirdly, consider a mean-zero smooth Gaussian process X_t on $t_1 \leq t \leq t_2$ such that var $X_t = \sigma^2(t)$ is maximized at some $t_0 \in (t_1, t_2)$. Let $\theta(t) = EV_t^2$. We can use the heuristic to estimate the tail of $M \equiv \max_{t_1 \leq t \leq t_2} X_t$. Indeed, writing $\rho_b(t)$ for the upcrossing rate over a high level b, and identifying ρ with the clump rate and using (A10f),

$$\boldsymbol{P}(M > b) \approx \int_{t_1}^{t_2} \rho_b(t)\,dt. \qquad \text{(C27d)}$$

Rice's formula (C12f) gives

$$\begin{aligned} \rho_b(t) &= E(V_t^+ \mid X_t = b) f_{X_t}(b) \\ &\approx \left(\frac{\theta(t_0)}{2\pi}\right)^{\frac{1}{2}} \sigma^{-1}(t_0)\phi\left(\frac{b}{\sigma(t)}\right) \qquad \text{for } t \approx t_0, \end{aligned}$$

after some simple calculations. Evaluating the integral via (C21e) gives

$$\boldsymbol{P}(M > b) \approx \left(\frac{\theta(t_0)\sigma(t_0)}{-2\pi\sigma''(t_0)}\right)^{\frac{1}{2}} b^{-1} \exp\left(\frac{-b^2/2}{\sigma^2(t_0)}\right); \qquad b \text{ large.} \qquad \text{(C27e)}$$

C28 Example: Smooth $\mathcal{X}^2$ processes. For $1 \leq i \leq n$ let $X_i(t)$ be independent stationary mean-zero Gaussian processes as in Section C23, that is with

$$R_i(t) \sim 1 - \frac{1}{2}\theta t^2 \quad \text{as } t \to 0.$$

Let $Y^2(t) = \sum_{i=1}^n X_i^2(t)$. Then Y^2 is a stationary process with smooth paths and with $\mathcal{X}^2$ marginal distribution: this is sometimes calls a $\mathcal{X}^2$ *process*. Studying extremes of Y^2 is of course equivalent to studying extremes of Y, and the latter is more convenient for our method. Y has marginal distribution

$$f_Y(y) = \left(\frac{1}{2}\right)^a y^{n-1} \frac{e^{-\frac{1}{2}y^2}}{a!}; \qquad a = \frac{1}{2}(n-2).$$

Regard $X(t)$ as a n-dimensional Gaussian process. As in the 1-dimensional case (C22c), the velocity $\boldsymbol{V}(t) = (V_1(t), \ldots, V_n(t))$ is Gaussian and independent of $X(t)$ at fixed t. Now $Y(t)$ is the radial distance of $X(t)$; let $V(t)$ be the velocity of $Y(t)$. By rotational symmetry, the distribution of V given $Y = b$ is asymptotically ($b \to \infty$) the same as the distribution of V_1 given $X_1 = b$. Thus we can use Rice's formula for upcrossings of Y over b:

$$\begin{aligned} \lambda_b &\approx \rho_b = f_Y(b)E(V^+ \mid Y = b) \approx f_Y(b)E(V_1^+ \mid X_1 = b) \\ &\approx \theta^{\frac{1}{2}}(2\pi)^{-\frac{1}{2}} f_Y(b) \qquad \text{using (C22c).} \end{aligned}$$

Then as usual

$$P(\sup_{s \le t} Y(s) \le b) \approx \exp(-\lambda_b t).$$

We can also use the heuristic to study the minimum of $Y(t)$, but we leave this as an exercise for the reader.

COMMENTARY

C29 General references. The natural theoretical reference is the monograph by Leadbetter et al. (1983) supplemented by the survey article of Leadbetter and Rootzen (1988). A different theoretical perspective is given by the survey paper Berman (1982b) and subsequent papers of Berman (see references). Resnick (1987) emphasizes the point process approach. On the applied side, the survey by Abrahams (1984b) has an extensive bibliography; the monograph by Vanmarcke (1982) uses the heuristic. The conference proceedings ed. de Oliveira (1984) and Gani (1988) give an overview of current interests in theory and applications.

C30 Proving limit theorems. As for Markov chains (Section B24), to formalize the "asymptotic Poisson" property (Section C3) of extrema is easier than justifying explicit estimates of the normalizing constants (clump rates λ_b). This "asymptotic Poisson" property can be proved

(i) under mixing hypotheses: Leadbetter et al (1983), Berman (1982b);

(ii) by representing the process as a function of a general-space Markov chain, and exploiting the regenerative property: O'Brien (1987), Rootzen (1988);

(iii) by exploiting special structure (e.g. Gaussian).

The known general methods of computing normalizing constants are essentially just formalizations of various forms of the heuristic.

Under regularity conditions, the classical extremal distributions (C1c) are the only possible limits (under linear rescaling) for maxima of dependent stationary sequences: see Leadbetter (1983).

C31 Approximate independence of tails. The idea that (C7a) implies $EC \approx 1$ and thus that maxima behave as if the X's were i.i.d. is easy to formalize: this is essentially condition D' of Leadbetter et al. (1983).

C32 Moving average and autoregressive processes. Extremes for these processes have been studied in some detail in the theoretical literature. See Rootzen (1978) and Davis and Resnick (1985) for the polynomial-tail case; and Rootzen (1986) for the case $\boldsymbol{P}(Y > y) \sim \exp(-y^a)$, $a > 0$.

C33 Additive processes. The G/G/1 queue has been studied analytically in much detail; a central fact being that its stationary distribution is related to M at (C11c). Prabhu (1980) and Asmussen (1987) are good introductions. Iglehart (1972) treats its extremal behavior.

The point of the heuristic approach is to make clear that the relation (C11e) between the stationary distribution and the extremal behavior is true for the more general class of additive processes, and has nothing to do with the special structure of the G/G/1 queue. There seems no literature on this, not even a proof of the tail behavior (C11b). Informally, (C11b) holds because

1. the stationary probability $\boldsymbol{P}(X_0 > b)$ is proportional to the mean sojourn time above b in an "excursion" from the boundary;

2. this mean sojourn time is proportional to the probability that an excursion hits $[b, \infty)$;

3. this probability is proportional to $e^{-\theta b}$ because (C11a) implies $M_n = e^{\theta X_n}$ is a martingale when X is away from the boundary.

Martingale arguments are readily used to establish rigorous bounds on mean hitting times to high levels; see e.g. Hajek (1982), Yamada (1985).

C34 Rice's formula. Leadbetter et al. (1983), Theorem 7.2.4, give a precise formulation. Vanmarcke (1982) gives applications.

The "local" argument for Rice's formulation goes as follows. For t small,

$$\begin{aligned} \boldsymbol{P}(X_0 < b, X_t > b) &= \int_{x<b}\int_{v>0} \boldsymbol{P}(X_t > b \mid X_0 = x, V_0 = v) f(x,v)\,dx\,dv \\ &\approx \int_{x<b}\int_{v>0} 1_{(x+vt>b)} f(x,v)\,dx\,dv \\ &= \int_{v>0} vtf(b,v)\,dv. \end{aligned}$$

So the upcrossing rate is $\rho_b = \frac{d}{dt}\boldsymbol{P}(X_0 < b, X_t > b) = \int_{v>0} vf(b,v)\,dv$.

Our "uniform distribution" example (Example C14) is surely well known, but I do not know a reference. The large deviation results used can be deduced from more general results in Varadhan (1984).

C35 Drift-jump processes. Asmussen (1987) Chapter 13 contains an account of these processes. The special case at (C18d) is treated by Tsurui and Osaki (1976).

C36 Single-server queues. There are innumerable variations on single-server queueing models. Even for complicated models, our heuristic can relate hitting times to the stationary distribution; the difficulty is to approximate the stationary distribution. Some analytic techniques are developed in Knessl et al (1985; 1986b; 1986a).

C37 The ergodic-exit form of the heuristic. The method (C10a) of estimating clump rate as $\boldsymbol{P}(X_0 \geq b)f_b$ has been formalized, in the context of functions of general-space Markov chains, by O'Brien (1987). In the continuous-time setting, see Section D38.

C38 Smooth Gaussian processes. Most of the material in Sections C22–C26 is treated rigorously in Leadbetter et al. (1983).

Processes with a point of maximum variance (C27e) are treated by Berman (1987).

Smooth $\mathcal{X}^2$ processes have been studied in some detail: see Aronowich and Adler (1986) for recent work.

C39 Normalizing constants. In the context of Gaussian processes, our heuristic conclusions correspond to limit assertions of the form

$$\sup_x \left| \boldsymbol{P}(M_t \leq x) - \exp(-tax^b\phi(x)) \right| \to 0 \quad \text{as } t \to \infty, \tag{C39a}$$

with the usual monotonicity convention (Section C24). Putting this in the classic form gives

$$\frac{M_t - c_t}{s_t} \stackrel{\mathcal{D}}{\to} \xi_3 \quad \text{as } t \to \infty, \tag{C39b}$$

where ξ_3 has double-exponential extreme value distribution,

$$\begin{aligned} s_t &= (2\log t)^{-\frac{1}{2}} \\ c_t &= (2\log t)^{\frac{1}{2}} + (2\log t)^{-\frac{1}{2}}\left(\log(a/\sqrt{2\pi}) + \frac{1}{2} b \log(2\log t)\right). \end{aligned}$$

C40 Multivariate extremes. There is some theoretical literature on d-dimensional extreme value distributions, generalizing (C1c), and the corresponding classical limit theorems: see e.g. Leadbetter and Rootzen (1988). But I don't know any interesting concrete examples. Chapter I treats multidimensional processes, but with a different emphasis.

D Extremes of Locally Brownian Processes

This chapter looks at extrema and boundary crossings for stationary and near-stationary 1-dimensional processes which are "locally Brownian". The prototype example is the Ornstein-Uhlenbeck process, which is both Gaussian and Markov. One can then generalize to non-Gaussian Markov processes (diffusions) and to non-Markov Gaussian processes; and then to more complicated processes for which these serve as approximations. In a different direction, the Ornstein-Uhlenbeck process is a time and space-change of Brownian motion, so that boundary-crossing problems for the latter can be transformed to problems for the former: this is the best way to study issues related to the law of the iterated logarithm.

D1 Brownian motion. We assume the reader has some feeling for standard Brownian motion B_t and for Brownian motion X_t with constant drift μ and variance σ^2:

$$X_t = \mu t + \sigma B_t. \tag{D1a}$$

In fact, most of our calculations rest upon one simple fact, as follows. For $A \subset R$, let $\Gamma(A)$ be the total length of time that $(X_t; t \geq 0)$ spends in A. Then

$$\lim_{\delta \downarrow 0} \delta^{-1} E_0 \Gamma(0, \delta) = \frac{1}{|\mu|}; \qquad \mu \neq 0. \tag{D1b}$$

In words: the sojourn density of X at its starting point equals 1/drift.

Occasionally we use facts about sojourn times in the half-line. Let $\mu < 0$. Given $X_0 = 0$, $\Gamma(0, \infty) \stackrel{\mathcal{D}}{=} \sigma^2/\mu^2 \cdot \Gamma$, where Γ is a standardized random variable satisfying

$$\text{mean:} \qquad E\Gamma = \frac{1}{2} \tag{D1c}$$

$$\text{second moment:} \qquad E\Gamma^2 = 1 \tag{D1d}$$

$$\text{transform:} \qquad E\exp(-\theta\Gamma) = 2((1+2\theta)^{\frac{1}{2}} + 1)^{-1} \tag{D1e}$$

$$\text{density:} \qquad f_\Gamma(x) = 2x^{-\frac{1}{2}}\phi(x^{\frac{1}{2}}) - 2\overline{\Phi}(x^{\frac{1}{2}}) \tag{D1f}$$

where ϕ and $\overline{\Phi}$ are the standard Normal density and tail probability. See Section D40 for derivations.

D2 The heuristic for locally Brownian processes. Let (X_t) be a stationary, continuous-path process with marginal density f_X. Take b such that $\boldsymbol{P}(X_t > b)$ is small. Suppose X has the property

$$\text{Given } X_t = x, \text{ where } x \approx b, \text{ the incremental process } (X_{t+u} - x; u \geq 0) \text{ evolves for small } u \text{ like Brownian motion with drift } -\mu(b) \text{ and some variance.} \tag{D2a}$$

This is our "locally Brownian" property. Now apply the heuristic to the random set $\mathcal{S} = \{\, t : X_t \in (b, b+\delta) \,\}$. By (D2a) the clumps behave like the corresponding clumps for Brownian motion with drift $-\mu(b)$, so by (D1b) the mean clump length is $EC = \delta/\mu(b)$. Since

$$p \equiv \boldsymbol{P}(t \in \mathcal{S}) = \delta f_X(b),$$

the fundamental identity $\lambda_b = p/EC$ gives the clump rate

$$\lambda_b = f_X(b)\mu(b). \tag{D2b}$$

So if we define

$$M_t = \sup_{0 \leq s \leq t} X_s; \qquad T_b = \inf\{\, t : X_t = b \,\}$$

then the heuristic approximation (Section A4) is

$$\boldsymbol{P}(M_t \leq b) = \boldsymbol{P}(T_b > t) \approx \exp(-\lambda_b t). \tag{D2c}$$

This simple result, and its analogues for non-stationary processes and curved boundaries $b(t)$, covers most of the examples in this chapter.

We can also get the compound Poisson approximation for sojourn times above a fixed level. Let $\sigma^2(b)$ be the local variance in (D2a). Let $S(t, b)$ be the total sojourn time of $(X_u : 0 \leq u \leq t)$ above level b. Then as at Section A19

$$S(t,b) \stackrel{\mathcal{D}}{\approx} \mathrm{POIS}(\nu) \qquad \text{for } \nu = t\lambda_b \mathrm{distribution}(C_b). \tag{D2d}$$

Here C_b has the distribution, described in Section D1, of $\Gamma(0, \infty)$ for Brownian motion with drift $-\mu(b)$ and variance $\sigma^2(b)$: that is,

$$C_b \stackrel{\mathcal{D}}{=} \frac{\sigma^2(b)}{\mu^2(b)} \cdot \Gamma, \qquad \text{for } \Gamma \text{ as in Section D1.}$$

Using the transform formula for compound Poisson (Section A19) and the transform formula for Γ (D1e), we can write

$$E\exp(-\theta S(t,b)) \quad \approx \quad \exp(-tf_X(b)\mu(b)\psi(b,\theta)), \tag{D2e}$$

where

$$\psi(b,\theta) = 1-2((1+2\theta\mu^2(b)/\sigma^2(b))^{\frac{1}{2}}+1)^{-1}.$$

In the examples we will usually state only the simpler results corresponding to (D2c).

Two final comments on the heuristic assumption (D2a). First, if we interpret "locally Brownian" in the asymptotic ($u \to 0$) sense, we need to specify that the local drift $\mu(X_s)$ should not change much during a typical excursion above level b. Second, we do not need to explicitly assume the Markov property: "given $X_t = x \ldots$" really does mean "given $X_t = x \ldots$" and not "given $X_t = x$ and the past X_s, $s \le t \ldots$". The point is that we could use the ergodic-exit form (Section A9) of the heuristic in place of the renewal-sojourn form (Section A8), and this requires no explicit Markov property.

D3 One-dimensional diffusions. One-dimensional diffusions are a tractable class of processes for which explicit calculations are feasible. For this reason, heuristic arguments are somewhat redundant. However, to illustrate the heuristic it seems sensible to start out with the simplest cases; and we need some diffusion results later. So below we give a concise listing of some basic results about diffusions, and in (D4) give the heuristic estimate of hitting time distributions. Karlin and Taylor (1982) Chapter 15 is an excellent introduction to diffusions, and this book [KT] may be consulted for the results stated below.

By a *diffusion* $X_t, t \ge 0$, I mean a continuous path Markov process such that, writing $\Delta X_t = X_{t+\Delta t} - X_t$,

$$\begin{aligned} E(\Delta X_t \mid X_t = x) &\approx \mu(x)\Delta t \\ \mathrm{var}(\Delta X_t \mid X_t = x) &\approx \sigma^2(x)\Delta t \end{aligned} \quad \text{as } \Delta t \to 0 \qquad \text{(D3a)}$$

where $\mu(x)$ and $\sigma^2(x)$ are nice functions (continuous will do), and $\sigma^2(x) > 0$. The role of these functions is the same as the role of the transition matrix in a discrete chain — they, together with the initial distribution, determine the distribution of the whole process. By a version of the central limit theorem (applied to infinitesimal increments), (D3a) is equivalent to the stronger property

$$\text{distribution}(\Delta X_t \mid X_t = x) \sim \text{Normal}(\mu(x)\Delta t, \sigma^2(x)\Delta t) \quad \text{as } \Delta t \to 0. \qquad \text{(D3b)}$$

Call μ the *drift* function, σ^2 the *variance* function. Regard these as given; we are interested in computing probabilities associated with the corresponding diffusion X_t.

Notation: $\boldsymbol{P}_x()$, $E_x()$ mean "given $X_0 = x$",
T_a is the first hitting time on a;

$T_{a,b}$ is the first hitting time on $\{a, b\}$. Define, for $-\infty < x < \infty$,

$$\begin{aligned} s(x) &= \exp(-\int_0^x \frac{2\mu(y)}{\sigma^2(y)}\, dy) \\ S(x) &= \int_0^x s(y)\, dy \\ m(x) &= (\sigma^2(x)s(x))^{-1} \\ M(a,b) &= \int_a^b m(x)\, dx. \end{aligned} \tag{D3c}$$

$S(\cdot)$ is the *scale function*, $m(\cdot)$ the *speed density*, $M(dx)$ the *speed measure*. In the integrations defining s and S, we could replace "0" by any x_0 without affecting the propositions below.

Proposition D3.1 (KT p.195) *Let $a < x < b$.*

$$\boldsymbol{P}_x(T_b < T_a) = \frac{S(x) - S(a)}{S(b) - S(a)}.$$

Next, consider the diffusion X_t started at $x \in (a, b)$ and run until time $T_{a,b}$. The mean total time spent in (a, y) in this period is

$$\Gamma_{a,b}(x,y) = E_x \int_0^{T_{a,b}} 1_{(X_t<y)}\, dt.$$

The derivative

$$G_{a,b}(x,y) = \frac{d}{dy}\Gamma_{a,b}(x,y); \qquad a < y < b \tag{D3d}$$

is the *mean occupation density* at y; informally, $G_{a,b}(x,y)\Delta y$ is the mean time spent in $(y, y + \Delta y)$ by the process X_t started at x and run until it exits (a, b). One reason this quantity is useful is

$$E_x T_{a,b} = \int_a^b G_{a,b}(x,y)\, dy \tag{D3e}$$

Proposition D3.2 (KT p.198) *$G_{a,b}(x,y)$ is given by the formulas*

$$\frac{2(S(x) - S(a))(S(b) - S(y))}{S(b) - S(a)} m(y) \qquad a < x \le y < b$$

$$\frac{2(S(b) - S(x))(S(y) - S(a))}{S(b) - S(a)} m(y) \qquad a < y \le x < b.$$

Call a diffusion *positive-recurrent* if $E_x T_y < \infty$ for all x, y. As with discrete-space chains, this is equivalent to the existence of a stationary distribution π.

Proposition D3.3 (KT p.221) *A diffusion is positive-recurrent if and only if $M(-\infty,\infty) < \infty$. The stationary distribution has density $\pi(dx) = m(x)\,dx/M(-\infty,\infty)$.*

Finally, we specialize some of these results to the Brownian case.

Proposition D3.4 *Let X_t be Brownian motion with drift $-\mu$ and variance σ^2.*

$$\boldsymbol{P}_x(T_b < T_a) = \frac{\exp(2\mu x/\sigma^2) - \exp(2\mu a/\sigma^2)}{\exp(2\mu b/\sigma^2) - \exp(2\mu a/\sigma^2)}, \qquad a < x < b \qquad \text{(D3f)}$$

$$\boldsymbol{P}_x(\sup_{t\geq 0} X_t > x + z) = \exp(-2\mu z/\sigma^2), \qquad z \geq 0 \qquad \text{(D3g)}$$

$$E_x T_a = \frac{x-a}{\mu}, \qquad a < x \qquad \text{(D3h)}$$

$$\begin{aligned} G_{-\infty,\infty}(x,y) &= \frac{1}{\mu}, & y \leq x \\ &= \frac{\exp(-2\mu(y-x)/\sigma^2)}{\mu}, & x \leq y \end{aligned} \qquad \text{(D3i)}$$

$$E_x(\textit{total sojourn time of } X_t,\ t \geq 0,\ \textit{in } [x,\infty)) = \frac{\sigma^2}{2\mu^2}. \qquad \text{(D3j)}$$

D4 First hitting times for positive-recurrent diffusions. There are explicit conditions for positive-recurrence (D3.3) and an explicit form for the stationary density π. For such a diffusion, fix b such that the stationary probability of (b,∞) is small. Then our heuristic (Section D2) should apply, and says that the first hitting time T_b satisfies

$$T_b \overset{\mathcal{D}}{\approx} \text{exponential}(\lambda_b); \qquad \lambda_b = -\mu(b)\pi(b). \qquad \text{(D4a)}$$

Equivalently, we have an approximation for maxima:

$$\boldsymbol{P}(\max_{s\leq t} X_s \leq b) \approx \exp(-\lambda_b t), \qquad b \text{ large}, \qquad \text{(D4b)}$$

and (D2d) gives the compound Poisson approximation for sojourn time above b. These approximations should apply to the stationary process, or to the process started with any distribution not near b. How good are these approximations? The exponential limit law

$$\frac{T_b}{ET_b} \overset{\mathcal{D}}{\to} \text{exponential}(1) \quad \text{as } b \to \infty \qquad \text{(D4c)}$$

is easy to prove (Section D37) under no assumptions beyond positive-recurrence. So the issue is the mean ET_b. Now for a diffusion we can improve the heuristic (D2b) as follows. The factor "$\mu(b)$" in (D2b) arises as (occupation density at $b)^{-1}$, where we used the occupation density for an approximating Brownian motion with drift. But for a diffusion we can use the true occupation density (D3.2) for the diffusion (killed on reaching some point x_0 in the middle of the stationary distribution, say). Then (D2b) becomes

$$\lambda_b = \frac{\pi(b)}{G_{x_0,\infty}(b,b)}. \tag{D4d}$$

Using $ET_b = \lambda_b^{-1}$, these two approximations (D4a),(D4d) become

$$ET_b \approx M(\infty,\infty)\sigma^2(b)\frac{s(b)}{|\mu(b)|} \tag{D4e}$$

$$ET_b \approx 2M(\infty,\infty)(S(b)-S(x_0)). \tag{D4f}$$

Let us compare these heuristic estimates with the exact formula given by (D3e,D3.2), which is

$$E_xT_b = 2(S(b)-S(x))M(-\infty,x) + 2\int_x^b (S(b)-S(y))m(y)\,dy. \tag{D4g}$$

Taking limits in this exact formula,

$$E_xT_b \sim 2M(-\infty,\infty)S(b) \quad \text{as } b\to\infty \tag{D4h}$$

which certainly agrees asymptotically with the heuristic (D4f). To relate (D4h) to (D4e), if $\sigma(\cdot)$ and $\mu(\cdot)$ are not changing rapidly around b then the definition of $s(x)$ gives

$$\frac{s(b-u)}{s(b)} \approx \exp\left(\frac{u\cdot 2\mu(b)}{\sigma^2(b)}\right) \qquad \text{for small } u \geq 0.$$

Then if $\mu(b) < 0$ the definition of $S(\cdot)$ gives

$$\frac{S(b)}{s(b)} \approx \frac{1}{2}\frac{\sigma^2(b)}{-\mu(b)}$$

which reconciles (D4e) and (D4h). From a practical viewpoint, it is more convenient to use (D4e), since it avoids the integral defining $S(b)$; from the theoretical viewpoint, (D4f) is always asymptotically correct, whereas (D4e) depends on smoothness assumptions.

D5 Example: Gamma diffusion. The diffusion on range $(0,\infty)$ with drift and variance

$$\mu(x) = a - bx, \qquad \sigma^2(x) = \sigma^2 x$$

occurs in several contexts, e.g. as a limit of "birth, death and immigration" population processes. From (D3.3) the stationary distribution works out to be a gamma distribution

$$\pi(x) = cx^{-1+2a/\sigma^2} \exp\left(\frac{-2bx}{\sigma^2}\right).$$

So for x in the upper tail of this distribution, (D4a) says

T_x has approximately exponential distribution, rate $(bx - a)\pi(x)$. (D5a)

As discussed above, this is an example where it is helpful to be able to avoid calculating the integral defining $S(x)$.

D6 Example: Reflecting Brownian motion. This is the diffusion on range $[0, \infty)$ with drift $-\mu$ and variance σ^2 and with a reflecting boundary (see Karlin and Taylor [KT] p.251) at 0. This diffusion arises e.g. as the heavy traffic limit for the M/M/1 queue. The stationary distribution is exponential:

$$\pi(x) = 2\mu\sigma^{-2} \exp\left(\frac{-2\mu x}{\sigma^2}\right). \tag{D6a}$$

So for x in the upper tail of this distribution, (D4a) says

T_x has approximately exponential distribution, rate $2\mu^2\sigma^{-2}\exp(-2\mu x/\sigma^2)$. (D6b)

D7 Example: Diffusions under a potential. Given a smooth function $H(x)$, we can consider the diffusion with

$$\mu(x) = -H'(x); \qquad \sigma^2(x) = \sigma^2.$$

Call H the *potential function.* Such diffusions arise as approximations in physics or chemical models: the process X_t might represent the energy of a molecule, or the position of a particle moving under the influence of a potential and random perturbations (in the latter case it is more realistic to model the potential acting on velocity; see Section I13). Provided $H(x) \to \infty$ not too slowly, as $|x| \to \infty$, (D3.3) gives the stationary density

$$\pi(x) = c\exp\left(\frac{-2H(x)}{\sigma^2}\right); \qquad c \text{ the normalizing constant.} \tag{D7a}$$

Now suppose H has a unique minimum at x_0. Then for b in the upper tail of the stationary distribution, (D4a) gives

$$T_b \overset{\mathcal{D}}{\approx} \text{exponential}, \quad \text{rate } H'(b)\pi(b). \tag{D7b}$$

If H is sufficiently smooth, we can get a more explicit approximation by estimating c using the quadratic approximation to H around x_0:

$$\begin{aligned} c^{-1} &= \int \exp\left(\frac{-2H(x)}{\sigma^2}\right) dx \\ &= \exp\left(\frac{-2H(x_0)}{\sigma^2}\right) \int \exp\left(\frac{-2(H(x_0+u)-H(x))}{\sigma^2}\right) du \\ &\approx \exp\left(\frac{-2H(x_0)}{\sigma^2}\right) \int \exp\left(\frac{-u^2 H''(x_0)}{\sigma^2}\right) du \\ &= \sigma \left(\frac{\pi}{H''(x_0)}\right)^{\frac{1}{2}} \exp\left(\frac{-2H(x_0)}{\sigma^2}\right). \end{aligned}$$

So (D7a) becomes

$$\pi(x) \approx \sigma^{-1} \left(\frac{H''(x_0)}{\pi}\right)^{\frac{1}{2}} \exp\left(\frac{-2(H(x)-H(x_0))}{\sigma^2}\right). \qquad \text{(D7c)}$$

This makes explicit the approximation (D7b) for T_b.

D8 Example: State-dependent M/M/1 queue. Take service rate $= 1$ and arrival rate $= a(i)$ when i customers are present, where $a(x)$ is a smooth decreasing function. This models e.g. the case where potential arrivals are discouraged by long waiting lines. We could consider this process directly by the methods of Chapter B; let us instead consider the diffusion approximation X_t. Let x_0 solve

$$a(x_0) = 1 \qquad \text{(D8a)}$$

so that x_0 is the "deterministic equilibrium" queue length: for a continuous-space approximation to be sensible we must suppose x_0 is large. The natural diffusion approximation has

$$\mu(x) = a(x) - 1; \qquad \sigma(x) = a(x) + 1.$$

We shall give an approximation for the first hitting time T_b on a level $b > x_0$ such that

$$1 - a(b) \text{ is small;} \qquad \int_{x_0}^{b} (1 - a(x))\, dx \text{ is not small.} \qquad \text{(D8b)}$$

This first condition implies we can write $\sigma^2(x) \approx 2$ over the range we are concerned with. But then we are in the setting of the previous example, with potential function

$$H(x) = \int^{x} (1 - a(y))\, dy.$$

The second condition of (D8b) ensures that b is in the tail of the stationary distribution, and then (D7b, D7c) yield

T_b has approximately exponential distribution, rate

$$\lambda_b = (1-a(b))\left(\frac{-a'(x_0)}{2\pi}\right)^{\frac{1}{2}} \exp\left(-\int_{x_0}^{b} (1-a(x))\,dx\right) \tag{D8c}$$

D9 Example: The Ornstein-Uhlenbeck process. The (general) Ornstein-Uhlenbeck process is the diffusion with

$$\mu(x) = -\mu x, \qquad \sigma^2(x) = \sigma^2.$$

The *standard* Ornstein-Uhlenbeck process is the case $\mu = 1$, $\sigma^2 = 2$:

$$\mu(x) = -x, \qquad \sigma^2(x) = 2.$$

The stationary distribution is, in the general case, the Normal$(0, \sigma^2/2\mu)$ distribution; so in particular, the *standard* Ornstein-Uhlenbeck process has the *standard* Normal stationary distribution.

From (D4a) we can read off the exponential approximation for hitting times T_b. It is convenient to express these in terms of the standard Normal density $\phi(x)$. In the general case

$$T_b \stackrel{\mathcal{D}}{\approx} \text{exponential}, \quad \text{rate } \lambda_b = \left(\frac{2\mu^3}{\sigma^2}\right)^{\frac{1}{2}} b\phi(\sigma^{-1}b\sqrt{2\mu}). \tag{D9a}$$

provided that b is large compared to $\sigma^2/(2\mu)$. For the standard Ornstein-Uhlenbeck process, this takes the simple form

$$T_b \stackrel{\mathcal{D}}{\approx} \text{exponential}, \quad \text{rate } \lambda_b = b\phi(b), \tag{D9b}$$

and this approximation turns out to be good for $b \geq 3$, say.

The Ornstein-Uhlenbeck process is of fundamental importance in applications, because almost any stochastic system which can be regarded as "a stable deterministic system plus random fluctuations" can be approximated (for small random fluctuations, at least) by an Ornstein-Uhlenbeck process. For instance, it arises as the heavy-traffic limit of the M/M/∞ queue, in stable population models, as well as numerous "small noise" physics settings. In such settings, (D9b) gives the chance of large deviations (in the non-technical sense!) from equilibrium during a time interval, using the equivalent form

$$M_t \equiv \sup_{0\leq s\leq t} X_s \text{ satisfies } \boldsymbol{P}(M_t \leq b) \approx \exp(-b\phi(b)t); \qquad t \text{ large} \tag{D9c}$$

for the standard process.

For later use, we record that by (D1c) the mean clump size of $\{ t : X_t \geq b \}$ for the general Ornstein-Uhlenbeck process is

$$EC_b = \frac{1}{2}\sigma^2\mu^{-2}b^{-2}. \tag{D9d}$$

D10 Gaussian processes. The stationary Ornstein-Uhlenbeck process is not only a diffusion but also a Gaussian process (recall Section C22 for discussion) with mean zero and covariance function

$$\begin{aligned} R(t) &= (2\mu)^{-1}\sigma^2\exp(-\mu|t|) \quad \text{(general)} \\ &= \exp(-|t|) \quad \text{(standard)} \end{aligned} \tag{D10a}$$

In particular,

$$\begin{aligned} R(t) &\sim (2\mu)^{-1}\sigma^2 - \tfrac{1}{2}\sigma^2|t| \quad \text{as } t \to 0 \quad \text{(general)} \\ &\sim 1 - |t| \quad \text{as } t \to 0 \quad \text{(standard)} \end{aligned} \tag{D10b}$$

Consider now a stationary mean-zero Gaussian process X_t which is not Markovian. In Section C23 we treated the case where $R(t) \sim R(0) - \theta t^2$ as $t \to 0$, which is the case where the sample paths are differentiable. Consider now the case

$$R(t) \sim v - \theta|t| \quad \text{as } t \to 0. \tag{D10c}$$

This is the "locally Brownian" case. For we can directly calculate that, given $X_0 = x$, then $X_t - x \overset{\mathcal{D}}{\approx} \text{Normal}(-\frac{\theta}{v}xt, 2\theta t)$ for small $t \geq 0$; more generally, given $X_0 = x$ then for small t the process $X_t - x$ is approximately Brownian motion with drift $-\theta x$ and variance $2\theta v$. Thus we can apply our heuristic (Section D2) in this setting. Rather than repeat calculations, for a Gaussian process satisfying (D10c) we simply "match" with the corresponding Ornstein-Uhlenbeck process via

$$\sigma^2 = 2\theta; \qquad \mu = \frac{\theta}{v}. \tag{D10d}$$

Then (D9a) shows that, for a Gaussian process of form (D10c),

$$T_b \overset{\mathcal{D}}{\approx} \text{exponential, rate } \lambda_b = \theta v^{-3/2} b\phi(bv^{-1/2}). \tag{D10e}$$

It is worth recording also that, by (D9d), the mean clump size for $\{ t : X_t \geq b \}$ is

$$EC_b = v^2\theta^{-1}b^{-2}. \tag{D10f}$$

In practice, when studying a stationary Gaussian process it is natural to scale so that the variance is 1, so let us explicitly state:

> For a stationary mean-zero Gaussian process with $R(t) \sim 1 - \theta|t|$ as $t \to 0$, we have $T_b \overset{\mathcal{D}}{\approx}$ exponential, rate $\lambda_b = \theta b\phi(b)$, $b \geq 3$. (D10g)

Of course, the heuristic also requires some "no long-range dependence" condition, which essentially means that $R(t)$ must go to 0 not too slowly as $t \to \infty$.

D11 Example: System response to external shocks. Suppose a unit "shock" at time t_0 causes response $h(t-t_0)$ at times $t \geq t_0$, where h is a smooth function with $h(t) \downarrow 0$ as $t \to \infty$. At Example C13 we considered the case of Poisson shocks; now let us consider the case of "white noise", in which case the total response at time t may be written as

$$X_t = \int_{-\infty}^{t} f(t-s)\,dB_s. \qquad \text{(D11a)}$$

Then X is a stationary mean-zero Gaussian process with covariance function

$$R(t) = \int_0^\infty f(s)f(s+t)\,ds.$$

Suppose we normalize so that $\text{var}\, X_t \equiv \int_0^\infty f^2(s)\,ds = 1$. Then (D10g) gives the heuristic approximation for T_b, or equivalently for $\max_{s \leq t} X_s$, in terms of

$$\theta = -R'(0) = -\int_0^\infty f(s)f'(s)\,ds = \frac{1}{2}f^2(0).$$

More interesting examples involve non-stationary forms of the heuristic, arising from non-stationary processes or from curved boundary crossing problems. Here is a simple example of the former.

D12 Example: Maximum of self-normalized Brownian bridge. Let B_t^0 be the standard Brownian bridge and, for small $a > 0$, consider

$$M_a = \max_{a \leq t \leq 1-a} \left(\frac{B_t^0}{\sigma(t)} \right),$$

where $\sigma^2(t) = \text{var}\, B_t^0 = t(1-t)$. We can re-write this as

$$M_a = \max_{a \leq t \leq 1-a} X_t; \qquad X_t = \frac{B_t^0}{\sigma(t)}$$

and now X is a mean-zero Gaussian process with variance 1. But X is not stationary; instead we can calculate that the covariance function satisfies

$$R(t, t+s) \sim 1 - \theta_t |s| \quad \text{as } s \to 0; \quad \text{where } \theta_t = (2t(1-t))^{-1}. \qquad \text{(D12a)}$$

For any Gaussian process of the form (D12a) (for any θ_t), we can argue as follows. We want to apply the heuristic to the random set $\mathcal{S} = \{\, t : X_t \geq b \,\}$. Around any fixed t_0 the process behaves like the stationary process with

$R(s) \sim 1 - \theta_{t_0}|s|$ as $s \to 0$, so that the clump rate around t_0 is given by (D10g) as

$$\lambda_b(t_0) = \theta_{t_0} b\phi(b).$$

Thus the non-stationary form (A4g) of the heuristic gives

$$\begin{aligned} \boldsymbol{P}(\max_{t_1 \le t \le t_2} X_t \le b) &= \boldsymbol{P}(\mathcal{S} \cap [t_1, t_2] \text{ empty}) \\ &\approx \exp\Big(-\int_{t_1}^{t_2} \lambda_b(t)\,dt\Big) \\ &\approx \exp\Big(-b\phi(b)\int_{t_1}^{t_2} \theta_t\,dt\Big). \end{aligned} \qquad \text{(D12b)}$$

In our particular case, the integral is

$$\int_a^{1-a} (2t(1-t))^{-1}\,dt = \log(a^{-1} - 1)$$

and so we get

$$\boldsymbol{P}(M_a \le b) \approx \exp(-b\phi(b)\log(a^{-1} - 1)). \qquad \text{(D12c)}$$

D13 Boundary-crossing. For a locally Brownian process X_t we can use the heuristic to study the first hitting (or crossing) times

$$T = \min\{\, t : X_t = b(t)\,\} \qquad \text{(D13a)}$$

where $b(t)$ is a smooth boundary or barrier. The essential requirement is that the boundary be *remote* in the sense

$$\boldsymbol{P}(X_t \ge b(t)) \text{ is small for each } t. \qquad \text{(D13b)}$$

(our discussion treats upper boundaries, but obviously can be applied to lower boundaries too). Recall now the discussion (Section D2) of the heuristic for a stationary locally Brownian process X_t crossing the level b. There we used the random set $\mathcal{S}_1 = \{\, t : X_t \in (b, b+\delta)\,\}$ for δ small. So here it is natural to use the random set $\mathcal{S} = \{\, t : X_t \in (b(t), b(t)+\delta)\,\}$. In estimating the clump size, a crude approximation is to ignore the slope of the boundary and replace it by the level line; thus estimating the mean clump size for a clump near t_0 as the mean clump size for $\{\, t : X_t \in (b(t), b(t)+\delta)\,\}$. And this is tantamount to estimating the clump rate for $\mathcal{S}$ as

$$\begin{aligned} \lambda(t) &= \lambda_b(t); \quad \lambda_b \text{ the clump rate for } X_t \text{ crossing level } b &\text{(D13c)} \\ &= f_X(b(t))\mu(b(t)) \qquad \text{by (D2b).} &\text{(D13d)} \end{aligned}$$

Naturally, for this "level approximation" to be sensible we need $b(t)$ to have small slope:

$$b'(t) \text{ small for all } t. \qquad \text{(D13e)}$$

Taking account of the slope involves some subtleties, and is deferred until Section D29: the asymptotics of the next few examples are not affected by the correction for slope.

As usual, given the clump rate $\lambda(t)$ for $\mathcal{S}$ we estimate

$$\boldsymbol{P}(T > t) \approx \exp(-\int_0^t \lambda(s)\,ds) \tag{D13f}$$

$$\boldsymbol{P}\begin{pmatrix}X \text{ does not cross bound-}\\ \text{ary between } t_1 \text{ and } t_2\end{pmatrix} \approx \exp(-\int_{t_1}^{t_2} \lambda(s)\,ds). \tag{D13g}$$

Similarly, we can adapt (D2d, D2e) to this setting to obtain a "non-homogeneous compound Poisson" approximation for the length of time X_t spends above the boundary $b(t)$, but the results are rather complicated.

D14 Example: Boundary-crossing for reflecting Brownian motion. As the simplest example of the foregoing, consider reflecting Brownian motion X_t as in Example D6. For a remote barrier $b(t)$ with $b'(t)$ small, we can put together (D6b) and (D13c, D13g) to get

$$\begin{aligned}&\boldsymbol{P}(X \text{ does not cross } b(t) \text{ between } t_1 \text{ and } t_2)\\ &\approx \exp\left(\frac{-2\mu^2}{\sigma^2}\cdot\int_{t_1}^{t_2} \exp\left(\frac{-2\mu b(t)}{\sigma^2}\right)dt\right).\end{aligned} \tag{D14a}$$

We can use these estimates to study asymptotic sample path questions such as: is $X_t \le b(t)$ ultimately (i.e. for all sufficiently large t)? Indeed (D14a) gives

$$\boldsymbol{P}(X_t \le b(t) \text{ ultimately}) = \begin{cases} 1 & \text{if } \int^\infty \exp\left(\frac{-2\mu b(t)}{\sigma^2}\right)dt < \infty \\ 0 & \text{if } \int^\infty \exp\left(\frac{-2\mu b(t)}{\sigma^2}\right)dt = \infty \end{cases}. \tag{D14b}$$

In particular, if we consider $b(t) = c\log t$ for large t,

$$\boldsymbol{P}(X_t \le c\log t \text{ ultimately}) = \begin{cases} 1 & \text{if } c > \frac{\sigma^2}{2\mu} \\ 0 & \text{if } c < \frac{\sigma^2}{2\mu} \end{cases}.$$

In other words,

$$\limsup_{t\to\infty} \frac{X_t}{\log t} = \frac{\sigma^2}{2\mu}. \quad \text{a.s.} \tag{D14c}$$

There is an important conceptual point to be made here. The initial approximation (D14a) is already somewhat rough, because e.g. of the "level approximation" in Section D13. Then the "integral test" result (D14b) is much cruder, being purely asymptotic and throwing away the constant factor "$2\mu^2/\sigma^2$" as irrelevant for convergence of the integral. Finally the "lim sup" result (D14c) is much cruder again, a kind of 1-parameter version of (D14b). The general point is that a.s. limit results may look deep or sharp at first sight, but are always merely crude corollaries of distributional approximations which tell you what's really going on.

D15 Example: Brownian LIL. Let X_t be the standard Ornstein-Uhlenbeck process, and let $b(t)$ be a remote barrier with $b'(t)$ small. We can argue exactly as in the last example. Putting together (D9b) and (D13c, D13g) gives

$$\boldsymbol{P}\begin{pmatrix} X_t \text{ does not cross } b(t) \\ \text{between } t_1 \text{ and } t_2 \end{pmatrix} \approx \exp(-\int_{t_1}^{t_2} b(t)\phi(b(t))\,dt). \tag{D15a}$$

Then

$$\boldsymbol{P}(X_t \leq b(t) \text{ ultimately}) = \begin{cases} 1 & \text{if } \int^\infty b(t)\exp(-\frac{1}{2}b^2(t))\,dt < \infty \\ 0 & \text{if } \int^\infty b(t)\exp(-\frac{1}{2}b^2(t))\,dt = \infty \end{cases}. \tag{D15b}$$

Considering $b(t) = (2c\log t)^{1/2}$ gives

$$\limsup_{t\to\infty} \frac{X_t}{(2\log t)^{\frac{1}{2}}} = 1 \quad \text{a.s.} \tag{D15c}$$

The significance of these results is that standard Brownian motion and standard Ornstein-Uhlenbeck process are deterministic space-and-time-changes of each other. Specifically,

> If $B(t)$ is standard Brownian motion then $X(t) \equiv e^{-t}B(e^{2t})$ is standard Ornstein-Uhlenbeck process; if $X(t)$ is standard Ornstein-Uhlenbeck process then $B(t) \equiv t^{1/2}X(\frac{1}{2}\log t)$ is standard Brownian motion. (D15d)

Using this transformation, it is easy to see that (D15c) is equivalent to the usual law of the iterated logarithm for a standard Brownian motion:

$$\limsup_{t\to\infty} \frac{B_t}{(2t\log\log t)^{\frac{1}{2}}} = 1 \quad \text{a.s.} \tag{D15e}$$

And (D15b) is equivalent to the integral test

$$\boldsymbol{P}(B_t \leq t^{\frac{1}{2}}c(t) \text{ ultimately}) = \begin{cases} 1 & \text{if } \int^\infty t^{-1}c(t)\exp(-\frac{1}{2}c^2(t))\,dt < \infty \\ 0 & \text{if } \int^\infty t^{-1}c(t)\exp(-\frac{1}{2}c^2(t))\,dt = \infty \end{cases}. \tag{D15f}$$

Again, it is important to understand that these a.s. limit results are just crude consequences of distributional approximations for boundary-crossings. It is straightforward to use the heuristic (D15a) to obtain more quantitative information, e.g. approximations to the distributions of the times of

$$\text{the last crossing of } B_t \text{ over } (ct\log\log t)^{1/2},\ (c > 2) \tag{D15g}$$

the first crossing after t_0 of B_t over $(ct \log\log t)^{1/2}$, $(c < 2)$: (D15h)

since the calculus gets messy we shall leave it to the reader!

D16 Maxima and boundary-crossing for general Gaussian processes. Given a Gaussian process X_t with $m(t) \equiv EX_t$ non-zero, and a boundary $b(t)$, then we can simplify the boundary-crossing problem in one of two ways: replace X_t by $X_t - m(t)$ and $b(t)$ by $b(t) - m(t)$, to get a boundary-crossing problem for a mean-zero process; or replace X_t by $X_t - b(t)$ to get a level-crossing problem for a non-zero-mean process. The former is useful when the transformed boundary is only slowly sloping, as the examples above show. The latter is useful when $\boldsymbol{P}(X_t > b(t))$ is maximized at some point t^* and falls off reasonably rapidly on either sider of t^*. In such cases, the transformed problem can often be approximated by the technique in the following examples. The technique rests upon the fact: if $f(t)$ has its minimum at t_0, if $f(t_0) > 0$ and $\phi(f(t_0))$ is small, and if f and g are smooth and $g > 0$, then

$$\int g(t)\phi(f(t)) \approx (f(t_0)f''(t_0))^{-\frac{1}{2}} \exp(-\frac{1}{2}f^2(t_0))g(t_0). \tag{D16a}$$

This is a restatement of (C21e), obtained by Laplace's method.

D17 Example: Maximum of Brownian bridge. Let B_t^0, $0 \leq t \leq 1$ be Brownian bridge, that is the non-homogeneous Gaussian diffusion with

$$\begin{aligned} B_t^0 &\stackrel{\mathcal{D}}{=} \text{Normal}(0, \sigma^2(t)), \qquad \sigma^2(t) = t(1-t) \\ \text{cov}(B_s^0, B_t^0) &= s(1-t), \qquad s \leq t \end{aligned} \tag{D17a}$$

$$\text{drift rate } \mu(x,t) = -\frac{x}{1-t}; \qquad \text{variance rate } \equiv 1. \tag{D17b}$$

Let $M = \sup_{0 \leq t \leq 1} B_t^0$. It turns out that M has a simple exact distribution:

$$\boldsymbol{P}(M > b) = \exp(-2b^2), \qquad 0 \leq b < \infty. \tag{D17c}$$

Let us see what the heuristic gives. Fix a high level b. Let S be the random set $\{\, t : B_t^0 \in (b, b+\delta) \,\}$. By (D17a),

$$p_b(t) \equiv \frac{\boldsymbol{P}(B_t^0 \in (b, b+\delta))}{\delta} = \sigma^{-1}(t)\phi\left(\frac{b}{\sigma(t)}\right).$$

Now the non-homogeneous version of our heuristic (D2b) is

$$\lambda_b(t) = p_b(t)\mu_b(t) \tag{D17d}$$

where $-\mu_b(t)$ is the drift of the incremental process $(X_{t+u} - b \mid X_t = b)$, $u \geq 0$. By (D17b), $\mu_b(t) = b/(1-t)$ and hence

$$\lambda_b(t) = b(1-t)^{-1}\sigma^{-1}(t)\phi\left(\frac{b}{\sigma(t)}\right). \tag{D17e}$$

Since we are interested in the tail of M rather than the whole distribution, we write (A10f)

$$\boldsymbol{P}(M > b) \approx \int_0^1 \lambda_b(t)\,dt; \qquad b \text{ large.} \tag{D17f}$$

We estimate the integral using (D16a), with

$$f(t) = \frac{b}{\sigma(t)} = \frac{b}{\sqrt{t(1-t)}}; \qquad g(t) = b(1-t)^{-1}\sigma^{-1}(t).$$

We easily find

$$t_0 = \frac{1}{2}; \qquad f(t_0) = 2b; \qquad f''(t_0) = 8b; \qquad g(t_0) = 4b$$

and so (D16a) gives

$$\boldsymbol{P}(M > b) \approx \int \lambda_b(t)\,dt \approx \exp(-2b^2); \qquad b \text{ large.} \tag{D17g}$$

It is purely fortuitous, though reassuring, that the heuristic approximation gives the actual exact answer in this example. In the examples below the heuristic estimates of $\boldsymbol{P}(M > b)$ are only asymptotically correct. The maximum of the Brownian bridge arises as the limiting null distribution of the Kolmogorov-Smirnov test statistic in 1 dimension; d-dimensional analogs lead to the study of maxima of Gaussian fields, treated in Chapter J.

D18 Maxima of non-stationary Gaussian processes. We can abstract the last example as follows. Let X_t, $t_1 \leq t \leq t_2$ be Gaussian with

$$EX_t = m(t), \qquad \operatorname{var} X_t = v(t), \tag{D18a}$$

$$\begin{aligned} &E(X_{t+u} - b \mid X_t = b) \sim -u\mu(b,t) \\ &\text{and } \operatorname{var}(X_{t+u} \mid X_t = b) \sim u\sigma^2(t) \text{ as } u \downarrow 0. \end{aligned} \tag{D18b}$$

Let $M = \max_{t_1 \leq t \leq t_2} X_t$. Fix b and let $f_b(t) = (b - m(t))/v^{1/2}(t)$. Suppose $f_b(t)$ is minimized at some $t_b^* \in (t_1, t_2)$ and $\phi(f(t_b^*))$ is small. Then

$$\boldsymbol{P}(M > b) \approx v^{-\frac{1}{2}}(t_b^*)\mu(b, t_b^*)\left(f_b(t_b^*)f_b''(t_b^*)\right)^{-\frac{1}{2}}\exp(-f_b^2(t_b^*)/2). \tag{D18c}$$

This is exactly the same argument as in Example D17; as at (D17d), the random set $\mathcal{S} = \{\, t : X_t \in (b, b+\delta) \,\}$ has clump rate

$$\lambda_b(t) = p_b(t)\mu(b,t) = v^{-\frac{1}{2}}(t)\phi(f_b(t))\mu(b,t),$$

and writing $\boldsymbol{P}(M > b) \approx \int \lambda_b(t)\,dt$ and estimating the integral via (D16a) gives the result (D18c).

Here are three examples.

D19 Example: Maximum of Brownian Bridge with drift. Consider

$$\begin{aligned} X_t &= B_t^0 + ct, \quad 0 \le t \le 1, \qquad \text{where } B^0 \text{ is Brownian bridge;} \\ M &= \max X_t. \end{aligned}$$

We study $\boldsymbol{P}(M > b)$ for $b > \max(0, c)$; this is equivalent to studying the probability of B^0 crossing the sloping line $b - ct$.

In the notation of Section D18 we find

$$\begin{aligned} m(t) &= ct \\ v(t) &= t(1-t) \\ \mu(b,t) &= \frac{b-c}{1-t} \\ f_b(t) &= (b-ct)(t(1-t))^{-\frac{1}{2}} \\ f_b(t_b^*) &= 2(b(b-c))^{\frac{1}{2}} \\ t_b^* &= \frac{b}{2b-c} \\ v(t_b^*) &= b(b-c)(2b-c)^{-2} \\ f_b''(t_b^*) &= \frac{1}{2}(2b-c)v^{-\frac{3}{2}}(t_b^*) \end{aligned}$$

and (D18c) gives

$$\boldsymbol{P}(M > b) \approx \exp(-b(b-c)); \qquad b \text{ large.} \tag{D19a}$$

D20 Example: Brownian motion and quadratic boundary. Let

$$X_t = B_t - t^2, \qquad t \ge 0.$$

We study $M = \sup_{t \ge 0} X_t$. More generally, for $a, \sigma > 0$ we could consider

$$M_{\sigma,a} = \sup_{t \ge 0} \sigma B_t - at^2$$

but a scaling argument shows $M_{\sigma,a} \stackrel{\mathcal{D}}{=} a^{1/5}\sigma^{4/5} M$. Studying M is equivalent to studying crossing probabilities for quadratic boundaries:

$$\boldsymbol{P}(B_t \text{ crosses } b + at^2) = \boldsymbol{P}(M_{1,a} > b) = \boldsymbol{P}(M > ba^{-1/5}).$$

In the notation of Section D18,

$$\begin{aligned}
m(t) &= -t^2 \\
v(t) &= t \\
\mu(b,t) &= 2t \\
f_b(t) &= (b+t^2)t^{-\frac{1}{2}} \\
t_b^* &= \left(\frac{b}{3}\right)^{1/3} \\
f_b(t_b^*) &= 4\left(\frac{b}{3}\right)^{3/4} \\
f_b''(t_b^*) &= 3^{5/4}b^{-1/4}
\end{aligned}$$

and (D18c) gives

$$\boldsymbol{P}(M > b) \approx 3^{-\frac{1}{2}} \exp\left(-\left(\frac{4b}{3}\right)^{\frac{3}{2}}\right), \qquad b \text{ large.} \tag{D20a}$$

D21 Example: Ornstein-Uhlenbeck quadratic boundary. Let Y_t be the standard Ornstein-Uhlenbeck process (Example D9), let

$$X_t = Y_t - at^2, \qquad -\infty < t < \infty.$$

and let $M_a = \sup X_t$. For small a, we shall estimate the distribution of M_a. This arises in the following context. If $Z(t)$ is a process which can be regarded as "a deterministic process $z(t)$ + small noise", then one can often model $Z(t) - z(t)$ as an Ornstein-Uhlenbeck process $\widehat{Y}_t$, say, with parameters (σ^2, μ) as at Example D9, with σ^2 small. Suppose $z(t)$ is maximized at t_0, and suppose we are interested in $\max Z(t)$. Then we can write

$$\begin{aligned}
\max_t Z(t) - z(t_0) &\approx \max_t \widehat{Y}_t - \frac{1}{2}z''(t_0)t^2 \\
&\stackrel{\mathcal{D}}{=} 2^{-\frac{1}{2}}\sigma M_a; \quad \text{where } a = -\frac{1}{2}\mu^{-2}z''(t_0).
\end{aligned}$$

(The last relation is obtained by scaling the general Ornstein-Uhlenbeck process $\widehat{Y}$ into the standard Ornstein-Uhlenbeck process Y).

In the notation of Section D18,

$$\begin{aligned}
m(t) &= -at^2 \\
v(t) &= 1 \\
f_b(t) &= b + at^2 \\
t_b^* &= 0 \\
f_b(t_b^*) &= b \\
f_b''(t_b^*) &= 2a \\
\mu(b,0) &= b
\end{aligned}$$

and (D18c) gives

$$\boldsymbol{P}(M_a > b) \approx \lambda_{a,b} = \left(\frac{b}{2a}\right)^{\frac{1}{2}} \exp(-\frac{1}{2}b^2); \qquad b \text{ large} \tag{D21a}$$

In this setting, as $a \to 0$ we can use the full form of the heuristic to approximate the whole distribution of M_a:

$$\boldsymbol{P}(M_a \leq b) \approx \exp(-\lambda_{a,b}); \qquad a \text{ small.} \tag{D21b}$$

Remarks All of our examples so far in this chapter rest upon the simple form (Section D2) of the heuristic. This is certainly the most natural form of the heuristic to use for locally Brownian processes; and we could give more examples in the same spirit. Instead, it seems more interesting to give applications of other forms of the heuristic. Example D23 is a nice application of the "conditioning on semi-local maxima" approach: here are some preliminaries.

D22 Semi-local maxima for the Ornstein-Uhlenbeck process. For the standard Ornstein-Uhlenbeck process, our basic result is that the clump rate for $\{\, t : X_t \geq b \,\}$ is

$$\lambda_b = b\phi(b), \qquad b \text{ large.}$$

Recall from Sections A7 and C26 the notion of the point process (t^*, x^*) of times and heights of semi-local maxima of X_t. At high levels, this will approximate the Poisson point process of rate

$$L(x) = -\frac{d}{dx}\lambda_x = x^2\phi(x) \tag{D22a}$$

(to first order), by (C26b). It is interesting to derive this directly by the "conditioning on semi-local maxima" form of the heuristic.

Given $X_0 = x_0$, we know that $x_0 - X_t$ behaves (in the short term) as Brownian motion Y_t with drift x_0 and variance 2. Now condition on $X_0 = x_0$ and on $(0, x_0)$ being a semi-local maximum; then $x_0 - X_t \equiv Z_t^{x_0}$, say, will behave as Y_t conditioned to stay positive. It turns out (see Section D41) that such "conditioned Brownian motion" is a certain diffusion, and we can therefore calculate its mean sojourn density at y, which turns out to be

$$m(x_0, y) = x_0^{-1}(1 - \exp(-x_0 y)). \tag{D22b}$$

Now as at (C26e) the rate $L(x)$ of semi-local maxima satisfies

$$\phi(y) = 2\int_y^\infty L(x)m(x, x-y)\,dx \tag{D22c}$$

where the factor 2 appears because (C26e) uses two-sided occupation density. Anticipating a solution of the form $L(x) = a(x)\phi(x)$ for polynomial $a(x)$, (D22c) becomes

$$\begin{aligned} 1 &\approx 2a(y)\int_y^\infty \frac{\phi(x)}{\phi(y)} \cdot m(x, y-x)\,dx \\ &\approx 2a(y)y^{-1}\int_0^\infty e^{-yu}(1-e^{-yu})\,du \\ &= a(y)y^{-2}, \qquad \text{giving (D22a).} \end{aligned}$$

D23 Example: A storage/queuing process. Imagine a supermarket with a linear parking lot with spaces number $1, 2, 3, \ldots$. Cars arrive as a Poisson process of rate ρ, and park in the lowest numbered vacant space. Each car remains for a random time with exponential(1) distribution, independent of everything else, and then departs. This describes a certain Markov process whose states are subsets of the positive integers, representing the set of occupied spaces. This process has a stationary distribution; it seems hard to describe completely the stationary distribution, but some features can be studied. Consider the sub-processes

$$\begin{aligned} V_t &= \text{number of cars parked at time } t \\ R_t &= \text{right-most occupied space at time } t. \end{aligned}$$

The process V_t is just the M/M/∞ queue, whose stationary distribution is Poisson(ρ). R_t is a complicated non-Markov process, but we shall give a heavy-traffic ($\rho \to \infty$) approximation. The first idea is that the rescaled "number of cars" process $X_t = \rho^{-1/2}(V_t - \rho)$ approximates, for large ρ, the standard Ornstein-Uhlenbeck process (calculate conditional means and variances!). Set $D = \log(\rho^{1/2})$ and define $b = b(\rho)$ by

$$b\phi(b)D = 1. \tag{D23a}$$

The point process $\mathcal{N}$ of semi-local maxima (x, t) of X is approximately Poisson with rate

$$\begin{aligned} L(x) &= x^2\phi(x) \qquad \text{by (D22a)} \\ &\approx b^2\phi(b)\exp(-b(x-b)) \qquad \text{for } x \text{ around } b \\ &\approx \frac{b}{D}\cdot\exp(-b(x-b)) \qquad \text{by (D23a).} \end{aligned} \tag{D23b}$$

Now consider a semi-local maximum (v_0, t_0) of V, where v_0 is around $\rho + \rho^{1/2}b$, and consider how many of the K cars in places ρ thru v_0 are still present at time $t_0 + t$. Each car has chance e^{-t} of being present. So for $t = (1-\epsilon)\log K$ this chance is $K^{\epsilon-1}$ and so about K^ϵ cars will still remain;

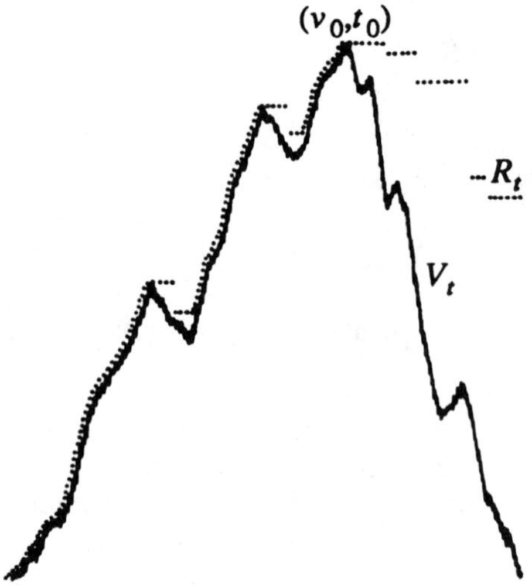

FIGURE D23a.

whereas at time $t = (1+\epsilon)\log K$ it is likely that no cars remain. Thus the position Y_t of the right-most of these cars is about

$$\begin{aligned} Y_t &\approx v_0 \qquad \text{for } t_0 \le t \le t_0 + \log(b\rho^{1/2}) \approx t_0 + D \\ &= \rho + O(\rho^{1/2}) \qquad \text{for } t > t_0 + D \end{aligned}$$

Except for times around semi-local maxima, V_t is small compared to $\rho + b\rho^{1/2}$, and arrivals at those times will not affect R_t. This argument leads to the approximation

$$R_t \approx \max\{\, r : (r,s) \text{ is a semi-local maximum of } V, \text{ for some } t - D < s < t\,\}.$$

Putting $R_t = \rho + \rho^{1/2} R_t^*$,

$$R_t^* \approx \max\{\, x : (x,s) \in \mathcal{N} \text{ for some } t - D < s < t\,\}.$$

Thus the point process description of maxima gives the description of the process R_t. We can specialize to get the stationary distribution:

$$\begin{aligned} \boldsymbol{P}(R_t^* \le r) &\approx \boldsymbol{P}(\text{no points of } \mathcal{N} \text{ in } [r,\infty) \times [t-D, t]) \\ &\approx \exp(-\int_r^\infty \int_{t-D}^t L(x')\, dt'\, dx') \\ &= \exp(-\exp(-b(r-b)) \qquad \text{using (D23b).} \end{aligned}$$

This translates to

$$R_t \stackrel{\mathcal{D}}{\approx} \rho + \rho^{1/2}(b + b^{-1}\xi_3) \tag{D23d}$$

where ξ_3 has the double-exponential distribution (C1c). From (D23a), we can calculate that $b = b(\rho)$ is asymptotic to $(2 \log \log \rho)^{1/2}$.

D24 Approximation by unstable Ornstein-Uhlenbeck process. Our earlier applications of the heuristic to hitting times for diffusions used the fact that around a high level b the diffusion could be approximated by Brownian motion with drift. Occasionally, as in the example below, we are interested in an "unstable equilibrium" point b, and here the natural approximating process is the unstable Ornstein-Uhlenbeck process Y_t, defined as the diffusion with drift and variance

$$\mu(y) = \mu y, \qquad \sigma^2(y) = \sigma^2; \quad \text{where } \mu > 0.$$

This is a transient process. We shall need the result, given by (D3.2), that its mean occupation density at 0 is

$$G_{-\infty,\infty}(0,0) = \frac{1}{2}\pi^{\frac{1}{2}}\mu^{-\frac{1}{2}}\sigma^{-1}. \tag{D24a}$$

D25 Example: Escape from a potential well. As in Example D7 let X_t be a diffusion controlled by a smooth potential H:

$$\mu(x_0) = -H'(x); \qquad \sigma^2(x) = \sigma^2.$$

Suppose H is a double-welled potential, as in the sketch, with the barrier height $H(b) - H(x_0)$ large compared to σ^2. Let T be the time, starting in the well near x_0, to cross into the other well near x_1. By (D3e) one can find an exact expression for ET, but the heuristic gives an informative approximation. First, we can say $ET \approx 2E_{x_0}T_b$, since after reaching b the process is equally likely to descend into either well, and the descent time is small compared to the ascent time. Next, to calculate $E_{x_0}T_b$ there is no harm is making the potential symmetric about b, by replacing the well around x_1 by the mirror image of the left well.

Write $\pi(x)$ for the stationary density and

$$G(b)\delta = E_b(\text{ sojourn time in } (b, b-\delta) \text{ before hitting } x_0 \text{ or } x_1).$$

By the "renewal-sojourn" form (Section A8) of the heuristic applied to $\{\, t : X_t \in (b, b-\delta)\,\}$, this random set has clump rate

$$\lambda_b = \frac{\pi(b)}{G(b)} \tag{D25a}$$

and $T_b \overset{\mathcal{D}}{\approx} \text{exponential}(\lambda_b)$. Now by (D7c),

$$\pi(b) \approx \frac{1}{2}\pi^{-1}\left(\frac{H''(x_0)}{\pi}\right)^{\frac{1}{2}} \exp\left(\frac{-2(H(b) - H(x_0))}{\sigma^2}\right),$$

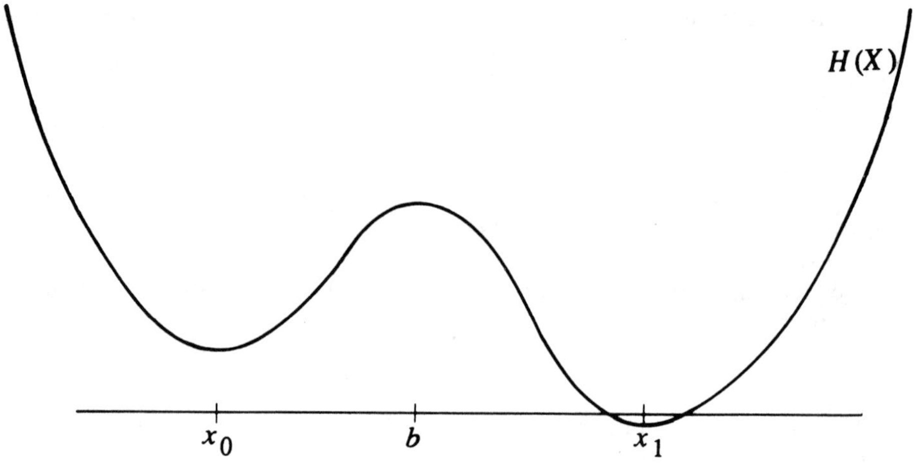

FIGURE D25a.

the factor $\frac{1}{2}$ arising from the fact we have two wells. Next, given $X_{t_0} = b$ the incremental process $Y_t = X_{t_0+t} - b$ behaves in the short run like the unstable Ornstein-Uhlenbeck process with drift

$$\mu(y) = -H'(b+y) \approx -yH''(b).$$

So (D24a) gives

$$G(b) \approx \frac{1}{2}\pi^{\frac{1}{2}}(-H''(b))^{-\frac{1}{2}}\sigma^{-1}.$$

So by (D25a)

$$ET \approx 2ET_b \approx \frac{2}{\lambda_b} \approx 2\pi(-H''(b)H''(x_0))^{-\frac{1}{2}}\exp(2\sigma^{-2}(H(b)-H(x_0))). \tag{D25b}$$

In chemical reaction theory, this is called the *Arrhenius formula.*

D26 Example: Diffusion in random environment. As above, consider a diffusion with variance rate σ^2 controlled by a potential H. Now regard H as a sample path from a stationary process $(H_x : -\infty < x < \infty)$ with smooth paths. For simplicity, suppose the random process H has distribution symmetric in the sense $(H_x) \stackrel{\mathcal{D}}{=} (-H_x)$, and that the clump rates for $\{\, x : H_x \geq b \,\}$ or $\{\, x : H_x \leq -b \,\}$ satisfy

$$\lambda_b \sim a(b)e^{-\theta b} \quad \text{as } b \to \infty; \quad \text{where } a(\cdot) \text{ is subexponential.} \tag{D26a}$$

Suppose also that around large negative local minima $(x^*, H_{x^*} = -b)$ the incremental process $H_{x^*+x} + b$ is locally like $\boldsymbol{\xi}_b x^2$, for (maybe random) $\boldsymbol{\xi}_b$ such that $\boldsymbol{\xi}_b$ does not go to 0 or ∞ exponentially fast as $b \to \infty$.

Consider a large interval $[-L, L]$ of the space line. On this interval $\boldsymbol{P}(\max H_x \leq b) \approx \exp(-L\lambda_b)$ by the heuristic for the extremes of H, and so by (D25a) $\max H_x \approx \theta^{-1}\log L$ and $\min H_x \approx -\theta^{-1}\log L$. Let Δ be the depth of the deepest "well" on the interval; that is, the maximum over $-L \leq x_1 < x_2 < x_3 \leq L$ of $\min(H_{x_1} - H_{x_2}, H_{x_3} - H_{x_4})$. Then $\Delta \approx 2\theta^{-1}\log L$. Now consider the diffusion X_t controlled by H. The arguments of Example D25 indicate that the time to escape from a well of depth Δ is of order $\exp(2\Delta/\sigma^2)$. Hence the time to escape from the deepest well in $[-L, L]$ is of order

$$L^{4/(\theta\sigma^2)} \tag{D26b}$$

Now consider the long-term behavior of X_t as a function of the variance rate σ^2. For $\sigma^2 < 2/\theta$, (D26b) shows that the time to exit $[-L, L]$ is of larger order than L^2, which is the exit time for Brownian motion without drift. In other words, X_t is "subdiffusive" in the sense that $|X_t|$ grows of slower order than $t^{1/2}$. On the other hand, if σ is large then (D26b) suggests that the influence of the potential on the long-term behavior of X_t is negligible compared with the effect of the intrinsic variance, so that X_t should behave roughly like Brownian motion without drift in the long term. (Although the potential will affect the variance).

D27 Interpolating between Gaussian processes. In studying maxima of stationary Gaussian processes, we have studied the two basic cases where the covariance function $R(t)$ behaves as $1 - \frac{1}{2}\theta t^2$ or as $1 - \theta|t|$ as $t \to 0$. But there are other cases. A theoretically interesting case is where there is a fractional power law:

$$R(t) \sim 1 - \theta|t|^\alpha \quad \text{as } t \to 0, \qquad \text{some } 0 < \alpha \leq 2.$$

In this case it turns out that the clump rate λ_b for $\{\,t : X_t \geq b\,\}$, b large, has the form

$$\lambda_b = K_{1,\alpha}\theta^{1/\alpha}b^{-1+2/\alpha}\phi(b)$$

where $K_{1,\alpha}$ is a constant depending only on α. The best way to handle this case is via the "harmonic mean" form of the heuristic; the argument is the same for d-parameter fields, and we defer it until Section J18.

The next example treats a different kind of interpolation.

D28 Example: Smoothed Ornstein-Uhlenbeck. Let X_t be the Ornstein-Uhlenbeck process with covariance $R(t) = \exp(-\mu t)$, $t > 0$. For fixed, small $T > 0$ let Y_t be the smoothed "local average" process

$$Y_t = T^{-1}\int_{t-T}^{t} X_s\, ds.$$

The idea is that a physical process might be modeled adequately on the large scale by an Ornstein-Uhlenbeck process, but on the small scale its paths should be smooth; the averaging procedure achieves this. We calculate

$$\begin{aligned} EY_0 &= 0 \\ EY_0^2 &\approx 1-\frac{\mu T}{3} && \text{(D28a)} \\ EY_0Y_t &\approx EY_0^2-\frac{\mu t^2}{T} \qquad 0\le t\ll T && \text{(D28b)} \\ &\approx \exp(-\mu t) \qquad t\ge T. && \text{(D28c)} \end{aligned}$$

We want to estimate the clump rates $\lambda_{b,T}$ for high levels b. Of course, if we fix T and let $b\to\infty$ we can apply the result for smooth processes. What we want is an estimate which for fixed b interpolates between the Ornstein-Uhlenbeck result for $T=0$ and the smooth result for T non-negligible. This is much harder than the previous examples: there our estimates can be shown to be asymptotically correct, whereas here we are concerned with a non-asymptotic question. Our method, and result, is rather crude.

Let C be the size of the clumps of $\{t : Y_t > b\}$. Let $\rho_{b,T}$ be the rate of downcrossing of Y over b, which we will calculate later from Rice's formula. Now consider the processes X_t, Y_t conditioned on Y making a downcrossing of b at time 0. Under this conditioning, define

$$\begin{aligned} q &= \boldsymbol{P}(Y_t=b \text{ for some } T\le t\le t_0) \\ \alpha &= E(\text{duration of time } t \text{ in } [T,t_0] \text{ that } Y_t>b). \end{aligned}$$

Here t_0, representing "the short term" in the heuristic, is such that $T \ll t_0 \ll$ mean first hitting time on b. We will calculate α later by approximating by Brownian motion. I assert

$$\begin{aligned} \alpha &\approx qEC && \text{(D28d)} \\ \lambda_{b,T} &\approx \rho_{b,T}(1-q). && \text{(D28e)} \end{aligned}$$

The idea is that a downcrossing is unlikely to be followed by an upcrossing within time T; and the distributions of (X,Y) at successive upcrossings within a clump should be roughly i.i.d. Thus the number N of upcrossings in a clump should be roughly geometric: $\boldsymbol{P}(N=n)=(1-q)q^{n-1}$. So (D28e) follows from (A9f) and (D28d) is obtained by conditioning on some upcrossing occurring. Now the fundamental identity and the Normal tail estimate give

$$\lambda_{b,T}EC = \boldsymbol{P}(Y_0>b)\approx\frac{\phi_0(b)}{b}, \text{ where } \phi_0 \text{ is the density of } Y_0 \qquad \text{(D28f)}$$

Solving (D28d), (D28e), (D28f) for $\lambda_{b,T}$ gives

$$\lambda_{b,T} = \left(\frac{1}{\rho_{b,T}}+\frac{\alpha b}{\phi_0(b)}\right)^{-1}. \qquad \text{(D28g)}$$

We calculate $\rho_{b,T}$ from (C23b), using $\theta = 2\mu/T$ because of (D28b):

$$\rho_{b,T} = \phi_0(b) \cdot \left(\frac{\mu}{T\pi}\right)^{\frac{1}{2}}. \qquad \text{(D28h)}$$

Estimating α is harder. Conditional on $X_0 = x_0 \approx b$, we can calculate from (C22a) that for $T \leq t \ll t_0$,

$$E(Y_t - X_0) \approx -b\mu(t - T/2); \qquad \mathrm{var}(Y_t - X_0) \approx 2\mu(t - T/2). \qquad \text{(D28i)}$$

Now α depends only on the conditional means and variances of Y_t, not on the covariances. By (D28i), these means and variances are approximately those of $Z_{t-T/2}$, where Z is Brownian motion with drift $-\widehat{\mu} = -\mu b$ and variance $\widehat{\sigma}^2 = \sigma^2 = 2\mu$, and $Z_0 = b$. So

$$\alpha \approx \widehat{\alpha} \equiv E(\text{duration of time } t \text{ in } [T/2, \infty) \text{ that } Z_t > b).$$

A brief calculation gives

$$\widehat{\alpha} = \frac{\widehat{\sigma}^2}{\widehat{\mu}^2} \cdot \overline{\Phi}(\widehat{\mu} T^{\frac{1}{2}} 2^{-\frac{1}{2}} \widehat{\sigma}^{-1}); \qquad \text{where } \overline{\Phi} \text{ is the Normal tail d.f.}$$

Substituting this and (D28h) into (D28g) and rearranging, we get

$$\lambda_{b,T} = \mu b \phi_0(b)(c\pi^{\frac{1}{2}} + 2\overline{\Phi}(\frac{1}{2}c))^{-1}; \qquad \text{where } c = (\mu b^2 T)^{\frac{1}{2}}. \qquad \text{(D28j)}$$

Note that by (D28a),

$$\phi_0(b) \approx \phi(b) e^{\mu b T/3}; \qquad \text{where } \phi \text{ is the Normal density.}$$

We can check that (D28j) agrees with the known limits. If $T = 0$ then $\overline{\Phi}(0) = \frac{1}{2}$ implies $\lambda_{b,T} = \mu b \phi(b)$, the Ornstein-Uhlenbeck result. If T is fixed and $b \to \infty$, then $\lambda_{b,T} \sim \mu^{1/2} T^{-1/2} \pi^{-1/2} \phi_0(b)$, which is the result for the smooth Gaussian process satisfying (D28b).

D29 Boundary-crossing revisited. Let us return to the setting of Section D2, the heuristic for a locally Brownian process X_t, and suppose now that we have a smooth barrier $b(t)$, which is remote in the sense

$$\boldsymbol{P}(X_t > b(t)) \text{ is small, for each } t.$$

As in Section D2, we can apply the heuristic to $\mathcal{S} = \{\, t : X_t \in (b(t), b(t) + \delta)\,\}$. Given $X_{t_0} = b(t_0)$, the incremental process $X_{t_0+u} - b(t_0)$ behaves for small u like Brownian motion with drift $-\mu(b(t_0))$, and hence $X_{t_0+u} - b(t_0 + u)$ behaves like Brownian motion with drift $-(\mu(b(t_0)) + b'(t_0))$. Using this latter Brownian motion as in Section D2 to estimate the clump size of $\mathcal{S}$, we get the non-homogeneous clump rate

$$\lambda(t) = f_{X_t}(b(t))(\mu(b(t)) + b'(t)). \qquad \text{(D29a)}$$

As usual, this is used in

$$\boldsymbol{P}(X_s \text{ crosses } b(s) \text{ during } [0,t]) \approx 1 - \exp(-\int_0^t \lambda(s)\,ds).$$

Previously (Section D13) we treated boundary-crossing by ignoring the $b'(t)$ term in (D29a); that gave the correct first-order approximation for the type of slowly-increasing boundaries in those examples.

It is natural to regard (D29a) as the "second-order approximation" for boundary-crossing. But this is dangerous: there may be other second-order effects of equal magnitude. Let us consider a standard Ornstein-Uhlenbeck process, and look for a clump rate of the form

$$\lambda(t) = \phi(b(t))a(t); \qquad a(\cdot) \text{ varying slowly relative to } \phi. \tag{D29b}$$

Fix t, and compute the density of X_t at $b(t)$ by conditioning on the time $t-s$ at which the clump $\{\, u : X_u = b(u) \,\}$ started:

$$\begin{aligned}\phi(b(t))\,dt &= \int_{0+} \lambda(t-s)\boldsymbol{P}(X_t \in db(t) \mid X_{t-s} = b(t-s))\,ds \qquad \text{(D29c)}\\ &\approx a(t)\,dt \int_{0+} \phi(b(t-s))(4\pi s)^{-\frac{1}{2}} \exp\left(\frac{-(b(t)+b'(t))^2 s^2}{4s}\right) ds\end{aligned}$$

approximating $X_{t-s+u} - X_{t-s}$ for u small by Brownian motion with drift $-b(t)$ and variance 2. The usual Normal tail estimate (C21d) gives $\phi(b(t-s))/\phi(b(t)) \approx \exp(sb(t)b'(t))$, and so

$$1 \approx a(t) \int_0^\infty (4\pi s)^{-\frac{1}{2}} \exp\left(\frac{-(b(t)-b'(t))^2 s}{4}\right) ds$$

Now the integral also occurs as the mean sojourn density at 0 for Brownian motion with constant drift $(b(t) - b'(t))$ and variance 2; and this mean sojourn density is $1/(b(t) - b'(t))$. Thus $a(t) = b(t) - b'(t)$ and hence

$$\lambda(t) = \phi(b(t))(b(t) - b'(t)). \tag{D29d}$$

This is different from the approximation (D29a): the "+" has turned into a "−".

What's going on here is difficult to say in words. Essentially it is an effect caused by the rapid decrease of the Normal density $\phi(x)$. For a process X_t with marginal density $f(x)$ which decreases exponentially or polynomially (or anything slower than all e^{-ax^2}) as $x \to \infty$, the original approximation (D29a) is correct; so (D29a) applies to reflecting Brownian motion or the Gamma process, for instance. The Normal density is a critical case, and (D29a) does indeed change to (D29d) for the Ornstein-Uhlenbeck process.

D30 Tangent approximation for Brownian boundary-crossing. As in (D15), a time-change argument transforms results for the Ornstein-Uhlenbeck process into results for standard Brownian motion B_t. Here is the transform of (D29d). Let $g(t)$ be smooth with $t^{-1/2}g(t)$ large. Let T be the first crossing time of B_t over $g(t)$. Let

$$\lambda(s) = s^{-3/2}(g(s) - sg'(s))\phi(s^{-\frac{1}{2}}g(s)). \qquad \text{(D30a)}$$

Then under weak conditions

$$\boldsymbol{P}(T \le t) \approx \int_0^t \lambda(s)\,ds \qquad \text{provided this quantity is small;} \qquad \text{(D30b)}$$

and under more stringent conditions

$$\boldsymbol{P}(T > t) \approx \exp(-\int_0^t \lambda(s)\,ds) \qquad \text{for all } t. \qquad \text{(D30c)}$$

This has a nice interpretation: let L_s be the line tangent to g at s, then there is an explicit density $h_{L_s}(t)$ for the first hitting time of B on L_s, and $\lambda(s) = h_{L_s}(s)$. So (D30a) is the *tangent approximation.*

To justify (D30c) one needs roughly (see Section D33 for references) that $g(s)$ grows like $s^{1/2}$. For instance, if

$$g_a(s) = (ct\log(a/t))^{\frac{1}{2}}; \qquad c > 0 \text{ fixed}$$

or

$$g_a(t) = (c + at)^{\frac{1}{2}}; \qquad c > 0 \text{ fixed}$$

then the approximations $\widehat{T}_a$ given by (D30b) are asymptotically correct as $a \to \infty$;

$$\sup_t |\boldsymbol{P}(\widehat{T}_a \le t) - \boldsymbol{P}(T_a \le t)| \to 0 \quad \text{as } a \to \infty.$$

On the other hand, if

$$g_a(t) = at^c; \qquad 0 < c < \frac{1}{2} \text{ fixed}$$

then only the weaker result (D30b) holds: precisely, the approximation $\widehat{T}_a$ satisfies

$$\boldsymbol{P}(\widehat{T}_a \le t_a) \sim \boldsymbol{P}(T_a \le t_a) \quad \text{as } a \to \infty \quad \text{for all } (t_a) \text{ s.t. one side } \to 0.$$

COMMENTARY

D31 General references. The remarks at Section C29 on general stationary processes apply in part to the special processes considered in this chapter. There is no comprehensive account of our type of examples. Leadbetter et al. (1983) Chapter 8 treats Gaussian processes which behave locally like the Ornstein-Uhlenbeck process; papers of Berman (1982a; 1982b; 1983a; 1988) cover diffusions. But these rigorous treatments make the results look hard; the point of the heuristic is to show they are mostly immediate consequences of the one simple idea in Section D2.

D32 Diffusion background. Karlin and Taylor [KT] is adequate for our needs; another good introduction is Oksendal (1985). Rigorous treatments take a lot of time and energy to achieve useful results: Freedman (1971) is a concise rigorous introduction. Rogers and Williams (1987) is the best theoretical overview.

D33 Boundary crossing for Brownian motion. There is a large but somewhat disorganized literature on this topic. Jennen (1985) and Lerche (1986) discuss the tangent approximation (D30), the latter in relation to the LIL. Karatzas and Shreve (1987) sec. 4.3C give an introduction to Brownian motion boundary crossing via differential equations and martingales. Siegmund (1985; 1986) discusses boundary crossing from the statistical "sequential analysis" viewpoint, where the errors in the diffusion approximation need to be considered.

Another set of references to Brownian motion and Ornstein-Uhlenbeck process boundary crossing results can be found in Buonocore et al. (1987).

D34 References for specific examples not mentioned elsewhere. More general queueing examples in the spirit of Example D8 are in Knessl et al. (1986b). For Brownian motion and a quadratic boundary (Example D20) the exact distribution has been found recently by Groeneboom (1988). The problem can be transformed into the maximum of a mean-zero Gaussian process, and relates to bounds on the size of Brownian motion stopped at random times — see Song and Yor (1987). I don't know any discussion of the analogous Ornstein-Uhlenbeck problem (Example D21). The argument in Example D23 (queueing/storage) is from Aldous (1986); an exact expression is in Coffman et al. (1985). Example D26 (diffusion in random environment): this precise example does not seem to have been discussed, although the "phase change" behavior it exhibits is theoretically interesting. Schumacher (1985) treats some related 1-dimensional examples. Bramson and Durrett (1988) treat some discrete d-dimensional models which are subdiffusive. The literature on random walks and diffusions in random environments mostly deals with the case where the drift, not the potential, is stationary (in other words, the potential has stationary increments) — see e.g. Durrett (1986). Example D28 (smoothed Ornstein-Uhlenbeck) is treated differently in Vanmarcke (1982) and Naess (1984).

D35 Slepian model processes. At various places we have used the idea that a process, looked at in a short time interval after an upcrossing, has a simple form. In Gaussian theory this is called the Slepian model process: Lindgren (1984b) gives a nice survey.

D36 Long-range dependence. It must be kept in mind that all our heuristic results require a background assumption of "no long-range dependence". Results for some processes with long-range dependence are given in Taqqu (1979), Maejima (1982), Berman (1984).

D37 Exponential limit distribution for hitting times. For positive-recurrent diffusions, the exponential limit distribution (D4c) is trivial to prove, by adapting the "regeneration" argument of Section B24.1. For non-Markov processes some explicit mixing condition is required.

In principle, for diffusions one can find the exact hitting time distribution by analytic methods (Karlin and Taylor [KT] p. 203), but one rarely gets an explicit solution. The regeneration argument and the argument for (D4h) combine to make a simple rigorous proof of

Proposition D37.1 *For a positive-recurrent diffusion on* (a, ∞)*,*

$$\sup_t \left| P(\max_{0 \le s \le t} X_s \le b) - \exp\left(-\frac{t}{2S(b)M(a,\infty)} \right) \right| \to 0 \quad \textit{as } b \to \infty.$$

Various complicated proofs appeared in the past — see Davis (1982) for discussion. Berman (1983a) discusses smoothness assumptions on μ, σ^2 leading to the simpler form (D4a), and treats sojourn time distributions.

D38 Berman's method. The ergodic-exit form (A9) of the heuristic for continuous process, using clump distribution C^+ for clumps of time spent above b, has been formalized by Berman (1982b). Here are his results, in our heuristic language.

Let X_t be stationary with only short-range dependence. We study

$$\begin{aligned} M_t &= \max_{0 \le s \le t} X_s \\ L_{t,b} &= \text{sojourn time of } X_s,\ 0 \le s \le t, \text{ in } [b, \infty). \end{aligned}$$

Define $X_b^*(t)$ to be the process $X(t)$ conditioned on $\{X(0) > b\}$. Suppose that as $b \to \infty$ we can rescale X^* so that it approximates some limit process Z; that is

$$w(b)\left(X_b^*\left(\frac{t}{v(b)}\right) - y\right) \xrightarrow{\mathcal{D}} Z(t) \quad \text{as } b \to \infty, \tag{D38a}$$

where $v(b) \to \infty$. Let D^+ be the sojourn time of $Z(t)$, $t \geq 0$ in $[0, \infty)$, and let $h(x)$ be the density of D^+.

With those assumptions, here is the heuristic analysis. Let C_b be the distribution of the clumps of time that X spends in $[b, \infty)$, and let λ_b be the clump rate. The fundamental identity is

$$\lambda_b EC_b = \boldsymbol{P}(X_0 > b). \tag{D38b}$$

Think of D^+ as the distribution of the clump of time that Z spends above 0 during time $[0, \infty)$ conditional on $\{Z(0) > 0\}$. The corresponding unconditioned clump distribution D, obtained from the relation (A9d), satisfies

$$\boldsymbol{P}(D > x) = h(x)ED \tag{D38c}$$

$$ED = \frac{1}{h(0)}. \tag{D38d}$$

And from assumption (D38a),

$$v(b)EC_b \to ED \quad \text{as } b \to \infty. \tag{D38e}$$

Solving (D38b,D38d,D38e) for λ_b gives

$$\lambda_b \sim h(0)v(b)\boldsymbol{P}(X_0 > b), \tag{D38f}$$

which we use in our usual estimate form M_t:

$$\boldsymbol{P}(M_t \leq b) \approx \exp(-t\lambda_b). \tag{D38g}$$

One way to make a limit theorem is to fix t and let $b \to \infty$; then

$$\boldsymbol{P}(M_t > b) \sim th(0)v(b)\boldsymbol{P}(X_0 > b) \tag{D38h}$$

which is Berman's Theorem 14.1. Now consider sojourn times in the same setting of t fixed, $b \to \infty$. Ultimately there will be at most one clump, occurring with chance $\sim t\lambda_b$, whose duration C_b satisfies $v(b)C_b \xrightarrow{\mathcal{D}} D$ by the approximation (D38a) of X by Z. So

$$\begin{aligned}\boldsymbol{P}(L_{t,b} \cdot v(b) > x) &\sim t\lambda_b \boldsymbol{P}(D > x) \\ &\sim tv(b)\boldsymbol{P}(X_0 > b)h(x) \quad \text{by (D38c,D38f)}\end{aligned} \tag{D38i}$$

and this is Berman's Theorem 3.1.

Now consider t and b both large. The Compound Poisson form (Section A19) of the heuristic gives

$$L_{t,b} \overset{\mathcal{D}}{\approx} \text{POIS}(t\lambda_b \mu_{C_b}(\cdot)),$$

where μ_{C_b} is the distribution of C_b. Since $v(b)C_b \overset{\mathcal{D}}{\to} D$, this scales to

$$v(b)L_{t,b} \overset{\mathcal{D}}{\approx} \text{POIS}(t\lambda_b \mu_D(\cdot)). \tag{D38j}$$

Now think of $h(\cdot)$ as the measure $h(dx) = h'(x)\,dx$. Then (D38c) says

$$\mu_D(\cdot) = EDh(\cdot)$$

and using (D38f,D38d) we find that (D38j) becomes

$$v(b)L_{t,b} \overset{\mathcal{D}}{\approx} \text{POIS}(tv(b)\boldsymbol{P}(X_0 > b)h(\cdot)). \tag{D38k}$$

This is the "natural" compound Poisson approximation for sojourn time, just as (D38g,D38f) is the "natural" approximation for M_t. To make a limit theorem, define $b = b(t)$ by

$$tv(b)\boldsymbol{P}(X_0 > b) = 1. \tag{D38l}$$

Then (D38k) gives

$$v(b)L_{t,b} \overset{\mathcal{D}}{\to} \text{POIS}(h(\cdot)) \quad \text{as } t \to \infty \tag{D38m}$$

which is Theorem 4.1 of Berman (1983b). A final result, Theorem 19.1 of Berman (1982b), is:

for $b = b(t)$ defined at (D38j),

$$\boldsymbol{P}(w(b)(M_t - b) < x) \to \exp(-h(0)e^{-x}) \quad \text{as } t \to \infty. \tag{D38n}$$

For $x = 0$, this is just (D38f,D38g); establishing this for general x involves a clever rescaling argument for which the reader is referred to the original paper.

This is one of the most wide-ranging formalizations of any version of the heuristic which has been developed. But in several ways it is not completely satisfactory. The reader will notice that we didn't use this form of the heuristic in any of the examples. I do not know any continuous-path example in which this ergodic-exit form is easiest; for locally Brownian process the renewal-sojourn form (Section D2) is easier to use. Thus for ease of application one would like to see a wide-ranging formalization of Section D2. From an opposite viewpoint, any formalization of this ergodic-exit method will require smoothness hypotheses to ensure the density h of D^+ exists; the "harmonic mean" form of the heuristic does not require so much smoothness, and I suspect it can be formalized in greater generality.

D39 Durbin's formula. For a discrete-time, integer-valued, skip-free process X_n, the first hitting time T on a skip-free increasing barrier $b(n)$ satisfies (trivially)

$$\boldsymbol{P}(T=n)=\boldsymbol{P}(X_n=b(n))\boldsymbol{P}(X_m<b(m) \text{ for all } m<n \mid X_n=b(n)). \tag{D39a}$$

Now let X_t be a continuous-time continuous-path process with marginal density f_t; let $b(t)$ be a smooth barrier; let T be the first hitting time of X_t on $b(t)$; and let $g(t)$ be the density of T. Then one expects a formula analogous to (D39a):

$$g(t)=f_t(b(t))\theta(t) \tag{D39b}$$

where $\theta(t)$ is some continuous analogue of the final term of (D39a). For a process with smooth paths it is easy to give a variant of Rice's formula in form (D39b). For locally Brownian processes, it is rather less easy to guess that the formula for $\theta(t)$ in (D39b) is

$$\theta(t)=\lim_{\delta\downarrow 0}\delta^{-1}E((b(t-\delta)-X_{t-\delta})1_{A_{t-\delta}} \mid X_t=b(t)) \tag{D39c}$$

where $A_t=\{X_s<b(s) \text{ for all } 0\le s\le t\}$. Durbin (1985) developed this in the context of Gaussian processes, so we name it *Durbin's formula.* Once written down, it is not so hard to *verify* the formula: heuristically, the essential condition seems to be that $\mathrm{var}(X_{t+\delta}-X_t \mid X_t=x)\sim\sigma^2(x,t)\delta$ as $\delta\downarrow 0$ for smooth σ^2.

Thus another approach to approximations for boundary-crossing probabilities is to start from the exact formula (D39b,D39c) and then approximate. Durbin (1985) develops the tangent approximation for Brownian motion boundary-crossing, and several ingenious and more refined approximations, in this way. Both the theory (exactly what type of processes does (D39c) work for?) and applications seem worthy of further study: a start is made in Rychlik (1987).

D40 Sojourn distribution for Brownian motion. For Brownian motion X_t with drift -1 and variance 1, the sojourn time Γ in $[0,\infty)$ has distribution given by (D1f). This may be derived by setting up and solving a differential equation. A more elegant probabilistic approach is as follows. Let L be the last time t that $X_t=0$. Informally "each time X is at 0 has the same chance to be the last time", so the density $f_L(t)$ is proportional to the density of X_t at 0, giving

$$f_L(t)=(2\pi t)^{-\frac{1}{2}}e^{-\frac{1}{2}t}.$$

Given $L=t_0$, the process X behaves during $[0,t_0]$ as rescaled Brownian bridge, so $(\Gamma,L)=(UL,L)$ where U is the sojourn time in $[0,\infty)$ for Brownian bridge. But U is uniform on $(0,1)$ by a symmetry argument and this specifies the distribution $UL\overset{\mathcal{D}}{=}\Gamma$. See Imhof (1986) for details.

D41 Conditioned diffusion. Consider a diffusion X_t with drift and variance $\mu_(x)$, $\sigma^2(x)$, and let $X_0 = 0$. For $a < 0 < b$, let $\widehat{X}_t$ be X_t conditioned on $\{T_b < T_a\}$, and killed at b. Then $\widehat{X}$ is again a diffusion, and its drift and variance can be calculated explicitly (Karlin and Taylor [KT] p. 261). In some cases we can let $a \to 0$ and $b \to \infty$ and get a limit diffusion $\widehat{X}_t$ which we interpret as "X_t conditioned on $X_t > 0$ for all $t > 0$". In particular, let X_t be Brownian motion with drift $\mu > 0$ and variance σ^2; then $\widehat{X}_t$ has

$$\widehat{\mu}(x) = \mu + 2\mu\left(\exp\left(\frac{2\mu x}{\sigma^2}\right) - 1\right)^{-1}; \qquad \widehat{\sigma}^2 = \sigma^2$$

and the mean occupation density $G(0, x)$ at x is

$$G(0, x) = \mu^{-1}\left(1 - \exp\left(-\frac{2\mu x}{\sigma^2}\right)\right).$$

This result is used in the discussion of semi-local maxima at (D22).

D42 The quasi-Markov estimate of clump size. The argument of Example D28 can be abstracted as follows. Consider a sparse random mosaic $\mathcal{S}$ on $\boldsymbol{R}^1$, where the clumps consist of N component intervals. Condition on 0 being the right endpoint of some component interval of a clump $\mathcal{C}$; let

$$\mathcal{C}^+ = \mathcal{C} \cap (0, \infty), \qquad C^+ = \text{length}(\mathcal{C}^+).$$

Now assume N has a geometric distribution:

$$\boldsymbol{P}(N = n) = q(1-q)^{n-1}, \quad n \geq 1 \text{ (for some } q)$$

and suppose the lengths of the component intervals are i.i.d. This implies

$$EC^+ = (1-q)EC. \tag{D42a}$$

Using the notation of Section A9, write $p = \boldsymbol{P}(x \in \mathcal{S})$ and let ψ be the rate of component intervals of clumps. Then

$$\lambda = \psi q \quad \text{by (A9f)} \tag{D42b}$$

$$p = \lambda EC \quad \text{by the fundamental identity.} \tag{D42c}$$

Eliminating q and EC from these equations gives *the quasi-Markov estimate* of the clump rate λ:

$$\lambda = \left(\frac{EC^+}{p} + \frac{1}{\psi}\right)^{-1}. \tag{D42d}$$

We should emphasize that (D42d) is unlikely to give the "correct" value of λ. Rather, it is a crude method to use only when no better method can be found.

E Simple Combinatorics

E1 Introduction. Here are four classic elementary problems, and approximate solutions for large N.

For the first three problems, imagine drawing at random with replacement from a box with N balls, labeled 1 through N.

E1.1 Waiting time problem. What is the number T of draws required until a prespecified ball is drawn?

Solution: $T/N \overset{\mathcal{D}}{\approx} \text{exponential}(1)$.

E1.2 Birthday problem. What is the number T of draws required until some (unspecified) ball is drawn which had previously been drawn?

Solution: $T/N^{1/2} \overset{\mathcal{D}}{\approx} \mathcal{R}$, where $\boldsymbol{P}(\mathcal{R} > x) = \exp(-x^2/2)$.

E1.3 Coupon-collector's problem. What is the number T of draws required until every ball has been drawn at least once?

Solution: $T \approx N \log N$, or more precisely

$$N^{-1}(T - N\log N) \overset{\mathcal{D}}{\approx} \boldsymbol{\xi}, \qquad \text{where } \boldsymbol{P}(\boldsymbol{\xi} \le x) = \exp(-e^{-x}).$$

For the fourth problem, imagine two well-shuffled decks of cards, each deck having cards labeled 1 through N. A *match* occurs at i if the i'th card in one deck is the same (i.e., has the same label) as the i'th card in the other deck.

E1.4 Matching problem. What is the total number T of matches between the two decks?

Solution: $T \overset{\mathcal{D}}{\approx} \text{Poisson}(1)$.

By stretching our imagination a little, we can regard almost all the problems discussed in these notes as generalizations of these four elementary problems. For instance, problem E1.1 concerns the time for a certain process X_1, X_2, X_3, (which happens to be i.i.d. uniform) to first hit a value i; Chapters B, C, D were mostly devoted to such first hitting time problems for more general random processes. Chapter F will give extensions of problems E1.2–E1.4 to Markov chains, and Chapter H will treat geometrical problems, such as the chance of randomly-placed discs covering the unit square, which are generalizations of the coupon-collector's problem.

Of course these basic problems, and simple extensions, can be solved exactly by combinatorial and analytic techniques, so studying them via our heuristic seems silly. But for more complicated extensions it becomes harder to find informative combinatorial solutions, or to prove asymptotics analytically, whereas our heuristics allow us to write down approximations with little effort. The aim of this chapter is to discuss the immediate extensions of Examples E1.2, E1.3, E1.4. First, this is a convenient time to discuss

E2 Poissonization. Let $1 \geq p(1) \geq p(2) \geq \cdots \geq p(n) \to 0$ as $n \to \infty$. Think of $p(n)$ as the probability of some given event happening, in the presence of n objects (balls, particles, random variables, etc.). Sometimes it is easier to calculate, instead of $p(n)$, the chance $q(\theta)$ of this same event happening with a random Poisson(θ) number of objects. Then

$$q(\theta) = \sum_{n \geq 0} p(n) \frac{e^{-\theta} \theta^n}{n!}.$$

Given $q(\theta)$, one might try to invert analytically to find $p(n)$; instead, let us just ask the obvious question "when is $q(n)$ a reasonable estimate of $p(n)$?" I assert that the required condition for q to be a good approximation to p in mid-range (i.e., when $q(n)$ is not near 1 or 0) is

$$-\theta^{1/2} q'(\theta) \text{ is small;} \qquad \text{for } \theta \text{ such that } q(\theta) = 1/2, \text{ say.} \tag{E2a}$$

For consider the extreme case where $p(n)$ jumps from 1 to 0 at n_0, say. Then $q(\theta)$ is a "smoothed" version of $p(n)$, and needs the interval $(n_0 - 2n_0^{1/2}, n_0 + 2n_0^{1/2})$ to go from near 1 to near 0, so the derivative $q'(n_0)$ will be of order $n_0^{-1/2}$. Condition (E2a) stops this happening; it ensures that $q(\theta)$ and thence $p(n)$ do not alter much over intervals of the form $(n \pm n^{1/2})$ in mid-range.

Note that our heuristic usually gives estimates in the form

$$q(\theta) \approx \exp(-f(\theta)).$$

In this case (E2a) becomes, replacing 1/2 by e^{-1} for convenience:

$$\begin{array}{l} q(n) \text{ is a reasonable approximation for } p(n) \text{ in mid-range} \\ \text{provided } \theta^{1/2} f'(\theta) \text{ is small, for } \theta \text{ such that } f(\theta) = 1. \end{array} \tag{E2b}$$

The reader may check this condition is satisfied in the examples where we use Poissonization.

The situation is somewhat different in the tails. By direct calculation,

$$\begin{aligned}&\text{if } p(n) \sim an^j \text{ as } n \to \infty, \text{ then } q(\theta) \sim a\theta^j \text{ as } \theta \to \infty\\&\text{if } p(n) \sim an^j x^n \text{ as } n \to \infty, \text{ then } q(\theta) \sim ax^j\theta^j e^{-(1-x)\theta}.\end{aligned} \qquad \text{(E2c)}$$

Thus when $q(\theta)$ has polynomial tail it is a reasonable estimate of $p(n)$ in the tail, whereas when q has exponential tail we use in the tail the estimate of p obtained from (E2c):

$$\text{if } q(\theta) \sim a\theta^j e^{-s\theta} \text{ as } \theta \to \infty, \text{ then } p(n) \sim a(1-s)^{-j} n^j (1-s)^n. \qquad \text{(E2d)}$$

E3 Example: The birthday problem. Poissonization provides a simple heuristic for obtaining approximations in the birthday problem. Instead of drawing balls at times $1, 2, 3, \ldots$, think of balls being drawn at times of a Poisson process of rate 1. Say a "match" occurs at t if a ball is drawn at t which has previously been drawn. I assert that, for t small compared with N,

the process of matches is approximately a non-homogeneous Poisson process of rate $\lambda(t) = t/N$. (E3a)

Then T = time of first match satisfies

$$\boldsymbol{P}(T > t) \approx \exp\Big(-\int_0^t \lambda(u)\,du\Big) = \exp\left(-\frac{t^2/2}{N}\right). \qquad \text{(E3b)}$$

In other words $T \stackrel{\mathcal{D}}{\approx} N^{1/2}\mathcal{R}$, as stated at (E1.2). To argue (E3a),

$$\begin{aligned}&\boldsymbol{P}(\text{match involving ball } i \text{ during } [t, t+\delta])\\&\quad\approx \delta N^{-1}\boldsymbol{P}(\text{ball } i \text{ drawn before time } t)\\&\quad\approx \delta N^{-1}\frac{t}{N},\end{aligned}$$

and so

$$\boldsymbol{P}(\text{some match during } [t, t+\delta]) \approx \delta\frac{t}{N},$$

which gives (E3a), since the probability is only negligibly affected by any previous matches.

Of course one can write down the exact distribution of T in this basic birthday problem; the point is that the heuristic extends unchanged to variations of the problem for which the exact results become more complex. Here are some examples.

E4 Example: K-matches. What is the distribution of T_K = number of draws until some ball is drawn for the K'th time? I assert that (compare (E3a))

$$\text{the process of } K\text{-matches is approximately a non-homogeneous Poisson process of rate } \lambda(t) = (t/N)^{K-1}/(K-1)! \tag{E4a}$$

So T_K satisfies

$$\boldsymbol{P}(T_K > t) \approx \exp\left(-\int_0^t \lambda(u)\,du\right) = \exp\left(\frac{-t^K}{K!}N^{1-K}\right).$$

That is,

$$T_K \overset{\mathcal{D}}{\approx} N^{1-K^{-1}}\mathcal{R}_K; \qquad \text{where } \boldsymbol{P}(\mathcal{R}_K > x) = \exp(-x^K/K!). \tag{E4b}$$

To argue (E4a),

$$\begin{aligned} &\boldsymbol{P}(K\text{-match involving ball } i \text{ during } [t, t+\delta]) \\ &\quad\approx\ \delta N^{-1}\boldsymbol{P}(\text{ball } i \text{ drawn } K-1 \text{ times before time } t) \\ &\quad\approx\ \delta N^{-1}e^{-t/N}\frac{(t/N)^{K-1}}{(K-1)!} \end{aligned}$$

since the times of drawing ball i form a Poisson process of rate $1/N$. Now $e^{-t/N} \approx 1$ since t is supposed small compared to N; so

$$\boldsymbol{P}(\text{some } K\text{-match during } [t, t+\delta]) \approx \delta\frac{(t/N)^{K-1}}{(K-1)!}$$

giving (E4a).

E5 Example: Unequal probabilities. Suppose at each draw, ball i is drawn with chance p_i, where $\max_{i \le N} p_i$ is small. In this case, T = number of draws until some ball is drawn a second time satisfies

$$T \overset{\mathcal{D}}{\approx} \left(\sum p_i^2\right)^{-\frac{1}{2}} \mathcal{R}. \tag{E5a}$$

For the process of times at which ball i is drawn is approximately a Poisson process of rate p_i. So

$$\begin{aligned} &\boldsymbol{P}(\text{match involving ball } i \text{ during } [t, t+\delta]) \\ &\qquad\approx\ \delta p_i \boldsymbol{P}(\text{ball } i \text{ drawn before time } t) \\ &\qquad\approx\ \delta p_i t p_i; \end{aligned}$$

and so

$$\begin{aligned}&\boldsymbol{P}(\text{some match during } [t, t+\delta]) \\ &\qquad \approx \ \delta \sum p_i^2 t.\end{aligned}$$

Thus we have (E3a) with $1/N$ replaced by $\sum p_i^2$, and this leads to

$$\boldsymbol{P}(T > t) \approx \exp\left(-\frac{1}{2}t^2 \sum p_i^2\right) \tag{E5b}$$

which is equivalent to (E5a).

E6 Example: Marsaglia random number test. Pick K integers i.i.d. uniformly from $\{1, \dots, N\}$, and arrange in increasing order $1 \leq X_1 \leq X_2 \leq \cdots \leq X_K$. Form the successive differences $D_j = X_j - X_{j-1}$, and consider the chance that all these numbers $D_1, D_2, \dots, D_K$ are different. I assert

$$\boldsymbol{P}(\text{all } D\text{'s different}) \approx \exp\left(\frac{-K^3}{4N}\right). \tag{E6a}$$

This has been proposed as a test for computer random number generators. To argue (E6a), note that the D's are approximately i.i.d. geometric with mean $\mu = N/K$:

$$\boldsymbol{P}(D_j = i) \approx p_i = \mu^{-1}(1-\mu^{-1})^i, \qquad i \geq 1.$$

Thus (E5b) says

$$\boldsymbol{P}(D\text{'s all different}) \approx \exp\left(-\frac{1}{2}K^2 \sum p_i^2\right).$$

and (E6a) follows by calculating $\sum p_i^2 \approx \frac{1}{2}\mu^{-1} = \frac{1}{2}K/N$.

E7 Several types of coincidence. Another direction for generalization of the basic birthday problem can be stated abstractly as follows. Let (X_j) be i.i.d. with distribution μ on some space S. Let $(C_1, C_2, \dots)$ be a finite or countable collection of subsets of S. Let

$$\begin{aligned} p_i &= \boldsymbol{P}(X_1 \in C_i, X_2 \in C_i) = \mu^2(C_i) \\ p &= \boldsymbol{P}(X_1 \in C_i \text{ and } X_2 \in C_i, \text{ for some } i) \end{aligned}$$

and suppose

$$p \text{ is small; } \max p_i/p \text{ is small.} \tag{E7a}$$

For $j < k$ let $A_{j,k}$ be the event "$X_j \in C_i$ and $X_k \in C_i$, for some i". From (E7a) we can argue heuristically that the events $A_{j,k}$ are roughly independent, and then that

$$\begin{aligned}&\boldsymbol{P}(X_j \in C_i \text{ and } X_k \in C_i; \text{ for some } i \text{ and some} \\ &1 \leq j < k \leq N) \approx 1 - \exp(-\tfrac{1}{2}N^2 p).\end{aligned} \tag{E7b}$$

For a concrete example, suppose we pick N people at random and categorize each person in two ways, e.g., "last name" and "city of birth", which may be dependent. What is the chance that there is some coincidence, i.e., that some pair of people have either the same last name or the same city of birth? Let

$$\begin{aligned} q_{i,\widehat{\imath}} &= \boldsymbol{P}(\text{last name} = i, \text{ city of birth} = \widehat{\imath}) \\ q_{i,\cdot} &= \boldsymbol{P}(\text{last name } = i); \quad q_{\cdot,\widehat{\imath}} = \boldsymbol{P}(\text{ city of birth} = \widehat{\imath}) \\ p &= \sum_i q_{i,\cdot}^2 + \sum_{\widehat{\imath}} q_{\cdot,\widehat{\imath}}^2 - \sum_i \sum_{\widehat{\imath}} q_{i,\widehat{\imath}}^2 \end{aligned}$$

Then p is the chance of a coincidence involving a specified pair of people, and (E7b) says

$$\boldsymbol{P}(\text{some coincidence amongst } N \text{ people}) \approx \exp(-\tfrac{1}{2}N^2 p) \qquad \text{(E7c)}$$

is a reasonable approximation provided

$$p; \quad \max_i q_{i,\cdot}^2/p; \quad \max_{\widehat{\imath}} q_{\cdot,\widehat{\imath}}^2/p \qquad \text{all small.} \qquad \text{(E7d)}$$

For the simplest case, suppose there are K_1 (resp. K_2) categories of the first (second) type, and the distribution is uniform over categories of each type and independent between types: that is, $q_{i,\widehat{\imath}} \equiv 1/K_1K_2$. Then the number T of people one needs to sample until finding some coincidence is

$$T \overset{\mathcal{D}}{\approx} \left(\frac{1}{K_1} + \frac{1}{K_2}\right)^{-1} \mathcal{R}; \qquad \boldsymbol{P}(\mathcal{R} > t) = \exp(-\frac{1}{2}t^2). \qquad \text{(E7e)}$$

E8 Example: Similar bridge hands. As another example in this abstract set-up, the chance that two sets of 13 cards (dealt from different decks) have 8 or more cards in common is about 1/500. So if you play bridge for a (long) evening and are dealt 25 hands, the chance that some two of your hands will have at least 8 cards in common is, by (E7b), about

$$1 - \exp\left(-\frac{1}{2}\frac{25^2}{500}\right):$$

this is roughly a 50–50 chance. Bridge players often have remarkable memories of their past hands — this would make a good test of memory!

We now turn to matching problems. Fix N large, and let $X_1, \ldots, X_N$; $Y_1, \ldots, Y_N$ be independent uniform random permutations of $\{1, 2, \ldots, N\}$. A trite variation of the basic matching problem is to consider $M_j = \#\{\, i : |X_i - Y_i| \le j \,\}$. For fixed j, as $N \to \infty$ we have $\boldsymbol{P}(|X_i - Y_i| \le j) \sim (2j+1)/N$ and these events are asymptotically independent, so

$$M_j \overset{\mathcal{D}}{\approx} \text{Poisson}(2j+1).$$

Here is a more interesting variation.

E9 Example: Matching K-sets. For fixed small K, let I be the set of K-element subsets $\underset{\sim}{i} = \{i_1, \ldots, i_K\}$ of $\{1, \ldots, N\}$. Say $\underset{\sim}{i}$ is a K-match if the sets $\{X_{i_1}, \ldots, X_{i_K}\}$ and $\{Y_{i_1}, \ldots, Y_{i_K}\}$ are identical, but $\{X_{j_1}, \ldots, X_{j_k}\}$ and $\{Y_{j_1}, \ldots, Y_{j_k}\}$ are not identical for any proper subset $\{j_1, \ldots, j_k\}$ of $\{i_1, \ldots, i_K\}$. Let M_K be the number of K-matches (so M_1 is the number of matches, as in problem E1.4). So $M_K = \#(\mathcal{S} \cap I)$, where $\mathcal{S}$ is the random set of K-matches. We want to apply the heuristic to $\mathcal{S}$. Observe that if $\underset{\sim}{i}$ and $\underset{\sim}{\hat{\imath}}$ are K-matches, then either $\underset{\sim}{i} = \underset{\sim}{\hat{\imath}}$ or $\underset{\sim}{i}$ and $\underset{\sim}{\hat{\imath}}$ are disjoint (else the values of X and Y match on $\underset{\sim}{i} \cap \underset{\sim}{\hat{\imath}}$, which is forbidden by definition). Thus the clump size $C \equiv 1$. For each $\underset{\sim}{i}$, the chance that $\{X_{i_1}, \ldots, X_{i_K}\}$ and $\{Y_{i_1}, \ldots, Y_{i_K}\}$ are identical is $1/\binom{N}{K}$. Suppose they are identical. Define $u_1 = X_{i_1}$, $u_r =$ the Y_{i_j} for which $X_{i_j} = u_{r-1}$. Then $\underset{\sim}{i}$ is a K-match iff $u_2, u_3, \ldots, u_K$ are all different from u_1, and this has chance $(K-1)/K \times (K-2)/(K-1) \times \cdots \times 1/2 = 1/K$. So

$$p = \boldsymbol{P}(\underset{\sim}{i} \text{ is a } K\text{-match}) = \left(K\binom{N}{K}\right)^{-1}$$

and our heuristic clump rate is $\lambda = p/EC = p$, since $C \equiv 1$. So

$$\begin{aligned} M_K = \#(\mathcal{S} \cap I) &\overset{\mathcal{D}}{\approx} \text{Poisson}(\lambda \# I) \qquad \text{by the heuristic (Section A4)} \\ &\overset{\mathcal{D}}{\approx} \text{Poisson}(1/K) \qquad \text{since } \#I = \binom{N}{K}. \end{aligned} \tag{E9a}$$

This example is simple enough to solve exactly (see Section E21). Here's another example in the same setting where we really use clumping.

E10 Example: Nearby pairs. Let D be the smallest $L \geq 2$ such that for some i,

$$|\{X_i, X_{i+1}, \ldots, X_{i+L-1}\} \cap \{Y_i, Y_{i+1}, \ldots, Y_{i+L-1}\}| \geq 2 \tag{E10a}$$

We shall estimate the distribution of D. Fix L, and let $\mathcal{S}$ be the random set of i for which (E10a) holds. For each i the cardinality of the intersection in (E10a) is approximately Binomial$(L, L/N)$ because each Y has chance L/N of matching some X. So

$$p = p(i \in \mathcal{S}) \approx \frac{1}{2}\left(\frac{L^2}{N}\right)^2$$

provided this quantity is small. To estimate EC, fix i and condition on $i \in \mathcal{S}$. Then $\boldsymbol{P}(i+1 \notin \mathcal{S}) \approx \boldsymbol{P}(X_i \text{ or } Y_i \text{ is one of the matched values}) \approx 4/L$ provided this quantity is small. We can now use the ergodic-exit technique

(A9h) to estimate the clump rate

$$\lambda \approx p(4/L) \approx \frac{2L^3}{N^2},$$

and so

$$\begin{aligned} \boldsymbol{P}(D > L) &\approx \boldsymbol{P}(\mathcal{S} \cap \{1, 2, \ldots, N\} \text{ empty}) \\ &\approx \exp(-\lambda N) \\ &\approx \exp(-2\frac{L^3}{N}) \end{aligned} \tag{E10b}$$

E11 Example: Basic coupon-collectors problem. We now turn to other versions of the coupon-collector's problem. The basic example in Section E1.3 can be rephrased as follows. Suppose we have a large number N of boxes. Put balls independently uniformly into these boxes; what is the number T of balls needed until every box has at least one ball? As usual, we get a simple estimate by Poissonization. Imagine the placement times as a Poisson process of rate 1. Then

$$\boldsymbol{P}(\text{box } j \text{ empty at time } t) \approx \exp\left(-\frac{t}{N}\right),$$

for any particular box j. But Poissonization makes boxes independent. So Q_t = the number of empty boxes at time t satisfies

$$Q_t \overset{\mathcal{D}}{\approx} \text{Poisson, mean } N \exp\left(-\frac{t}{N}\right); \qquad t \text{ large.} \tag{E11a}$$

In particular

$$\begin{aligned} \boldsymbol{P}(T \le t) &= \boldsymbol{P}(Q_t = 0) \\ &\approx \exp(-N \exp(-t/N)). \end{aligned} \tag{E11b}$$

This can be rearranged to

$$N^{-1}(T - N \log N) \overset{\mathcal{D}}{\approx} \boldsymbol{\xi}; \qquad \text{where } \boldsymbol{P}(\boldsymbol{\xi} \le x) = \exp(-e^{-x}). \tag{E11c}$$

or more crudely to

$$T \approx N \log N. \tag{E11d}$$

Here are some simple variations on the basic problem.

E12 Example: Time until most boxes have at least one ball. Let $0 < \alpha < 1$. Let T_α be the time (= number of balls) until there are at most N^α empty boxes. By (E11a), T_α is approximately the solution t of $N \exp(-t/N) = N^\alpha$, and so the crude approximation analogous to (E11d) is

$$T_\alpha \approx (1 - \alpha) N \log N. \tag{E12a}$$

E13 Example: Time until all boxes have at least $(K+1)$ balls. For t large,

$$\begin{aligned}\boldsymbol{P}(\text{box } j \text{ has } < K+1 \text{ balls at time } t) \\ \approx \boldsymbol{P}(\text{box } j \text{ has } K \text{ balls at time } t) \\ = e^{-t/N}(t/N)^K/K!\end{aligned}$$

Write Q_t^K = number of boxes with $< K+1$ balls at time t. Then $Q_t^K \overset{\mathcal{D}}{\approx}$ Poisson, mean $Ne^{-t/N}(t/N)^K/K!$ So the time T_K until all boxes have at least $K+1$ balls satisfies

$$\begin{aligned}\boldsymbol{P}(T_K \leq t) &= \boldsymbol{P}(Q_t^K = 0) \\ &\approx \exp(-Ne^{-t/N}(t/N)^K/K!).\end{aligned}$$

This rearranges to

$$T_K \approx N\log N + KN\log\log N. \tag{E13a}$$

E14 Example: Unequal probabilities. Suppose each ball goes into box j with probability p_j, where $\max p_j$ is small. Then

$$\boldsymbol{P}(\text{box } j \text{ empty at time } t) \approx \exp(-p_j t),$$

and the crude result (E11c) becomes:

$$\begin{aligned}T &\approx \text{the solution } t \text{ of } \sum \exp(-p_j t) = 1 \qquad \text{(E14a)} \\ &= \Phi(\mu), \quad \text{say, where } \mu \text{ indicates the distribution } (p_j).\end{aligned}$$

This doesn't have an explicit solution in terms of the (p_j) — unlike the birthday problem (Example E5) — and this hampers our ability to handle more complicated extensions of the coupon-collector's problem.

E15 Abstract versions of CCP. Let (X_i) be i.i.d. with some distribution μ on a space S. Let (A_j) be subsets of S. Then a generalization of the coupon-collector's problem is to study

$$T \equiv \min\{\, n : \text{ for each } j \text{ there exist } m \leq n \text{ such that } X_m \in A_j \,\}. \tag{E15a}$$

In the case where (A_j) is a partition, we are back to the setting of Example E14 above, and $T \approx$ the solution t of

$$\sum_j \exp(-t\mu(A_j)) = 1. \tag{E15b}$$

This remains true for many more general families (A_j). For t defined at (E15b) let $\mathcal{S}$ be the random set of j such that A_j has not been hit by

$(X_1, \ldots, X_t)$. Using the heuristic, the essential condition for $T \approx t$ is that $\mathcal{S}$ should not form clumps, but instead consist of isolated points. In general (A6f) a sufficient condition for this is

$$\sum_{k \text{ near } j} \boldsymbol{P}(k \in \mathcal{S} \mid j \in \mathcal{S}) \qquad \text{is small, for all } j.$$

In the present setting,

$$\boldsymbol{P}(k \in \mathcal{S} \mid j \in \mathcal{S}) \approx \exp(-t\mu(A_k \setminus A_j)).$$

Thus the heuristic sufficient condition for $T \approx t$ at (E15b) is

$$\sum_k \exp(-t\mu(A_k \setminus A_j)) \qquad \text{is small, for each } j \qquad \text{(E15c)}$$

where the sum is taken over those A_k which overlap A_j significantly.

E16 Example: Counting regular graphs. We are familiar with the use of combinatorial counting results in probability theory. Here is an elegant example of the converse, a "counting" result whose only known proof is via an essentially probabilistic argument. Call a graph *improper* if we allow

1. loops (that is, an edge from a vertex to itself); and
2. multiple edges between the same pair of vertices.

Call a graph *proper* if these are not allowed. Let $A(N, d)$ be the number of proper graphs on N labeled vertices such that there are exactly d edges at each vertex (here $d \geq 3$ and dN is even). We shall argue

$$A(N,d) \sim \frac{(Nd)!}{(\frac{1}{2}Nd)!2^{Nd/2}(d!)^N} \exp(-\lambda - \lambda^2), \quad \text{as } N \to \infty,\ d \text{ fixed} \qquad \text{(E16a)}$$

where $\lambda = \frac{1}{2}(d-1)$.

Put Nd balls in a box: d balls marked "1", d balls marked "2", ... and d balls marked "N". Draw the balls two at a time, without replacement, and each time a pair is drawn create an edge between the corresponding vertices $1, 2, \ldots, N$. This constructs a random graph, which may be improper, but it is easy to see:

> conditional on the constructed graph being proper, it is equally likely to be any proper graph.

In other words,

$$\frac{1}{A(N,d)} = \frac{q}{\boldsymbol{P}(\text{graph is proper})} = \frac{q}{\boldsymbol{P}(X=0, Y=0)} \qquad \text{(E16c)}$$

where q is the chance the construction yields a specified proper graph,

$$X = \text{number of vertices } i \text{ such that there is a loop from } i \text{ to } i,$$
$$Y = \text{number of pairs of vertices } (i,j) \text{ linked by multiple edges.}$$

It is not hard to calculate $q = (\frac{1}{2}Nd)!(d!)^N 2^{Nd/2}/(Nd)!$. So the counting result (E16a) will be a consequence of the probabilistic result

$$(X,Y) \text{ are asymptotically independent Poissons, means } \lambda \text{ and } \lambda^2. \tag{E16d}$$

To argue this heuristically, let $B_{i,j}$ be the event that there are multiple edges between i and j. Then

$$P(B_{i,j}) \sim \frac{1}{2}\frac{(d-1)^2}{N^2}$$
$$\sum_{k \neq j} P(B_{i,k} \mid B_{i,j}) \to 0.$$

So applying the heuristic to $\mathcal{S} = \{\,(i,j) : B_{i,j} \text{ occurs}\,\}$, the clumps consist of single points (A6f) and so $Y = |\mathcal{S}|$ has approximately Poisson distribution with mean $E\,|\mathcal{S}| = \binom{N}{2} P(B_{i,j}) \to \lambda^2$. The argument for X is similar.

COMMENTARY

E17 General references. There is a vast literature on combinatorial problems. The basic problems are treated in David and Barton (1962), Feller (1968) and Johnson and Kotz (1977). Asymptotics of "balls in boxes" problems are treated in detail by Kolchin et al. (1978). Ivanov el al. (1984) discuss more recent Russian literature. The forthcoming work of Diaconis and Mosteller (1989) treats coincidence problems.

The references in Section A18 show how Stein's method can be used to give explicit bounds in certain approximations.

E18 Poissonization. Holst (1986) gives a careful account of Poisonization in several simple models. Presumably (E2a) can be formalized as follows:

> if $\sup_\theta -\theta^{1/2}q'(\theta) < \delta$ then $\sup|p(n) - q(n)| \leq \epsilon$ for some explicit function $\epsilon(\delta) \to 0$ as $\delta \to 0$.

Assertion (E2d) is really an easy Tauberian theorem.

E19 The random number test. Marsaglia (unpublished) gives a heuristic argument for Example E6; Diaconis (unpublished) has proved the corresponding limit theorem.

It is interesting to note that the sequential version of this problem is different. Fix N; pick $X_1, X_2, \ldots$ and for each K let A_K be the event "all D's different" for the D's formed from $X_1, \ldots, X_K$. Then Example E6 says $\boldsymbol{P}(A_K) \approx \exp(-K^3/4N)$. In this example, unlike our other "birthday" problems, it is possible for matches to be broken, so the (A_K) are not decreasing. So for

$$T = \min\{\, K : A_K \text{ does not occur} \,\}$$

$\boldsymbol{P}(T > K) \neq \boldsymbol{P}(A_K)$. My rough calculations give

$$\boldsymbol{P}(T > K) \approx \exp(-K^3/3N).$$

E20 Abstract forms of the birthday problem. Formalizations of Section E7 may be derived from the Poisson limit theorem for U-statistics (Silverman and Brown (1978)); explicit bounds on errors can be found using Stein's method.

E21 Cycles of random permutations. In the matching problems (Examples E9,E10) we could take the sequence (X_i) to be $(1, 2, \ldots, N)$ without changing the problem. In this setting, the r.v. M_K of Example E9 is just the number of cycles of length K in a uniform random permutation of $(1, \ldots, N)$. Properties of M_K have been studied extensively — see, e.g., Shepp and Lloyd (1966).

E22 Random regular graphs. The argument in Example E16 is from Bollobas (1985) Section 2.4.

F Combinatorics for Processes

In this chapter we study analogues of the birthday problem, the matching problem and the coupon-collector's problem for finite-valued stochastic processes $(X_n; n \geq 1)$ more complicated than i.i.d.

F1 Birthday problem for Markov chains. For an idea of the issues involved, consider a stationary Markov chain (X_n) with stationary distribution π, and consider the birthday problem, i.e. the distribution of

$$T = \min\{\, n \geq 2 : X_n = X_m \text{ for some } 1 \leq m < n \,\}. \qquad \text{(F1a)}$$

We seek approximations for T, in the case where ET is large. Let $1 \ll \tau \ll ET$, with τ representing "the short term". As in Section B2, suppose that at each state i we can approximate the distribution of $i = X_0, X_1, \ldots, X_\tau$ by the distribution of $i = X_0^*, X_1^*, \ldots, X_\tau^*$ for some transient chain X^*; write

$$q(i) = \boldsymbol{P}_i(X^* \text{ returns to } i).$$

We need to distinguish between "local" matches, where (F1a) holds for some $n - m \leq \tau$, and "long-range" matches with $n - m > \tau$. For local matches, consider the random set $\mathcal{S}_1$ of times m for which $X_n = X_m$ for some $m < n \leq m + \tau$. Then

$$p_1 \equiv \boldsymbol{P}(m \in \mathcal{S}_1) \approx \sum \pi(i)q(i).$$

Write c_1 for the mean clump size in $\mathcal{S}_1$; the heuristic says that the clump rate $\lambda_1 = p_1/c_1$. So

$$\begin{aligned}
\boldsymbol{P}(\text{no local matches before time } n) &\approx \boldsymbol{P}(\mathcal{S}_1 \cap [1,n] \text{ empty}) \\
&\approx \exp(-\lambda_1 n) \\
&\approx \exp(-nc_1^{-1} \textstyle\sum \pi(i)q(i)).
\end{aligned}$$

For long-range matches, consider the random set

$$\mathcal{S}_2 = \{\, (j,m) : m - j > \tau, \quad X_m = X_j \,\}.$$

Write c_2 for the mean clump size in $\mathcal{S}_2$. Assuming no long-range dependence,

$$p_2 \equiv \boldsymbol{P}((j,m) \in \mathcal{S}_2) \approx \sum \pi^2(i),$$

and the heuristic says that the clump rate $\lambda_2 = p_2/c_2$. So

$$\begin{aligned} \boldsymbol{P}(\text{no long-range matches before time } n) &\approx \boldsymbol{P}(\mathcal{S}_2 \cap [1,n]^2 \text{ empty}) \\ &\approx \exp(-\frac{1}{2}n^2\lambda_2) \\ &\approx \exp(-\frac{1}{2}n^2 c_2^{-1} \sum \pi^2(i)). \end{aligned}$$

Thus we get the approximation

$$\boldsymbol{P}(T > n) \approx \exp(-nc_1^{-1} \sum \pi(i)q(i) - \frac{1}{2}n^2 c_2^{-1} \sum \pi^2(i)). \tag{F1b}$$

In principle one can seek to formalize this as a limit theorem for a sequence of processes $X^{(K)}$. In this general setting it is not clear how to estimate the mean clump sizes c_1, c_2. Fortunately, in natural examples one type of match (local or long-range) dominates, and the calculation of c usually turns out to be easy, as we shall now show.

The simplest examples concern transient random walks. Of course these do not quite fit into the setting above: they are easy because they can have only local matches. Take T as in (F1a).

F2 Example: Simple random walk on $\boldsymbol{Z}^K$. Here $X_n = \sum_{m=1}^{n} \xi_m$, say. Now $T \leq R \equiv \min\{ m : \xi_m = -\xi_{m-1} \}$, and $R - 1$ has geometric distribution with mean $2K$. For large K it is easy to see that $\boldsymbol{P}(T = R) \approx 1$, so that $T/(2K)$ has approximately exponential(1) distribution.

F3 Example: Random walks with large step. Fix $d \geq 3$. For large K, consider the random walk $X_n = \sum_{m=1}^{n} \xi_m$ in $\boldsymbol{Z}^d$ whose step distribution ξ is uniform on $\{\underset{\sim}{i} = (i_1, \ldots, i_d) : |i_j| \leq K \text{ for all } j\}$. Write $\mathcal{S}_1$ for the random set of times n such that $X_m = X_n$ for some $m > n$. Let $q_K = \boldsymbol{P}(X_n$ ever returns to $0)$. Then $\boldsymbol{P}(n \in \mathcal{S}_1) = q_K$, and since the steps ξ are spread out there is no tendency for $\mathcal{S}_1$ to form clumps, so $\boldsymbol{P}(T > n) \approx \exp(-nq_K)$. Finally, we can estimate q_K by considering the random walk S_n on $\boldsymbol{R}^d$ whose steps have the continuous uniform distribution on $\{ \underset{\sim}{x} = (x_1, \ldots, x_d) : |x_j| \leq 1 \text{ for all } j \}$. Then $\boldsymbol{P}(X_n = 0) \approx K^{-d} f_n(0)$, where f_n is the density of S_n, and so

$$q_K \approx K^{-d} \sum_{n=1}^{\infty} f_n(0).$$

We turn now to random walks on finite groups, which are a special case of stationary Markov chains. Such random walks were studied at Examples B6,B7 in the context of first hitting times; here are two different examples where we study the "birthday" time T of (F1a).

F4 Example: Simple random walk on the K-cube. Take $I = \{0,1\}^K$, regarded as the vertices of the unit cube in K dimensions. Let X_n be simple random walk on I: at each step a random coordinate is chosen and changes parity. This is equivalent to the Ehrenfest urn model with K labeled balls. In the context of the birthday problem it behaves similarly to Example F2. Let R be the first time n such that the same coordinate is changed at $n-1$ and n. Then $T \leq R$, $\boldsymbol{P}(T = R) \approx 1$ for large K, and $R-1$ has geometric distribution with mean K. So for large K, T is approximately exponential with mean K.

F5 Example: Another card shuffle. Some effort is needed to construct an interesting example where the long-range matches dominate: here is one. As at Example B6 consider repeated random shuffles of a K-card deck (K even), and consider the shuffling method "take the top card and insert it randomly into the bottom half of the deck". I claim that for K large, T behaves as in the i.i.d. case, that is

$$\boldsymbol{P}(T > n) \approx \exp\left(-\frac{1}{2}\frac{n^2}{K!}\right).$$

Consider first the long-range matches. If $X_j = X_m$ for $m-j$ large then there is no tendency for nearby matches. So $c_2 = 1$, and then (F1b) shows that the contribution to $\boldsymbol{P}(T > n)$ from long-range matches is of the form stated. Thus the issue is to show that there are no local matches before time $O(\sqrt{K!})$. By (F1b) we have to show

$$q \equiv \boldsymbol{P}(\text{return to initial configuration in short term}) = o(K!)^{-\frac{1}{2}}. \quad \text{(F5a)}$$

Let i be the initial configuration. For $n < \frac{1}{2}K$ it is impossible that $X_n = i$. In any block of $\frac{1}{2}K$ shuffles there are $(K/2)^{K/2}$ possible series of random choices, and a little thought reveals these all lead to different configurations; it follows that

$$\begin{aligned} \boldsymbol{P}(X_n = i) &\leq \left(\frac{1}{2}K\right)^{-\frac{1}{2}K} \qquad \text{for } n \geq \frac{1}{2}K \\ &= o(K!)^{-\frac{1}{2}} \end{aligned}$$

and this leads to (F5a).

F6 Matching problems. Let us now shift attention away from birthday problems toward matching problems: we shall see later there is a connection. Let (X_n) and (Y_n) be stationary processes without long-range dependence, independent of each other. The matching time

$$T = \min\{\, n : X_n = Y_n \,\} \qquad \text{(F6a)}$$

is just the first time that the bivariate process (X_n, Y_n) enters the set $\{(x,x)\}$, and can be studied using the techniques of earlier chapters. The "shift-match" time

$$T = \min\{\, n : X_i = Y_j \text{ for some } i,\ j \le n \,\} \tag{F6b}$$

involves some new issues. To study T, we need some notation. Let $f_X(x) = \boldsymbol{P}(X_1 = x)$ and let $c_X(x)$ be the mean size of clumps of visits of (X_n) to x; define f_Y, c_Y similarly. Let $p = \boldsymbol{P}(X_1 = Y_1) = \sum_x f_X(x) f_Y(x)$, and let Z have the distribution of X_1 given $X_1 = Y_1$. To make T large, suppose p is small. We also make a curious-looking hypothesis: there exists $0 \le \theta < 1$ such that

$$\boldsymbol{P}(X_{i+1} = Y_{j+1} \mid X_i = Y_j,\ \text{past}) \approx \theta \text{ regardless of the past.} \tag{F6c}$$

We shall give an argument for the approximation

$$\boldsymbol{P}(T > K) \approx \exp(-K^2 p(1-\theta)); \tag{F6d}$$

to justify this we shall need the extra hypotheses

$$KEf_X(Z),\ KEf_Y(Z),\ E(c_X(Z)-1) \text{ and } E(c_Y(Z)-1) \text{ are all small.} \tag{F6e}$$

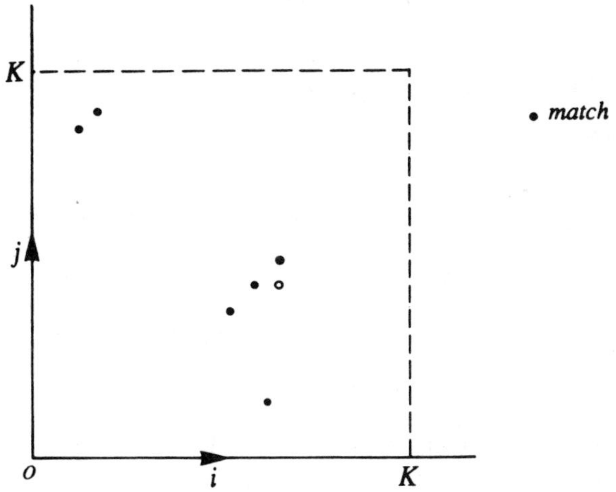

FIGURE F6a.

Consider the random set $\mathcal{S} = \{\,(i,j) : X_i = Y_j\,\}$. It is clear that, on the large scale, $\mathcal{S}$ does not look Poisson, since if a value x occurs as $X_{i_1}, X_{i_2}, \ldots$ and as $Y_{j_1}, Y_{j_2}, \ldots$ then all points of the irregular grid $\{i_v, j_w; v, w \ge 1\}$

occur in $\mathcal{S}$, and so $\mathcal{S}$ has this kind of long-range dependence. However, consider a large square $[1, K]^2$ and suppose we have the property

> when a match $X_i = Y_j = x$ occurs in the square, it is unlikely that x appears as any other X-value or Y-value in the square. (F6f)

In this case we can apply the heuristic to $\mathcal{S} \cap [1, K]^2$, and the clumps of matches will be diagonal lines only, as pictured.

For (F6f) implies first that the long-range dependence mentioned above does not affect the square; and second that points like $\circ$ can not be in $\mathcal{S}$, else the match-values at the points above and to the left of $\circ$ would be identical. Applying the heuristic, $\boldsymbol{P}((i,j) \in \mathcal{S}) = p$ and by (F6c) the clump sizes are approximately geometric, mean $(1-\theta)^{-1}$. Thus the clump rate is $\lambda = p(1-\theta)$, yielding (F6d). Finally, we want to show that (F6f) follows from (F6e). The value $Z = X_i = Y_j$ at match (i, j) is distributed as X_1 given $X_1 = Y_1$. So the mean number of extra times Z occurs locally in (X_n) is $E(c_X(Z) - 1)$; and the mean number of non-local times is $KEf_X(Z)$. So hypothesis (F6f) makes it unlikely that Z occurs as any other X-value or Y-value.

F7 Matching blocks. The most-studied aspect of these combinatorial topics is the problem of matching blocks of two sequences, motivated by DNA comparisons (see Section F19). Suppose we have underlying stationary sequences (ξ_n), (η_n), independent of each other, each without long-range dependence. Let M_K be the length of the longest block which occurs in the first K terms of each sequence:

$$M_K = \max\{\, m : (\xi_{i-m+1}, \ldots, \xi_i) = (\eta_{j-m+1}, \ldots, \eta_j) \text{ for some } i,\, j \le K \,\}.$$

To study M_K, fix m large. Let (X_n) be the process of m-blocks for (ξ_n), that is $X_n = (\xi_{n-m+1}, \ldots, \xi_n)$, and let (Y_n) be the process of m-blocks for (η_n). Then

$$\boldsymbol{P}(M_K < m) = \boldsymbol{P}(T > K) \tag{F7a}$$

where T is the shift-match time of (F6b). Let $u_\xi(s) = \boldsymbol{P}(\xi_1 = s)$, $u_\eta(s) = \boldsymbol{P}(\eta_1 = s)$.

F8 Example: Matching blocks: the i.i.d. case. Consider the case where (ξ_n) and (η_n) are each i.i.d. (with different distributions, perhaps). Then

$$p = \boldsymbol{P}(X_n = Y_n) = q^m \quad \text{where } q = \boldsymbol{P}(\xi_1 = \eta_1) = \sum_s u_\xi(s) u_\eta(s).$$

and (F6c) holds for $\theta = q$. So, if the extra hypotheses (F6e) hold, then (F6d) and (F7a) above imply the approximation

$$\boldsymbol{P}(M_K < m) \approx \exp(-K^2(1-q)q^m). \tag{F8a}$$

This distributional approximation implies the weaker approximation

$$M_K \approx \frac{2 \log K}{\log(1/q)}. \tag{F8b}$$

To verify the extra hypotheses, recall first the "patterns in coin-tossing" discussion (Example B5). Local matches are caused by a pattern overlapping itself; for m large, it is clear that a typical pattern of length m does not overlap itself substantially, so the hypothesis that $E(c_X(Z) - 1)$ be small is satisfied. For the other hypothesis, let ν have the distribution of ξ_1 given $\xi_1 = \eta_1$; then the Z in (F6e) is $Z = (\nu_1, \ldots, \nu_m)$ with i.i.d. entries. So $Ef_X(Z) = (Eu_\xi(\nu))^m$. Thus the condition we need to verify is

$$K(Eu_\xi(\nu))^m \to 0 \qquad \text{for } K,\, m \text{ related by } K^2(1-q)q^m \to c \in (0, \infty).$$

This reduces to the condition

$$\sum_s u_\xi^2(s) u_\eta(s) < q^{3/2} \tag{F8c}$$

and the similar condition with ξ, η interchanged.

Thus our heuristic arguments suggest (F8c) is a sufficient condition for the approximations (F8a,F8b). To see that (F8c) is a real constraint, consider the binary case where $u_\xi(0) = u_\xi(1) = \frac{1}{2}$, $u_\eta(0) = \alpha > \frac{1}{2}$, $u_\eta(1) = 1 - \alpha$. Then (F8c) is the condition $\alpha < 0.82$. This suggests that the limit theorems corresponding to (F8a,F8b) hold for some but not all α; and this turns out to be true (see Section F19) although our bound on α is conservative. In general it turns out that

$$M_K \sim \frac{C \log K}{\log(1/q)} \quad \text{a.s.} \quad \text{as } K \to \infty, \tag{F8d}$$

where B depends on the distributions ξ, η; and where $B = 2$ if these distributions are not too dissimilar.

F9 Example: Matching blocks: the Markov case. Now consider the setting above, but let (ξ_n) and(η_n) be stationary Markov chains with transition matrices $\boldsymbol{P}^\xi(s,t)$, $\boldsymbol{P}^\eta(s,t)$. Write

$$p_m = \boldsymbol{P}((\xi_1, \ldots, \xi_m) = (\eta_1, \ldots, \eta_m)).$$

Let Q be the matrix $Q(s,t) = \boldsymbol{P}^\xi(s,t)\boldsymbol{P}^\eta(s,t)$, that is with entrywise multiplication rather than matrix multiplication. Then

$$p_m \sim a\theta^m \quad \text{as } m \tag{F9a}$$

where θ is the largest eigenvalue of Q and a is related to the corresponding eigenvectors (see Example M4). Moreover, (F6c) holds for this θ, using

(F9a). So, as in the i.i.d. case, (F6d) yields the approximations

$$P(M_K < m) \approx \exp(-K^2 a(1-\theta)\theta^m) \tag{F9b}$$

$$M_K \sim \frac{2\log K}{\log(1/\theta)} \quad \text{as } K \to \infty. \tag{F9c}$$

Again, these require the extra hypotheses (F6e); as in the i.i.d case, these reduce to requirements that the transition matrices $\boldsymbol{P}^\xi$, $\boldsymbol{P}^\eta$ be not too different.

F10 Birthday problem for blocks. Given a single sequence (ξ_n), we can study the longest block which occurs twice:

$$M_K = \max\{\, m : (\xi_{j-m+1}, \ldots, \xi_j) = (\xi_{i-m+1}, \ldots, \xi_i) \text{ for some } i \leq j \leq K \,\}.$$

But long-range matches behave exactly like matches between two independent copies of (ξ_n). So if nearby matches can be neglected, then we can repeat the arguments for (F8a), (F9b) to get

$$\text{[i.i.d. case]} \qquad P(M_K < m) \approx \exp(-\tfrac{1}{2}K^2(1-q)q^m); \qquad q = \sum_s P^2(\xi_1 = s). \tag{F10a}$$

$$\text{[Markov case]} \qquad P(M_K < m) \approx \exp(-\tfrac{1}{2}K^2 a(1-\theta)\theta^m). \tag{F10b}$$

Note the "$\frac{1}{2}K^2$" here, being the approximate size of $\{\, (i,j) : 1 \leq i < j \leq K \,\}$. The condition under which nearby matches can be neglected is, in the notation of (F6e),

$$KE(c_X(X) - 1) \qquad \text{is small} \tag{F10c}$$

In the i.i.d. case this is automatic; the Markov case is less clear.

F11 Covering problems. Changing direction, let us consider for an I-valued sequence (X_n) the time

$$V = \min\left\{ t : \bigcup_{n=1}^{t} \{X_n\} = I \right\}.$$

In the i.i.d. setting (Example E14) this was the coupon collector's problem; in general let us call V the *covering time.* As observed in Example E14 even in the i.i.d. case the approximation for V is rather non-explicit for non-uniform distributions. Thus a natural class of dependent processes to study in this context is the class of random walks on finite groups (Examples F4,F5,B6,B7) since they have uniform stationary distribution.

F12 Covering problems for random walks. Let (X_n) be a stationary random walk on a finite group I and let i_0 be a reference state. Suppose as in Section B2 that, given $X_0 = i_0$, the process (X_n) can be approximated by a transient process (X_n^*). Then as at Section B2

$$T_{i_0} \text{ has approximately exponential distribution, mean } cN \tag{F12a}$$

where $N = |I|$ and c is the mean number of visits of X^* to i_0. By the symmetry of the random walk, (F12a) holds for all $i \in I$. Now let t, s be related by $t = cN(\log(N) + s)$. Then

$$\boldsymbol{P}(T_i > t) \approx \exp\left(-\frac{t}{cN}\right) \approx N^{-1}e^{-s}.$$

For fixed t let A_i be the event $\{T_i > t\}$. Under the condition

$$\sum_{j \text{ near } i_0} \boldsymbol{P}(A_j \mid A_{i_0}) \approx 0 \tag{F12b}$$

the heuristic says that the events $\{\, A_i; i \in I\,\}$ do not clump and so

$$\boldsymbol{P}(\bigcap_i A_i^c) \approx \prod_i \boldsymbol{P}(A_i^c)$$

That is

$$\begin{aligned} \boldsymbol{P}(V \le t) &\approx (\boldsymbol{P}(T_i > t))^N \\ &\approx \exp(-e^{-s}) \\ &= \boldsymbol{P}(\xi_3 \le s) \qquad \text{for the extreme value distribution } \xi_3. \end{aligned}$$

Thus we get the approximation

$$\frac{V - cN\log N}{cN} \overset{\mathcal{D}}{\approx} \xi_3, \tag{F12c}$$

or more crudely

$$\frac{V}{cN\log N} \approx 1. \tag{F12d}$$

These approximations depend not only on the familiar "local transience" property but also upon condition (F12b). To study this latter condition, write $q_j = \boldsymbol{P}_{i_0}(X^* \text{ hits } j)$, $q_j' = \boldsymbol{P}_j(X^* \text{ hits } i_0)$, $T_{j,i_0} = \min\{\, n : X_n^* = j \text{ or } i_0\}$. We can estimate the distribution of T_{j,i_0} using the heuristic method developed at Section B12. In the notation there , $E_{i_0}C_{i_0} = c = E_jC_j$, $E_{i_0}C_j = q_jc$, $E_jC_{i_0} = q_j'c$, and then (B12b) gives, after some algebra,

$$\boldsymbol{P}(T_{j,i_0} > t) \approx \exp(-\lambda_j t); \qquad \lambda_j = \frac{2 - q_j - q_j'}{cN(1 - q_jq_j')}.$$

Recall that $A_j = \{T_j > t\}$ for $t = cN(\log(N) + s)$, and that $\boldsymbol{P}(A_{i_0}) \approx N^{-1}e^{-s}$. So

$$\begin{aligned} \boldsymbol{P}(A_j \mid A_{i_0}) &= \frac{\boldsymbol{P}(T_{j,i_0} > t)}{\boldsymbol{P}(T_{i_0} > t)} \\ &\approx N^{-\alpha_j} \qquad \text{after some algebra,} \end{aligned}$$

where

$$\begin{aligned} \alpha_j &= \frac{1 - q_j - q'_j + q_j q'_j}{1 - q_j q'_j} \\ &= \frac{1 - q_j}{1 + q_j} \qquad \text{if } q_j = q'_j. \end{aligned}$$

Thus condition (F12b) is

$$\sum_{j \text{ near } i_0} N^{-\alpha_j} \approx 0. \tag{F12e}$$

Here are two explicit examples.

F13 Example: Random walk on Z^d modulo N. In this example (Example B7), for $d \geq 3$ the conclusion (F12c) is

$$\frac{V - R_d N^d \log(N^d)}{R_d N^d} \overset{\mathcal{D}}{\approx} \xi_3, \qquad N \text{ large.}$$

To verify condition (F12e), for the unrestricted transient walk X^* in $d \geq 3$ dimensions we have $q_j = q'_j \leq A|j|^{1-\frac{1}{2}d}$ for some $A < 1$. Writing $m = |j|$, the sum in (F12e) becomes

$$\sum_{m \geq 1} m^{d-1} (N^d)^{-1 + A(m^{1-\frac{1}{2}d})}$$

and the sum tends to 0 as $N \to \infty$.

F14 Example: Simple random walk on the K-cube. In this example (Example F4) there is an interesting subtlety. Here $N = 2^K$; take $i_0 = (0, \ldots, 0)$ and for $j = (j_1, \ldots, j_K)$ let $j = \sum_u |j_u|$. Then $q_j = q'_j = O(K^{-|j|})$ and so condition (F12e) is easily satisfied. To use (F12c) we need an estimate of the mean size c of clumps of visits to i_0, and the estimate has to be accurate to within $O(1/\log N)$. In this example, we take $c = 1 + 1/K + O(K^{-2})$, where the factor $1/K$ gives the chance of returning to i_0 after 2 steps. Then (F12c) gives

$$\frac{V - (K+1)2^K \log 2}{2^K} \overset{\mathcal{D}}{\approx} \xi_3.$$

The "1" makes this different from the result for i.i.d. uniform sampling on the K-cube. One might guess the difference is due to dependence between unvisited sites for the random walk, but our argument shows not; the difference is caused merely by the random walk being a little slower to hit specified points.

F15 Covering problem for i.i.d. blocks. Let (η_n) be i.i.d. finite-valued. For $m \geq 1$ let $X_n = (\eta_{n-m+1}, \ldots, \eta_n)$, $n \geq m$, and let $\mu_m =$ distribution(X_n). Let V_m be the coupon collectors time for (X_n). It is easy to see heuristically that, for Φ as at Example E14,

$$\frac{V_m}{\Phi(\mu_n)} \underset{p}{\to} 1 \quad \text{as } m \to \infty. \tag{F15a}$$

Except in the uniform case, it is not so clear how the sequence $\Phi(\mu_m)$ behaves in terms of distribution(η_1), nor how to prove the more refined results about convergence to ξ_3; this seems a natural thesis topic.

We end the chapter with some miscellaneous examples.

F16 Example: Dispersal of many walks. Consider N independent simple symmetric random walks on $\boldsymbol{Z}^d$, in continuous time (mean 1 holds), all started at the origin 0. For N large, what is the first time T that none of the walks is at the origin?

Consider first a single walk $X(t) = (X_1(t), \ldots, X_d(t))$. Each coordinate performs independent random walks on $\boldsymbol{Z}^1$ with variance $1/d$, so the CLT gives $\boldsymbol{P}(X_i(t) = 0) \approx (2\pi t/d)^{-1/2}$ for large t. So

$$\boldsymbol{P}(X(t) = 0) \approx \left(\frac{2\pi t}{d}\right)^{-d/2} \qquad \text{for } t \text{ large.}$$

Let $m(t) = N(2\pi t/d)^{-d/2}$. For large t, the number of walks at 0 at time t is approximately Poisson$(m(t))$. Moreover, the numbers $N^x(t)$ of walks at lattice points x near 0 will be approximately i.i.d. Poisson$(m(t))$. We want to apply the heuristic to the random set $\mathcal{S}$ of times t at which there are no walks at 0. So

$$p(t) \equiv \boldsymbol{P}(t \in \mathcal{S}) = \boldsymbol{P}(N^0(t) = 0) \approx \exp(-m(t)).$$

For the relevant values of t it turns out that $m(t)$ is large; given $t \in \mathcal{S}$ there will be many walks at neighbors x of 0, and the clump is likely to end when the first neighbor makes a transition to 0. So f_t, the rate of clump-ending given $t \in \mathcal{S}$, is about

$$(2d)^{-1} E\Bigg(\sum_{x \text{ neighbor } 0} N^x(t) \mid N^0(t) = 0 \Bigg) \approx m(t).$$

So the ergodic-exit form (A9c) of the heuristic gives clump rate

$$\begin{aligned}\lambda(t) &= p(t)f_t \\ &\approx m(t)\exp(-m(t)).\end{aligned}$$

So the non-stationary form of the heuristic gives

$$\begin{aligned}-\log \boldsymbol{P}(T>t) &\approx \int_0^t \lambda(s)\,ds \\ &\approx m(t)\int_0^t \exp(-m(s))\,ds \\ &\approx \frac{m(t)\exp(-m(t))}{-m'(t)} \\ &\approx 2d^{-1}t\exp(-m(t)) \\ &\approx 2d^{-1}t\exp\left(-N\left(\frac{2\pi t}{d}\right)^{-\frac{1}{2}d}\right). \qquad \text{(F16a)}\end{aligned}$$

More crudely, this gives

$$T \approx d(2\pi)^{-1}\left(\frac{N}{\log N}\right)^{2/d}. \qquad \text{(F16b)}$$

F17 Example: M/M/∞ combinatorics. A different class of "combinatorics for processes" examples arises as follows. The elementary combinatorial problems of Chapter E may be stated in terms of K draws from a box of N balls labeled $1,2,\ldots,N$. Now consider the M/M/∞ *ball process* whose states are (multi-)sets of labeled balls; new balls arrive as a Poisson (rate K) process and are given a random label (uniform on $1,\ldots,N$); balls stay for an exponential(1) time and are then removed. We consider the stationary processes, for which the number of balls present at time t has Poisson(K) distribution. Let B be a property applicable to labeled balls. There are associated events A_t = "B holds at time t for the M/M/∞ ball process". Suppose $p = \boldsymbol{P}(A_t) = \boldsymbol{P}(B \text{ holds at time } t)$ is small. Then we can estimate

$$T = \min\{\,t : B \text{ holds at time } t\,\}$$

by applying the main heuristic to $\mathcal{S} = \{\,t : B \text{ holds at time } t\,\}$. Here are the two basic examples.

F17.1 Birthday problems. Here T is the first time the M/M/∞ ball process contains two balls with the same label. Suppose K^2/N is small; then as at Example E3

$$p = \boldsymbol{P}(\text{some 2 balls have the same label at time } t) \approx \frac{K^2}{2N}.$$

We shall show that the clump size C for $\mathcal{S}$ has $EC \approx \frac{1}{2}$. Then the heuristic says T is approximately exponential with mean

$$ET \approx \frac{1}{\lambda} = \frac{EC}{p} \approx \frac{N}{K^2}.$$

To obtain the estimate for EC, suppose $0 \in \mathcal{S}$. Then at time 0 there are 2 balls present with the same label. These balls will be removed at times ξ_1, ξ_2 with exponential(1) distributions. So the C^+ of Section A9 has $C^+ \overset{\mathcal{D}}{\approx} \min(\xi_1, \xi_2) \overset{\mathcal{D}}{\approx}$ exponential(2), and then (A21c) gives $EC \approx \frac{1}{2}$.

F17.2 Coupon-collector's problem. Here T is the first time at which every label $1, 2, \ldots, N$ is taken by some ball. Suppose K is smaller than $N \log N - O(N)$, but not $o(N \log N)$. Write M_m for the number of labels l such that exactly m balls present at time t have label l. Consider the component intervals of the set $\mathcal{S} = \{t : \text{all labels present at time } t\}$. I assert that the rate of right end-points of such component intervals is

$$\psi = \boldsymbol{P}(M_1 = 1, M_0 = 0).$$

This follows by observing that an interval ends at t if some ball is removed at t whose label, l say, is not represented by any other ball present. Further, one can argue that it is unlikely that any new ball with label l will arrive before some other label is extinguished. So the clumps of $\mathcal{S}$ occur as isolated intervals. By Section A9

$$T \overset{\mathcal{D}}{\approx} \text{exponential}(\psi). \tag{F17a}$$

To estimate ψ, note that the number of balls with a specific label has Poisson(K/N) distribution. So

$$M_0 \overset{\mathcal{D}}{\approx} \text{Poisson, mean } Ne^{-K/N}$$
$$M_1 \overset{\mathcal{D}}{\approx} \text{Poisson, mean } Ke^{-K/N}$$

and M_0 and M_1 are approximately independent. Hence

$$\psi \approx Ke^{-K/N} \exp(-(K+N)e^{-K/N}). \tag{F17b}$$

COMMENTARY

I do not know any general survey of this area: results are scattered in research papers. One of the purposes of this chapter is to exhibit this field as an area of active current research.

F18 Birthday problems. Birthday problems for random walks on groups (Examples F2–F5) are studied in Aldous (1985), and the limit theorems corresponding to the heuristic are proved. Birthday problems for i.i.d. blocks (F10a) were studied in Zubkov and Mikhailov (1974); more recent work on the i.i.d. and Markov block cases is contained in references below.

F19 Block matching problems. DNA sequences can be regarded as sequences of letters from a 4-word alphabet. The occurrence of a long sequence in different organisms, or in different parts of the DNA of a single organism, has interesting biological interpretations; one can try to decide whether such matches are "real" or "just chance coincidence" by making a probability model for DNA sequences and seeing what the model predicts for long chance matches. This has been the motivation for recent work formalizing approximations (F8a,F8b), (F9b,F9c) as limit theorems. See Arratia et al. (1984; 1985b; 1985a; 1988; 1988) and Karlin and Ost (1987; 1988) for a variety of rigorous limit theorems in this setting. The main result which is not heuristically clear is the formula for B in (F8d), and the condition for $B = 2$ which justifies (F8b): see Arratia and Waterman (1985a). In the binary i.i.d. case discussed under (F8c), the critical value for the cruder "strong law" (F8b) is $\alpha \approx 0.89$. It seems unknown what is the critical value for the "extreme value distribution" limit (F8a) to hold: perhaps our heuristic argument for $\alpha \approx 0.82$ gives the correct answer.

F20 Covering problems. For random walks on groups, the weak result (F12d) holds under "rapid mixing" conditions: Aldous (1983a). Matthews (1988b) discusses the use of group representation theory to establish the stronger result (F12a), and treats several examples including Example F14, random walk on the k-cube. In Example F13 (random walk on Z^d modulo N) our heuristic results for $d \geq 3$ can presumably be formalized: thesis project! For $d = 1$ this is a classic elementary problem, and $EV \sim \frac{1}{2}N^2$ and the point visited last is distributed uniformly off the starting point. For $d = 2$ the problem seems hard: see (L9). Various aspects of the "i.i.d blocks" example, in the uniform case, have been treated in Mori (1988a; 1988b): the non-uniform case has apparently not been studied, and would make a good thesis project.

Covering problems for random walks on graphs have been studied in some detail: see Vol. 2 of Journal of Theoretical Probability.

F21 Miscellany. A simple example in the spirit of this chapter concerns runs in subsets for Markov chains: this is treated in Example M3 as the prototype example for the eigenvalue method.

Harry Kesten has unpublished work related to Example F16 in the discrete-time setting.

The examples on M/M/∞ combinatorics are artificial but cute.

G Exponential Combinatorial Extrema

G1 Introduction. We study the following type of problem. For each K we have a family $\{ X_i^K : i \in I_K \}$ of random variables which are dependent but identically distributed; and $|I_K| \to \infty$ exponentially fast as $K \to \infty$. We are interested in the behavior of $M_K = \max_{i \in I_K} X_i^K$. Suppose that there exists $c^* \in (0, \infty)$ such that (after normalizing the X's, if necessary)

$$|I_K|\boldsymbol{P}(X_i^K > c) \to 0 \quad \text{as } K \to \infty; \quad \text{all } c > c^*$$
$$|I_K|\boldsymbol{P}(X_i^K > c) \to \infty \quad \text{as } K \to \infty; \quad \text{all } c < c^*$$

Then Boole's inequality implies

$$\boldsymbol{P}(M_K > c) \to 0 \quad \text{as } K \to \infty; \quad \text{all } c > c^* \tag{G1a}$$

Call c^* the *natural outer bound* for M_K (for a minimization problem the analogous argument gives a lower bound c^*; we call these *outer* bounds for consistency).

Question: does $M_K \underset{p}{\to} \widehat{c}$, for some $\widehat{c}$ (with $\widehat{c} \leq c^*$ necessarily)?

In all natural examples, it does. Moreover we can divide the examples into three categories, as follows.

1. $M_K \underset{p}{\to} c^*$, and this can be proved using the simple "second moment method" described below.

2. $M_K \underset{p}{\to} c^*$, but the second moment method does not work.

3. $M_K \underset{p}{\to} \widehat{c}$ for some $\widehat{c} < c^*$.

We shall describe several examples in each category. These examples do not fit into the d-dimensional framework for the heuristic described in Chapter A. Essentially, we are sketching rigorous arguments related to the heuristic instead of using the heuristic itself.

The second-moment method was mentioned at Section A15. For asymptotics, we need only the following simple consequence of Chebyshev's inequality.

Lemma G1.1 *Let N_K be non-negative integer-valued random variables such that $EN_K \to \infty$ and $EN_K^2/(EN_K)^2 \to 1$. Then $\boldsymbol{P}(N_K = 0) \to 0$.*

Now consider a family $(A_i^K; i \in I_K)$ of events; typically we will have $A_i^K = \{X_i^K > c_K\}$ for some family of random variables X_i^K. Call (A_i^K) *stationary* if for each i_0, i_1 in I_K there exists a permutation π of I_K such that $\pi(i_0) = i_1$ and $(A_i^K; i \in I_K) \stackrel{\mathcal{D}}{=} (A_{\pi(i)}^K; i \in I_K)$. Then $p_K = \boldsymbol{P}(A_i^K)$ does not depend on i.

Lemma G1.2 *Suppose that (A_i^K) is stationary for each K, and suppose that $p_K|I_K| \to \infty$ as $K \to \infty$. If*

$$\sum_{i \neq i_0} \frac{\boldsymbol{P}(A_i^K \mid A_{i_0}^K)}{p_K|I_K|} \quad \to \quad 1 \quad \text{as } K \to \infty$$

then

$$\boldsymbol{P}\Big(\bigcup_{I_K} A_i^K\Big) \quad \to \quad 1.$$

This follows from Lemma G1.1 by considering $N_K = \sum_{I_K} 1_{A_i^K}$. Now the *second-moment method*, in the context (G1a), can be stated as follows. Take suitable $c_K \to c^*$, let $A_i^K = \{X_i^K \geq c_K\}$ and attempt to verify the hypotheses of Lemma G1.2; if so, this implies $\boldsymbol{P}(M_K \geq c_K) \to 1$ and hence $M_K \underset{p}{\to} c^*$.

We now start some examples; the first is the classic example of the method.

G2 Example: Cliques in random graphs. Given a graph G, a *clique* H is a subset of vertices such that (i, j) is an edge for every distinct pair i, $j \in H$; in other words, H is the vertex-set of a complete subgraph. The *clique number* of a graph G is

$$\text{cl}(G) = \max\{\, |H| : H \text{ is a clique of } G\,\}.$$

Let $\mathcal{G}(K, q)$ be the random graph on K labeled vertices obtained by letting $\boldsymbol{P}((i, j) \text{ is an edge}) = q$ for each distinct pair i, j, independently for different pairs. Let $\text{CL}(K, q)$ be the (random) clique number $\text{cl}(\mathcal{G}(K, q))$. It turns out that, as $K \to \infty$ for fixed $0 < q < 1$, the random quantity $\text{CL}(K, q)$ is nearly deterministic. Define $x = x(K, q)$ as the (unique, for large K) *real* number such that

$$\binom{K}{x} q^{\frac{1}{2}x(x-1)} = 1. \tag{G2a}$$

Then $x = (2 \log K)/(\log q^{-1}) + O(\log \log K)$; all limits are as $K \to \infty$ for fixed q. Let x^* be the nearest integer to x. We shall sketch a proof of

$$\boldsymbol{P}(\text{CL}(K, q) = x^* \text{ or } x^* - 1) \to 1. \tag{G2b}$$

First fix K and $1 \leq m \leq K$. Let $\mathcal{H} = \{ H \subset \{1,\ldots,K\} : |H| = m \}$. Let $\mathcal{G}(K,q)$ be the random graph on vertices $\{1,\ldots,K\}$. For $H \in \mathcal{H}$ let A_H be the event "H is a clique for $\mathcal{G}(K,q)$". Then $\boldsymbol{P}(A_H) = q^{m(m-1)/2}$ and so

$$\sum_H \boldsymbol{P}(A_H) = \binom{K}{m} q^{\frac{1}{2}m(m-1)}. \tag{G2c}$$

If $\mathrm{CL}(K,q) > x^*$ then there is some clique of size $x^* + 1$, so

$$\begin{aligned} \boldsymbol{P}(\mathrm{CL}(K,q) > x^*) &\leq \binom{K}{x^*+1} q^{\frac{1}{2}(x^*+1)x^*} \\ &\to 0, \quad \text{using the definition of } x \text{ and the fact that } x^*+1 \geq x + \tfrac{1}{2} \end{aligned} \tag{G2d}$$

For the other bound, note first that $\{ A_H : H \in \mathcal{H} \}$ is stationary. Put $m = x^* - 1 \leq x - \frac{1}{2}$ and let

$$\mu = \binom{K}{m} q^{\frac{1}{2}m(m-1)}. \tag{G2e}$$

Then $\mu \to \infty$ from the definition of x. Let $H_0 = \{1,\ldots,m\}$. If we can prove

$$\mu^{-1} \sum_{H \neq H_0} \boldsymbol{P}(A_H \mid A_{H_0}) \to 1 \tag{G2f}$$

then by Lemma G1.2 we have $\boldsymbol{P}(\mathrm{CL}(K,q) \geq x^* - 1) = \boldsymbol{P}(\bigcup A_H) \to 1$, establishing (G2b). To prove (G2f), note first that A_H is independent of A_{H_0} if $|H \cap H_0| \leq 1$. Since $\mu = \sum_H \boldsymbol{P}(A_H) \geq \sum_{|H \cap H_0| \leq 1} \boldsymbol{P}(A_H \mid A_{H_0})$, it will suffice to prove

$$\mu^{-1} \sum_{2 \leq |H \cap H_0| \leq m-1} \boldsymbol{P}(A_H \mid A_{H_0}) \to 0. \tag{G2g}$$

Now for $2 \leq i \leq K$ there are $\binom{m}{i}\binom{K-m}{m-i}$ sets H with $|H \cap H_0| = i$; for each such H there are $\binom{m}{2} - \binom{i}{2}$ possible edges $i, j \in H$ which are not in H_0, and so $\boldsymbol{P}(A_H \mid A_{H_0}) = q^{m(m-1)/2 - i(i-1)/2}$. So the quantity (G2g) is

$$\binom{K}{m}^{-1} \sum_{2 \leq i \leq m-1} \binom{m}{i}\binom{K-m}{m-i} q^{-\frac{1}{2}i(i-1)}.$$

Now routine but tedious calculations show this does indeed tend to zero.

G3 Example: Covering problem on the K-cube. The second-moment method provides a technique for seeking to formalize some of our heuristic results. Consider for instance Example F14, the time V_K taken for simple random walk on the vertices I_K of the K-cube to visit all 2^K vertices. For fixed s let $m_K(s) = (1 + K^{-1})2^K(s + \log 2^K)$. Let T_i be the first hitting time on i. For fixed K, s let $A_i = \{T_i > m_K(s)\}$. Suppose that the hitting time approximations used in Example F14 could be formalized to prove

$$2^K \boldsymbol{P}(A_i) \to e^{-s} \quad \text{as } K \to \infty; \tag{G3a}$$

$$\sum_{j \neq i} \boldsymbol{P}(A_j \mid A_i) \to e^{-s} \quad \text{as } K \to \infty. \tag{G3b}$$

Then for $s_K^- \to -\infty$ sufficiently slowly we can apply Lemma G1.2 and deduce

$$\begin{aligned} &\boldsymbol{P}\big((1 + K^{-1})2^K(s_K^- + \log 2^K) \\ &\qquad \le V_K \le (1 + K^{-1})2^K(s_K^+ + \log 2^K)\big) \\ &\qquad\qquad \to 1 \quad \text{for all } s_K^+ \to \infty,\ s_K^- \to -\infty. \end{aligned} \tag{G3c}$$

Of course this is somewhat weaker than the convergence in distribution assertion in Example F14.

G4 Example: Optimum partitioning of numbers. Let Y have a continuous density, $EY = 0$, $EY^2 = \sigma^2 < \infty$. Let $(Y_1, \ldots, Y_K)$, K even, be i.i.d. copies of Y and consider

$$M_K = \min_{\substack{H \subset \{1,\ldots,K\} \\ |H| = \frac{1}{2}K}} \left| \sum_{i \in H} Y_i - \sum_{i \notin H} Y_i \right|. \tag{G4a}$$

We shall estimate the size of M_K as $K \to \infty$. The set I_K of unordered partitions $\{H, H^C\}$ has $|I_K| = \frac{1}{2}\binom{K}{K/2} \sim (2\pi K)^{-1/2}2^K$. Write $X_H = \sum_{i \in H} Y_i - \sum_{i \notin H} Y_i$. The central limit theorem says $X_H \overset{\mathcal{D}}{\approx} \text{Normal}(0, K\sigma^2)$. Under suitable conditions on Y we can get a stronger "local" central limit theorem showing that the density of X_H at 0 approximates the Normal density at 0 and hence

$$\boldsymbol{P}(|X_H| \le x) \approx \frac{2x}{\sigma(2\pi K)^{\frac{1}{2}}} \quad \text{for small } x. \tag{G4b}$$

Now fix c and put

$$x_K = (\pi\sigma K)2^{-K}c.$$

Then (G4b) implies

$$|I_K| \boldsymbol{P}(|X_H| \le x_K) \to c \quad \text{as } K \to \infty. \tag{G4c}$$

Suppose we can formalize (G4c) and also

$$\sum_{G \neq H} \boldsymbol{P}(|X_G| \le x_K \mid |X_H| \le x_K) \to c \quad \text{as } K \to \infty. \tag{G4d}$$

Then (G4c) and (G4d) will hold for $c_K^+ \to \infty$ sufficiently slowly, and then Lemma G1.2 implies the following result:

$$\boldsymbol{P}(c_K^- \le (\pi\sigma K)^{-1} 2^K M_K \le c_K^+) \to 1 \quad \text{as } K \to \infty; \quad \text{all } c_K^- \to 0, c_K^+ \to \infty. \tag{G4e}$$

To sketch an argument for (G4d), let H_j have $|H_j \cap H| = \frac{1}{2}K - j$. The quantity at (G4d) is

$$\frac{1}{2} \sum_{j=1}^{\frac{1}{2}K-1} \binom{K/2}{j}^2 \boldsymbol{P}(|X_{H_j}| \le x_K \mid |X_H| \le x_K). \tag{G4f}$$

For fixed j we can argue that the conditional density of X_{H_j} given X_H is bounded as $K \to \infty$ and hence the contribution from j is $\binom{K/2}{j}^2 O(x_K) \to 0$. So it suffices to consider (G4f) with the sum taken over $j_0(K) \le j \le \frac{1}{2}K - j_0(K)$, for $j_0(K) \to \infty$ slowly. For j in this range we can appeal to a bivariate Normal approximation:

$$(X_{H_j}, X_H) \overset{\mathcal{D}}{\approx} \text{Normal, mean 0, variance } K\sigma^2, \text{ covariance } (K-4j)\sigma^2.$$

Putting $a_{j,K}^2 = 4j/K$, the corresponding local approximation is

$$\begin{aligned} \boldsymbol{P}(|X_{J_j}| \le x_K \mid |X_H| \le x_K) &\approx \frac{2x_K}{a_{j,K}\sigma(2\pi K)^{\frac{1}{2}}} \\ &\approx (2\pi K)^{-\frac{1}{2}} 2^{-K} a_{j,K}^{-1} c. \end{aligned}$$

Now the unconditional probability has the same form except with $a_{j,K}$ replaced by 1. And the sum of the unconditional probabilities tends to c, by (G4c). Thus the proof of (G4d) reduces to proving

$$\sum_{j=j_0(K)}^{\frac{1}{2}K-j_0(K)} \binom{K/2}{j}^2 2^{-K} K^{-\frac{1}{2}} (a_{j,K}^{-1} - 1) \to 0. \tag{G4g}$$

And this is routine calculus.

G5 Exponential sums. Several later examples involve sums of i.i.d. exponential random variables. It is convenient to record a lemma.

Lemma G5.1 *Let* (ξ_i) *be i.i.d.* exponential(1), *and let* $S_K = \sum_{i=1}^K \xi_i$. *Then as* $K \to \infty$,

$$\boldsymbol{P}(S_K \le a) \sim e^{-a}\frac{a^K}{K!}, \quad a > 0 \textit{ fixed}; \tag{G5a}$$

$$\boldsymbol{P}(S_K \le a_K) \sim (1-a)^{-1} e^{-a_K} \frac{(a_K)^K}{K!}, \quad K^{-1}a_K \to a < 1; \tag{G5b}$$

$$K^{-1} \log \boldsymbol{P}(K^{-1}S_K \leq a) \to 1 + \log a - a; \quad a < 1; \tag{G5c}$$

$$\boldsymbol{P}(S_K \geq b_K) \sim b(b-1)^{-1} e^{-b_K} \frac{(b_K)^K}{K!}, \quad K^{-1} b_K \to b > 1; \tag{G5d}$$

$$K^{-1} \log \boldsymbol{P}(K^{-1}S_K \geq b) \to 1 + \log b - b, \quad b > 1; \tag{G5e}$$

$$\boldsymbol{P}(S_K \geq K\psi\left(\frac{c}{K}\right)) \sim \frac{c^K}{K!}, \quad 0 < c < \infty, \tag{G5f}$$

where ψ is the inverse function of xe^{-x}, $x > 1$.

These estimates follow from the Poisson process representation: if $Z(\lambda)$ denotes a Poisson(λ) variable then $\boldsymbol{P}(S_K \leq a) = \boldsymbol{P}(Z(a) \geq K)$, and the results follow from the formula for the Poisson distribution. Our use of exponential ξ_i in examples is merely for convenience, to be able to use these explicit estimates, and often results extend to more general distributions. In particular, note that the basic large deviation theorem gives analogues of (G5c) and (G5e) for more general distributions.

We shall also use the simple result: if $n \to \infty$, $L/n \to a$, and $M/n \to b > a$ then

$$n^{-1} \log \binom{M}{L} \to b \log b - a \log a - (b-a) \log(b-a). \tag{G5g}$$

We now start "Category 2" examples: the following is perhaps the prototype.

G6 Example: First-passage percolation on the binary tree. Attach independent exponential(1) variables ξ_e to edges of the infinite rooted binary tree. For each K let I_K be the set of vertices at depth K. For $i \in I_K$ there is a unique path $\pi(i)$ from the root vertex to i; let $X_i = \sum_{e \in \pi(i)} \xi_e$. The intuitive story is that water is introduced at the root vertex at time 0 and percolates down edges, taking time $\xi_{i,j}$ to pass down an edge (i,j). So X_i is the time at which vertex i is first wetted.

Let $M_K = \max_{I_K} X_i$, $m_K = \min_{I_K} X_i$. We shall show

$$K^{-1}M_K \to c_2 \quad \text{a.s.}, \qquad K^{-1}m_K \to c_1 \quad \text{a.s., where } 0 < c_1 < 1 < c_2 < \infty \text{ are the solutions of } 2ce^{1-c} = 1. \tag{G6a}$$

We shall give the argument for M_K, and the same argument holds for m_K. Fix c. Since $|I_K| = 2^K$ and each X_i, $i \in I_K$ has the distribution of S_K in Lemma G5.1, it follows from (G5e) that

$$|I_K| \boldsymbol{P}(X_i^{(K)} \geq cK) \to \begin{cases} 0 & c > c_2 \\ \infty & c < c_2 \end{cases} \tag{G6b}$$

and the convergence is exponentially fast. Since

$$\boldsymbol{P}(M_K \geq cK) \leq |I_K| \boldsymbol{P}(X_i^{(K)} \geq cK),$$

by considering $c > c_2$ we easily see that $\limsup K^{-1}M_K \le c_2$ a.s.. The opposite inequality uses an "embedded branching process" argument. Fix $c < c_2$. By (G6b) there exists L such that

$$|I_L|\boldsymbol{P}(X_i^{(L)} \ge cL) > 1. \tag{G6c}$$

Consider the process $\mathcal{B}_j$, $j \ge 0$, defined as follows. $\mathcal{B}_0$ is the root vertex. $\mathcal{B}_j$ is the set of vertices i_j at level jL such that

$$\text{the ancestor vertex } i_{j-1} \text{ at level } (j-1)L \text{ is in } \mathcal{B}_{j-1}; \tag{G6d}$$

$$\sum_{e\in\sigma(i)} \xi_e \ge cL, \quad \text{where } \sigma(i) \text{ is the path from } i_{j-1} \text{ to } i_j. \tag{G6e}$$

Then the process $|\mathcal{B}_j|$ is precisely the branching process with offspring distribution $\eta = |\{i \in I_L : X_i^{(L)} \ge cL\}|$. By (G6c) $E\eta > 1$ and so $q \equiv \boldsymbol{P}(|\mathcal{B}_j| \to \infty \text{ as } j \to \infty) > 0$. Now if $\mathcal{B}_j$ is non-empty then $M_{jL}/jL \ge c$, so

$$\boldsymbol{P}(\liminf K^{-1}M_K > c) \ge q \tag{G6f}$$

Now consider a level d. Applying the argument above to initial vertex $i \in I_d$ instead of the root vertex, and using independence,

$$\boldsymbol{P}(\liminf K^{-1}M_K \ge c) \ge 1 - (1-q)^d.$$

Since d is arbitrary the probability must equal 1, completing the proof of (G6a).

Remarks

1. The second-moment method does not work here. Intuitively, if Z_n is the size of the n'th generation in a supercritical branching process then we expect $Z_n/EZ_n \to$ some non-degenerate Z, and hence we cannot have $EZ_n^2/(EZ_n)^2 \to 1$; so we cannot apply the second-moment method to counting variables which behave like populations of a branching process.

2. The result (G6a) generalizes to any distribution ξ satisfying an appropriate large deviation theorem. In particular, for the Bernouilli case $\boldsymbol{P}(\xi = 1) = p \ge \frac{1}{2}$, $\boldsymbol{P}(\xi = 0) = 1 - p$ we have

   $K^{-1}M_K \to 1$ a.s., $K^{-1}m_K \to c$ a.s., where c satisfies $\log 2 + c\log c + (1-c)\log(1-c) + c\log p + (1-c)\log(1-p) = 0$.

3. The other Category 2 examples known (to me) exploit similar branching process ideas. There are some open problems whose analysis involves "dependent branching processes", as in the next example.

G7 Example: Percolation on the K-cube. Consider the unit cube $\{0,1\}^K$ in K dimensions. To each edge attach independent exponential(1) random variables ξ. Write $\underset{\sim}{0} = (0,0,\ldots,0)$ and $\underset{\sim}{1} = (1,1,\ldots,1)$ for diametrically opposite vertices. Let I_K be the set of paths of length K from 0 to 1; each $i \in I_K$ is of the form $\underset{\sim}{0} = v_0, v_1, \ldots, v_k = \underset{\sim}{1}$. Let $X_i = \sum_{(v_j,v_{j-1})\in i} \xi_{v_{j-1},v_j}$ be the sum of the random edge-weights along path i. We shall consider $M_K = \max_{I_K} X_i$ and $m_K = \min_{I_K} X_i$.

Since $|I_K| = K!$ and each X_i is distributed as S_K in (G5.1), we obtain from (G5a,G5f) the following outer bounds:

$$\boldsymbol{P}(m_K \leq a) \to 0 \quad \text{as } K \to \infty; \qquad \text{each } a < 1 \tag{G7a}$$

$$\boldsymbol{P}(M_K \geq K\psi(c/K)) \to 0 \quad \text{as } K \to \infty; \qquad \text{each } c > 1. \tag{G7b}$$

Note (G7b) implies the cruder result

$$\boldsymbol{P}(M_K/(K \log K) \geq c) \to 0; \qquad c > 1. \tag{G7c}$$

It is easy to prove

$$M_K/(K \log K) \underset{p}{\to} 1. \tag{G7d}$$

For consider the greedy algorithm. That is, consider the random path G of vertices $\underset{\sim}{0} = V_0, V_1, V_2, \ldots, V_K = \underset{\sim}{1}$ chosen as follows: V_{j+1} is the neighbor v of V_j for which $\xi_{V_j,v}$ is maximal, amongst the $K-j$ allowable neighbors, that is those for which (V_j, v) is not parallel to any previous edge (V_u, V_{u+1}). For this path

$$X_G \overset{\mathcal{D}}{=} \sum_{j=1}^{K} \eta_j \qquad \text{where } (\eta_j) \text{ are independent, } \eta_j \overset{\mathcal{D}}{=} \max(\xi_1, \ldots, \xi_j). \tag{G7e}$$

Now $\eta_j \approx \log j$, so $X_G \approx \sum_{j=1}^{K} \log j \approx K(\log K - 1)$; it is easy to formalize this estimate (e.g. by considering variances and using Chebyshev's inequality) to prove

$$\boldsymbol{P}(X_G/(K \log K) \leq c) \to 0; \qquad c < 1.$$

Since $M_K \geq X_G$, this result and (G7c) proves (G7d).

This argument does not work so well for the minimum m_K. Using the greedy algorithm which chooses the minimal edge-weight at each step, we get the analogue of (G7e) with $\widehat{\eta}_j = \min(\xi_1, \ldots, \xi_j)$. Here $E\widehat{\eta}_j = 1/j$ and hence $EX_G = \sum_{j=1}^{K} 1/j \approx \log K$. Thus for m_K we get an upper bound of order $\log K$, which is a long way from the lower bound of 1 given by (G7a).

There is good reason to believe (Section G20) that m_K is in fact bounded as $K \to \infty$, and some reason to believe

Conjecture G7.1 $m_K \underset{p}{\to} 1$ *as* $K \to \infty$.

This seems a good thesis topic. A natural approach would be to try to mimic the argument in Example G6: fix $c > 1$ and $1 \ll L \ll K$ and consider the sets $\mathcal{B}_j$ of vertices at distance jL from 0 such that there is a path to some vertex of $\mathcal{B}_{j-1}$ with average ξ-value $\leq c/K$. Then $|\mathcal{B}_j|$ grows as a kind of non-homogeneous dependent branching process.

Turning to Category 3 examples, let us first consider an artificial example where it is easy to see what the correct limiting constant $\widehat{c}$ is.

G8 Example: Bayesian binary strings. Let $(L_m, m \geq 1)$ be i.i.d. random variables with $0 < L < 1$. For each $1 \leq i \leq K$ consider binary strings $\$_i(n) = X_i(1), X_i(2), \ldots, X_i(n)$ obtained as follows. Conditional on L_m, the m'th digits $(X_i(m), 1 \leq i \leq K)$ are i.i.d. Bernouilli, $\boldsymbol{P}(X_i(m) = 1) = L_m = 1 - \boldsymbol{P}(X_i(m) = 0)$. Let T_K be the smallest n for which the strings $(\$_i(n), 1 \leq i \leq K)$ are all distinct. We shall show that

$$\frac{T_K}{\log K} \underset{p}{\rightarrow} \widehat{c} \quad \text{as } K \rightarrow \infty \tag{G8a}$$

for a certain constant $\widehat{c}$.

Consider first a distinct pair i, j. Then

$$\begin{aligned} \boldsymbol{P}(X_i(1) = X_j(1)) &= E(L^2 + (1-L)^2) \\ &= a^2, \qquad \text{say,} \end{aligned} \tag{G8b}$$

and so $\boldsymbol{P}(\$_i(n) = \$_j(n)) = a^{2n}$. Writing N_n for the number of pairs $i, j \leq K$ such that $\$_i(n) = \$_j(n)$, we have

$$\boldsymbol{P}(T_K > n) = \boldsymbol{P}(N_n \geq 1) \leq EN_n = \binom{K}{2} a^{2n}.$$

By considering $n_K \sim c \log K$ we see that the natural outer bound is $c^* = -1/\log a$; that is

$$\boldsymbol{P}\left(\frac{T_K}{\log K} > c\right) \rightarrow 0 \quad \text{for } c > c^*.$$

But it turns out that c^* is *not* the correct constant for (G8a). To give the right argument, consider first the non-uniform birthday problem (Example E5) where (Y_i) are i.i.d. from a distribution with probabilities (q_j). The heuristic gave

$$\boldsymbol{P}(Y_i, 1 \leq i \leq r \text{ all distinct}) \approx \exp(-\frac{1}{2} r^2 \sum q_j^2). \tag{G8c}$$

and it can be shown that the error is bounded by $f(\max q_j)$ for some $f(x) \rightarrow 0$ as $x \rightarrow 0$. In the current example, conditional on $L_1, \ldots, L_n$ the strings $(\$_i(n); 1 \leq i \leq K)$ are i.i.d. from a distribution whose probabilities are the 2^n numbers of the form $\prod_{m=1}^n \widehat{L}_m$, where each $\widehat{L}_m = L_m$ or $1 - L_m$. Applying (G8c),

$$\left| \boldsymbol{P}(T_K \leq n \mid L_1, \ldots, L_n) - \exp(-\frac{1}{2}K^2 \prod_{m=1}^{n} (L_m^2 + (1 - L_m)^2)) \right|$$

$$\leq f(\prod_{m=1}^{n} \max(L_m, 1 - L_m)). \quad \text{(G8d)}$$

Define

$$\widehat{c} = -\frac{2}{E \log(L^2 + (1 - L)^2)}.$$

The strong law of large numbers says

$$n^{-1} \log(\prod_{m=1}^{n} (L_m^2 + (1 - L_m)^2) \to -\frac{2}{\widehat{c}} \quad \text{as } n \to \infty.$$

Consider $n_K \sim c \log K$. Then

$$\frac{1}{2}K^2 \prod_{m=1}^{n_K} (L_m^2 + (1 - L_m^2)) \to \begin{cases} 0 & \text{as } K \to \infty; \quad c > \widehat{c} \\ \infty & \text{as } K \to \infty; \quad c < \widehat{c} \end{cases}.$$

So from (G8d)

$$\boldsymbol{P}(T_K \leq n_K) \to \begin{cases} 1 & c > \widehat{c} \\ 0 & c < \widehat{c} \end{cases},$$

and this gives (G8a) for $\widehat{c}$.

G9 Example: Common cycle partitions in random permutations. A permutation π of $\{1, \ldots, N\}$ partitions that set into cycles of π. Consider K independent random uniform permutations (π_u) of $\{1, \ldots, N\}$ and let $Q(N, K)$ be the probability that there exists $i_1, i_2 \in \{1, \ldots, N\}$ such that i_1 and i_2 are in the same cycle of π_u for each $1 \leq u \leq K$. If $K_N \sim c \log N$ then one can show

$$Q(N, K_N) \to \begin{cases} 0 & \text{as } N \to \infty; \quad c > \widehat{c} \\ 1 & \text{as } N \to \infty; \quad c < \widehat{c} \end{cases} \quad \text{(G9a)}$$

for $\widehat{c}$ defined below. In fact, this example is almost the same as Example G8; that problem involved independent random partitions into 2 subsets, and here we have random partitions into a random number of subsets. For one uniform random permutation of $\{1, \ldots, N\}$ let $(Y_1, Y_2, \ldots) = N^{-1}$(vector of cycle lengths). Then as $N \to \infty$ it is known that

$$(Y_1, Y_2, \ldots) \xrightarrow{\mathcal{D}} (L_1, L_2, \ldots), \quad \text{where } \begin{array}{l} L_1 = (1 - U_1) \\ L_2 = U_1(1 - U_2) \\ L_3 = U_1 U_2 (1 - U_3) \end{array} \quad \text{(G9b)}$$

and so on, for i.i.d. (U_i) uniform on $(0,1)$. We can now repeat the arguments for Example G8 to show that (G9a) holds for

$$\widehat{c} = -2\left(E\log\sum L_i^2\right)^{-1}. \tag{G9c}$$

G10 Conditioning on maxima. One of our basic heuristic techniques is conditioning on semi-local maxima (Section A7). In hard combinatorial problems we can sometimes use the simpler idea of conditioning on the global maximum to get rigorous bounds. Here are two examples.

G11 Example: Common subsequences in fair coin-tossing. Let $(\xi_1,\ldots,\xi_K)$, $(\eta_1,\ldots,\eta_K)$ be i.i.d. with $\boldsymbol{P}(\xi_i=1)=\boldsymbol{P}(\xi_i=0)=\boldsymbol{P}(\eta_i=1)=\boldsymbol{P}(\eta_i=0)=\frac{1}{2}$. Let $L_K \le K$ be the length of the longest string of 0's and 1's which occurs as some increasing subsequence of $(\xi_1,\ldots,\xi_K)$ and which also occurs as some increasing subsequence of $(\eta_1,\ldots,\eta_K)$. For example, the starred valued below

ξ	0*	0	1*	1*	0*	1	0*	1*
η	0*	1*	1*	1	0*	0*	0	1*

indicate a common subsequence 011001 of length 6. The subadditive ergodic theorem (Section G21) implies

$$K^{-1}L_K \to c^* \quad \text{a.s.} \tag{G11a}$$

for some constant c^*. The value of c^* is unknown: we shall derive an upper bound.

Fix $1 \le m \le q \le K$. For specified $1 \le i_1 < i_2 < \cdots < i_m \le K$ and $1 \le j_1 < \cdots < j_m \le K$ we have $\boldsymbol{P}((\xi_{i_1},\ldots,\xi_{i_m}) = (\eta_{j_1},\ldots,\eta_{j_m})) = \frac{1}{2}^m$. So

$$E(\#\text{ of common subsequences of length } m) = \binom{K}{m}^2 \frac{1}{2}^m. \tag{G11b}$$

Now if there exists a common subsequence of length q, then by looking at any m positions we get a common subsequence of length m, and so the expectation in (G11b) is at least $\binom{q}{m}\boldsymbol{P}(L_K \ge q)$. Rearranging,

$$\boldsymbol{P}(L_K \ge q) \le \frac{1}{2}^m \binom{K}{m}^2 \Big/ \binom{q}{m}.$$

Now fix $0 < b < c < 1$ and take limits as $K \to \infty$, $m/K \to b$, $q/K \to c$, using (G5g).

$$\begin{aligned}\limsup_{K\to\infty} \log K^{-1}\log \boldsymbol{P}(L_K \ge q) &\le g(b,c)\\ &\equiv -b\log b - 2(1-b)\log(1-b) - c\log c + (c-b)\log(c-b) - b\log 2.\end{aligned}$$

Now the left side does not involve b, so we can replace the right side by its infimum over $0 < b < c$. Thus the constant in (G11a) must satisfy

$$c^* \leq \sup\left\{ c < 1 : \inf_{0<b<c} g(b,c) \geq 0 \right\} = 0.904 \qquad \text{(G11c)}$$

G12 Example: Anticliques in sparse random graphs. A subset H of vertices of a graph is an *anticlique* (or *independent set*) if there is no edge (i,j) with both endpoints in H. In other words, an anticlique is a clique of the complementary graph. The independence number $\mathrm{ind}(G)$ of a graph G is the size of the largest anticlique. Fix $1 < \alpha < \infty$ and consider the random graphs $\mathcal{G}(K, \alpha/K)$ (in the notation of Example G2): these graphs are *sparse* because the mean degree $\to \alpha$ as $K \to \infty$. It is believed that $K^{-1}\,\mathrm{ind}(\mathcal{G}(K,\alpha/K))$ tends to a constant limit as $K \to \infty$. We shall derive an upper bound $c^*(\alpha)$ such that

$$\boldsymbol{P}(\mathrm{ind}(\mathcal{G}(K,\alpha/K)) > cK) \to 0 \quad \text{as } K \to \infty; \qquad c > c^*(\alpha). \qquad \text{(G12a)}$$

The argument is very similar to the previous example. Fix $1 \leq m \leq q \leq K$. Then for a specified subset H of vertices of size m,

$$\boldsymbol{P}(H \text{ is an anticlique of } \mathcal{G}(K,\alpha/K)) = \left(1 - \frac{\alpha}{K}\right)^{\frac{1}{2}m(m-1)}.$$

So

$$E(\#\text{ of anticliques of size } m) = \binom{K}{m}\left(1 - \frac{\alpha}{K}\right)^{\frac{1}{2}m(m-1)}. \qquad \text{(G12b)}$$

If there exists an anticlique of size q, then all its subsets of size m are anticliques, so the expectation in (G12b) is at least $\binom{q}{m}\boldsymbol{P}(\mathrm{ind}\,\mathcal{G}(K,\alpha/K) \geq q)$. Rearranging,

$$\boldsymbol{P}(\mathrm{ind}\,\mathcal{G}(K,\alpha/K) \geq q) \leq (1-\alpha/K)^{\frac{1}{2}m(m-1)} \binom{K}{m} \bigg/ \binom{q}{m}.$$

Now fix $0 < b < c < 1$ and let $K \to \infty$, $m/K \to b$, $q/K \to c$.

$$\begin{aligned} &\limsup_{K\to\infty} K^{-1} \log \boldsymbol{P}(\mathrm{ind}\,\mathcal{G}(K,\alpha/K) \geq q) \leq f_\alpha(b,c) \\ &\equiv \; -(1-b)\log(1-b) - c\log c + (c-b)\log(c-b) - \frac{1}{2}\alpha b^2. \end{aligned}$$

Again the left side does not involve b, so we can take the infimum over b. Thus (G12a) holds for

$$c^*(\alpha) = \sup\{ c < 1 : \inf_{0<b<c} f_\alpha(b,c) \geq 0 \}$$

which can be calculated numerically.

G13 The harmonic mean formula. The examples above worked out easily because there is a deterministic implication: a long common subsequence or large anticlique implies there are at least a deterministic number of smaller common subsequence or anticliques. Where such deterministic relations are not present, the technique becomes impractical. The harmonic mean formula, in the exact form (Section A17), offers an alternative method of getting rigorous bounds. Let us sketch one example.

G14 Example: Partitioning sparse random graphs. For a graph G with an even number of vertices define

$$\text{part}(G) = \min_{H,H^C} \{\text{number of edges from } H \text{ to } H^C\},$$

the minimum taken over all partitions $\{H, H^C\}$ of the vertices with $|H| = |H^C|$. Consider now the sparse random graphs $\mathcal{G}(2K, \alpha/K)$ with $\alpha > \frac{1}{2}$. We shall argue

$$\boldsymbol{P}(\text{part}(\mathcal{G}(2K,\alpha/K)) \leq cK) \to 0 \quad \text{as } K \to \infty; \qquad c < \widehat{c}_\alpha \tag{G14a}$$

for $\widehat{c}_\alpha$ defined as follows. Let

$$\begin{aligned}
f_\alpha(\theta) &= -\theta \log 4 - \theta\log(\theta) - (1-\theta)\log(1-\theta) - \alpha\theta + (1-\theta)\log(1-e^{-\alpha}) \\
f_\alpha &= \sup_{0<\theta<1} f_\alpha(\theta) \\
h_\alpha(x) &= \log 4 + x - \alpha + x\log(\alpha/x) + f_\alpha. \\
\widehat{c}_\alpha &= \begin{cases} 0 & \text{if } h_\alpha(0) \geq 0 \\ \min\{\, x > 0 : h_\alpha(x) = 0\,\} & \text{otherwise} \end{cases}
\end{aligned}$$

To show this, fix $1 < m < K$. Let I be the set of partitions of $\{1, \ldots, 2K\}$ into sets H, H^c of size K. Let A_H be the event that $\mathcal{G}(2K, \alpha/K)$ has at most m edges from H to H^c. Then

$$\begin{aligned}
p \equiv \boldsymbol{P}(A_H) &= \boldsymbol{P}(Z \leq m) \quad \text{where } Z \stackrel{\mathcal{D}}{=} \text{Binomial}(K^2, \alpha/K) \\
&\approx \boldsymbol{P}(Z' \leq m) \quad \text{where } Z' \stackrel{\mathcal{D}}{=} \text{Poisson}(\alpha K); \\
|I| &= \binom{2K}{K}.
\end{aligned}$$

Write $X = \sum_H 1_{A_H}$. Applying Lemma A17.1,

$$\begin{aligned}
&\boldsymbol{P}(\text{part}(\mathcal{G}(2K,\alpha/K)) \leq m) \\
&\qquad = \boldsymbol{P}(\bigcup A_H) \\
&\qquad = \boldsymbol{P}(Z' \leq m)\binom{2K}{K} E(X^{-1} | A_{H_0}) \qquad \text{(G14b)}
\end{aligned}$$

where $H_0 = \{1, 2, \ldots, K\}$ say. Now let $K \to \infty$, $m/K \to c < \widehat{c}_\alpha$. Then

$$\begin{aligned} K^{-1} \log \boldsymbol{P}(Z' \leq m) &\to c - \alpha + c \log(\alpha/c) \\ K^{-1} \log \binom{2K}{K} &\to \log 4 \end{aligned}$$

and so to prove (G14a) is will suffice to show that

$$K^{-1} \log E(X^{-1} | A_{H_0}) \text{ is asymptotically } \leq f_\alpha. \tag{G14c}$$

To show this, let J_0 be the subset of vertices j in H_0 such that j has no edges to H_0; similarly let J_1 be the subset of vertices j in H_0^c such that j has no edges to H_0^c. Let $Y = \min(|J_0|, |J_1|)$. For any $0 \leq u \leq Y$ we can choose u vertices from J_0 and u vertices from J_1 and swap them to obtain a new partition $\{H, H^C\}$ which will have at most m edges from H to H^C. Thus

$$X \geq \sum_{u=0}^{Y} \binom{Y}{u}^2 = g(Y) \text{ say, on } A_{H_0}.$$

Now Y is independent of A_{H_0}, so

$$E(X^{-1} | A_{H_0}) \leq E(1/g(Y)) = \sum_{y=0}^{K} \frac{1}{g(y)} \boldsymbol{P}(Y = y).$$

Write $Y_0 = |J_0|$. Then $\boldsymbol{P}(Y = y) \leq 2\boldsymbol{P}(Y_0 = y)$. Since the sum is at most K times its maximum term, the proof of (G14c) reduces to

$$K^{-1} \log(\max_{0 \leq y \leq K} \frac{1}{g(y)} \boldsymbol{P}(Y_0 = y)) \to f_\alpha. \tag{G14d}$$

This in turn reduces to

$$K^{-1} \log(1/g(\theta K)) + K^{-1} \log \boldsymbol{P}(Y_0 = \theta K) \to f_\alpha(\theta); \qquad 0 < \theta < 1. \tag{G14e}$$

Now $g(y) = \sum_u \binom{y}{u}^2 \approx 2^{2y}$, and so the first term in (G14e) tends to $-\theta \log 4$. Next, suppose we had

$$Y_0 \stackrel{\mathcal{D}}{=} \text{Binomial}(K, e^{-\alpha}). \tag{G14f}$$

Then the second term in (G14e) would converge to the remaining terms of $f_\alpha(\theta)$, completing the proof. For a fixed vertex j in H_0, the chance that j has no edges to H_0 is $(1 - \alpha/K)^{K-1} \approx e^{-\alpha}$. If these events were independent as j varies then (G14f) would be exactly true; intuitively, these events are "sufficiently independent" that Y_0 should have the same large deviation behavior.

G15 Tree-indexed processes. We now describe a class of examples which are intermediate in difficulty between the (hard) "exponential maxima" examples of this chapter and the (easy) "d-parameter maxima" examples treated elsewhere in this book. Fix $r \geq 2$. Let I be the infinite $(r+1)$-tree, that is the graph without circuits where every vertex has degree $r+1$. Let $i_0 \in I$ be a distinguished vertex. Write I_K for the set of vertices within distance K of i_0. Then $|I_K|$ is of order r^K: precisely,

$$|I_K| = 1 + (r+1) + (r+1)r + \cdots + (r+1)r^{K-1} = \frac{(r+1)r^K - 2}{r-1}.$$

Consider a real-valued process $(X_i : i \in I)$ indexed by the $(r+1)$-tree. There are natural notions of *stationary* and *Markov* for such processes. Given any bivariate distribution (Y_1, Y_2) which is symmetric (i.e. $(Y_1, Y_2) \stackrel{\mathcal{D}}{=} (Y_2, Y_1)$) there exists a unique stationary Markov process $(X_i : i \in I)$ such that

$$(X_{i_1}, X_{i_2}) \stackrel{\mathcal{D}}{=} (Y_1, Y_2) \qquad \text{for each edge } (i_1, i_2).$$

For such a process, let us consider

$$M_K = \max_{I_K} X_i$$

as an "exponentially growing" analog of the maxima of 1-parameter stationary Markov chains discussed in Chapters B and C. Here are two examples

G16 Example: An additive process. Fix a parameter $0 < a < \infty$. Let (Y_1, Y_2) have a symmetric joint distribution such that

$$Y_1 \stackrel{\mathcal{D}}{=} Y_2 \stackrel{\mathcal{D}}{=} \text{exponential}(1) \tag{G16a}$$

$$\text{distribution}(Y_1 - y \mid Y_1 = y) \stackrel{\mathcal{D}}{\to} \text{Normal}(-a, 2a) \quad \text{as } y \to \infty. \tag{G16b}$$

Such a distribution can be obtained by considering stationary reflecting Brownian motion on $[0, \infty)$ with drift $-a$ and variance $2a$. Let (X_i) be the associated tree-indexed process and M_K its maximal process, as above. It turns out that

$$K^{-1} M_K \to \begin{cases} c(a) & a \leq \log r \\ \log r & a \geq \log r \end{cases} \quad \text{a.s.} \quad \text{as } K \to \infty \tag{G16c}$$

where $c(a) = (4a \log r)^{1/2} - a$. In other words, the behavior of M_K is different for $a > \log r$ than for $a < \log r$.

The full argument for (G16c) is technical, but let us just observe where the upper bound comes from. It is clear from (G16a) that $\log r$ is the natural upper bound (G1a) for M_K/K. Now consider

$$\widehat{M}_K = \max\{ X_i : \text{distance}(i, i_0) = K \}.$$

When $\widehat{M}_K$ is large, it evolves essentially like the rightmost particle in the following spatial branching process on $\boldsymbol{R}^1$. Each particle is born at some point $x \in \boldsymbol{R}^1$; after unit time, the particle dies and is replaced by r offspring particles at i.i.d. Normal$(x - a, 2a)$ positions. Standard theory for such spatial branching processes (e.g. Mollison (1978)) yields $\widehat{M}_K/K \to c(a)$ a.s. This leads to the asymptotic upper bound $c(a)$ for M_K/K.

G17 Example: An extremal process. The example above exhibited a qualitative change at a parameter value which was *a priori* unexpected. Here is an example which does not have such a change, although *a priori* one would expect it!

Fix a parameter $0 < q < 1$. We can construct a symmetric joint distribution (Y_1, Y_2) such that

$$Y_1 \stackrel{\mathcal{D}}{=} Y_2 \stackrel{\mathcal{D}}{=} \xi_3; \qquad \boldsymbol{P}(\xi_3 \le y) = \exp(-e^{-y}) \tag{G17a}$$

$$\boldsymbol{P}(Y_2 > z \mid Y_1 = y) = (1-q)\boldsymbol{P}(\xi_3 > z) \quad \text{for } z > y \tag{G17b}$$

$$\boldsymbol{P}(Y_2 = y \mid Y_1 = y) \to \frac{q}{1-q} \quad \text{as } y \to \infty. \tag{G17c}$$

We leave the reader to discover the natural construction. Let (X_i) be the associated tree-indexed process (Section G15) and M_K the maximal process. One might argue as follows. Suppose $X_{i_0} = y$, large. Then (G17c) implies that the size of the connected component of $\{\, i : X_i = y \,\}$ containing i_0 behaves like the total population size Z in the branching process whose offspring distribution is Binomial$(r+1, q/(1-q))$ in the first generation and Binomial$(r, q/(1-q))$ in subsequent generations. Clearly Z behaves quite differently in the subcritical $(rq/(1-q) < 1)$ and supercritical $(rq/(1-q) > 1)$ cases, and one might expect the difference to be reflected in the behavior of M_K, because the subcritical case corresponds to "finite-range dependence" between events $\{X_i \ge y\}$ while the supercritical case has infinite-range dependence. But this doesn't happen: using (G17b) it is easy to write down the exact distribution of M_K as

$$\boldsymbol{P}(M_K \le z) = (1 - (1-q)\boldsymbol{P}(\xi_3 > z))^{|I_K|-1}\boldsymbol{P}(\xi_3 \le z).$$

This distribution varies perfectly smoothly with the parameter q.

The paradox is easily resolved using our heuristic. The behavior of M_K depends on the mean clump size EC, and Z is the *conditioned* clump size $\widetilde{C}$ in the notation of Section A6. So the harmonic mean formula says that the behavior of M_K depends on $E(1/Z)$, and this (unlike EZ) behaves smoothly as q varies.

COMMENTARY

There seems no systematic account of the range of problems we have considered. Bollobas (1985) gives a thorough treatment of random graphs, and also has a useful account of different estimates for probabilities of unions (inclusion-exclusion, etc.) and of the second-moment method. Another area which has been studied in detail is first-passage percolation; see Kesten (1987).

G18 Notes on the examples. Example G2 (cliques) is now classic: see Bollobas (1985) XI.1 for more detailed results.

Example G4 (optimum partitioning) is taken from Karmarkar et al. (1986); some non-trivial analysis is required to justify the local Normal approximations.

Example G6 (percolation on binary tree) is well-known; see e.g. Durrett (1984) for the 0–1 case. The technique of looking for an embedded branching process is a standard first approach to studying processes growing in space.

Example G7 (percolation on the K-cube) seems a new problem. Rick Durrett has observed that, in the random subgraph of $\{0,1\}^K$ where edges are present with probability c/K, the chance that there exists a path of length K from 0 to 1 tends to 1 as $K \to \infty$ for $c > e$; it follows that an asymptotic upper bound for m_K is $\frac{1}{2}e$.

Examples G8 and G9 are artificial. The behavior of cycle lengths in random permutations (G9b) is discussed in Vershik and Schmidt (1977).

Examples G12 and G14 (sparse random graphs) are from Aldous (1988a), and improve on the natural outer bounds given by Boole's inequality.

G19 Longest common subsequences. Example G11 generalizes in a natural way. Let μ be a distribution on a finite set S and let (ξ_i), (η_i) be i.i.d. (μ) sequences. Let L_K be the length of the longest string $s_1, s_2, \ldots, s_L$ which occurs as some increasing subsequence of $(\xi_1, \ldots, \xi_K)$ and also as some increasing subsequence of $(\eta_1, \ldots, \eta_K)$. Then the subadditive ergodic theorem says

$$\frac{L_K}{K} \to c(\mu) \quad \text{a.s.}$$

for some constant $c(\mu)$. But the value of $c(\mu)$ is not known, even in the simplest case (Example G11) where $\mu(0) = \mu(1) = \frac{1}{2}$. For bounds see Chvatal and Sankoff (1975), Deken (1979). The argument for (G11c) is from Arratia (unpublished). A related known result concerns longest increasing sequences in random permutations. For a permutation π of $\{1, \ldots, K\}$ there is a maximal length sequence $i_1 < \cdots < i_m$ such that $\pi(i_i) < \pi(i_2) < \cdots < \pi(i_m)$. Let m_K be this maximal length for a uniform random permutation. Then

$$K^{-\frac{1}{2}} M_K \underset{p}{\to} 2;$$

see Logan and Shepp (1977), Vershik and Kerov (1977).

G20 Minimal matrix sums. For a random matrix $(\xi_{i,j} : 1 \leq i, j \leq K)$ with i.i.d. exponential(1) entries, consider

$$M_K = \min_{\pi} \sum_{i=1}^{K} \xi_{i,\pi(i)}$$

where the minimum is taken over all permutations π of $\{1, \ldots, K\}$. At first sight this problem is similar to Example G7; each involves the minimum of $K!$ sums with the same distribution, and the same arguments with Boole's inequality and the greedy algorithm lead to the same asymptotic upper and lower bounds (1 and $\log K$). But the problems are different. Here the asymptotic bounds can be improved to $1 + e^{-1}$ and 2; moreover simulation suggests

$$M_K \approx 1.6, \qquad K \text{ large}.$$

See Karp and Steele (1985).

G21 Abstract problems. In some hard (Category III) problems one can apply the subadditive ergodic theorem (Kingman (1976)) to show that a limiting constant c exists; this works in many first-passage percolation models (on a fixed graph) and in Example G11. But there is no known general result which implies the existence of a limiting constant in Examples G7,G12,G14 and G20; finding such a result is a challenging open problem.

There ought also to be more powerful general results for handling Category II examples. Consider the following setting. Let (ξ_i) be i.i.d. For each K let $(A_1^K, \ldots, A_{N_K}^K)$ be a family of K-element subsets of $\{1, 2, \ldots, Q_K\}$ satisfying some symmetry conditions. Suppose $K^{-1} \log N_K \to a > 0$. Let $\phi(\theta) = E \exp(\theta \xi_1)$ and suppose there exists c^* such that $a \inf_\theta e^{-\theta c^*} \phi(\theta) = 1$. Let $M_K = \max_{i \leq N_K} \sum_{j \in A_i^K} \xi_j$. Then the large deviation theorem implies that

$$\boldsymbol{P}(K^{-1} M_K > c) \to 0; \qquad c > c^*.$$

Can one give "combinatorial" conditions, involving e.g. the intersection numbers $|A_j^K \cap A_1^K|$, $j \geq 2$, which imply $K^{-1} M_K \underset{p}{\to} c^*$ and cover natural cases such as Example G6?

G22 Tree-indexed processes. Our examples are artificial, but this class of processes seems theoretically interesting as the simplest class of "exponential combinatorial" problems which exhibits a wide range of behavior.

There is a long tradition in probability -physics of using tree models as substitutes for lattice models: in this area (Ising models, etc.) the tree models are much simpler. For our purposes the tree processes are more complicated than d-parameter processes. For instance, there is no d-parameter analog of the behavior in Example G16.

H Stochastic Geometry

Readers who do not have any particular knowledge of stochastic geometry should be reassured to learn that the author doesn't, either. This chapter contains problems with a geometric (2-dimensional, mostly) setting which can be solved heuristically *without* any specialized geometric arguments. One exception is that we will need some properties of the Poisson line process at Section H2: this process is treated in Solomon (1978), which is the friendliest introduction to stochastic geometry.

Here is our prototype example.

H1 Example: Holes and clusters in random scatter. Throw down a large number θ of points randomly (uniformly) in the unit square. By chance, some regions of the square will be empty, while at some places several points will fall close together. Thus we can consider random variables

$$L = \text{radius of largest circle containing 0 points;} \quad \text{(H1a)}$$

$$M_K = \text{radius of smallest circle containing } K \text{ points } (K \geq 2) \quad \text{(H1b)}$$

where the circles are required to lie inside the unit square. We shall estimate the distributions of these random variables.

First consider a Poisson process of points (which we shall call *particles*) on the plane $\boldsymbol{R}^2$, with rate θ per unit area. Let $D(x, r)$ be the disc of center x and radius r. To study the random variable L of (H1a), fix r and consider the random set $\mathcal{S}$ of those centers x such that $D(x, r)$ contains 0 particles. Use the heuristic to suppose that $\mathcal{S}$ consists of clumps $\mathcal{C}$ of area C, occurring as a Poisson process of some rate λ per unit area. (This is one case where the appropriateness of the heuristic is intuitively clear.) Then

$$p = \boldsymbol{P}(D(x, r) \text{ contains 0 particles}) = \exp(-\theta\pi r^2)$$

and we shall show later that

$$EC \approx \pi^{-1} r^{-2} \theta^{-2}. \quad \text{(H1c)}$$

So the fundamental identity gives the clump rate for $\mathcal{S}$

$$\lambda = \frac{p}{EC} = \pi\theta^2 r^2 \exp(-\theta\pi r^2).$$

Now the original problem concerning a fixed number θ of particles in the unit square U may be approximated by regarding that process as the restriction to U of the Poisson process on $\boldsymbol{R}^2$ (this restriction actually produces a Poisson(θ) number of particles in U, but for large θ this makes negligible difference). The event "$L < r$" is, up to edge effects, the same as the event "$\mathcal{S}$ does not intersect $U^* = [r, 1-r] \times [r, 1-r]$", and so

$$\begin{aligned} \boldsymbol{P}(L < r) &\approx \boldsymbol{P}(\mathcal{S} \cap U^* \text{ empty}) \\ &\approx \exp(-\lambda(1-2r)^2) \\ &\approx \exp\left(-\pi\theta^2 r^2 \exp(-\theta\pi r^2)(1-2r)^2\right). \qquad \text{(H1d)} \end{aligned}$$

Consider the random variable M_K of (H1b). Fix r, and consider the random set $\mathcal{S}_K$ of those centers x such that $D(x, r)$ contains at least K particles. Use the heuristic:

$$\begin{aligned} p &= \boldsymbol{P}(D(x,r) \text{ contains at least } K \text{ particles}) \\ &\approx \exp(-\theta\pi r^2)\frac{(\theta\pi r^2)^K}{K!} \end{aligned}$$

since we are interested in small values of r. We will show later

$$EC_K \approx r^2 c_K, \qquad \text{(H1e)}$$

where c_K is a constant depending only on K. The fundamental identity yields λ_K. Restricting to the unit square U as above,

$$\begin{aligned} \boldsymbol{P}(M_K > r) &\approx \boldsymbol{P}(\mathcal{S}_K \cap U^* \text{ empty}) \\ &\approx \exp(-\lambda_K(1-2r)^2) \\ &\approx \exp\left(-r^{-2}c_K^{-1}\exp(-\theta\pi r^2)\frac{(\theta\pi r^2)^K}{K!}(1-2r)^2\right) \text{(H1f)} \end{aligned}$$

Remark: In both cases, the only effort required is in the calculation of the mean clump sizes. In both cases we estimate this by conditioning on a fixed point, say the origin $\underset{\sim}{0}$, being in $\mathcal{S}$, and then studying the area $\tilde{C}$ of the clump $\tilde{\mathcal{C}}$ containing $\underset{\sim}{0}$. Only at this stage do the arguments for the two cases diverge.

Let us first consider the clump size for $\mathcal{S}_K$. Given $\underset{\sim}{0} \in \mathcal{S}_K$, the distribution $D(\underset{\sim}{0}, r)$ contains at least K particles; since we are interested in small r, suppose exactly K particles. These particles are distributed uniformly i.i.d. on $D(\underset{\sim}{0}, r)$, say at positions $(X_1, \ldots, X_K)$. Ignoring other particles, a point x near $\underset{\sim}{0}$ is in $\tilde{\mathcal{C}}_K$ iff $|x - X_i| < r$ for all i. That is:

$$\tilde{\mathcal{C}}_K \text{ is the intersection of } D(X_i, r), \qquad 1 \le i \le K.$$

Thus the mean clump size is, by the harmonic mean formula (Section A6),

$$EC_K = \left(E\left(1/\operatorname{area}(\tilde{\mathcal{C}}_K)\right)\right)^{-1}. \qquad \text{(H1g)}$$

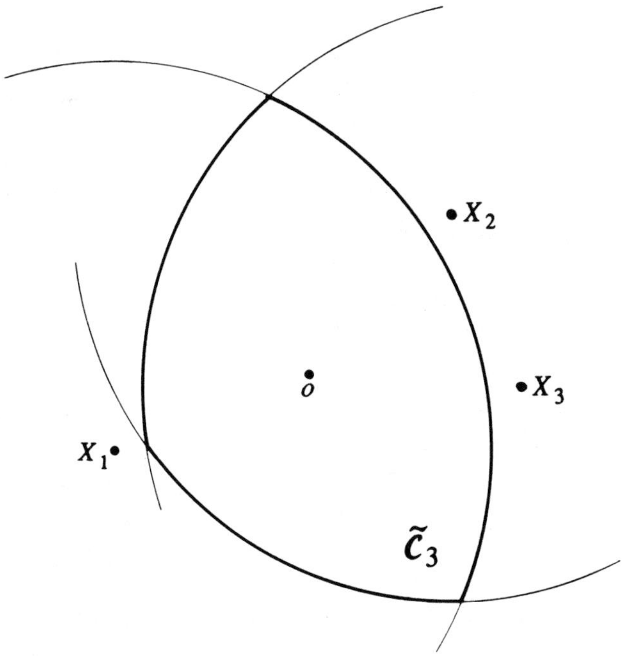

FIGURE H1a.

In particular, by scaling, we get (H1e):

$$EC_K = r^2 c_K,$$

where c_K is the value of EC_K for $r = 1$. I do not know an explicit formula for c_K — of course it could be calculated numerically — but there is some motivation for

Conjecture H1.1 $c_K = \pi/K^2$.

Indeed, this is true for $K = 1$ and 2 by calculation, and holds asymptotically as $K \to \infty$ by an argument in Section H3 below. An alternative expression for c_K is given at Section H16.

H2 The Poisson line process. A constant theme in using the heuristic is the local approximation of complicated processes by standard processes. To calculate mean clump size for $\mathcal{S}$ above, we shall approximate by the Poisson line process in the plane. A good account of this process is Solomon (1978), from which the following definition and result are taken.

A line can be described as a pair (d, ϕ), where d is the signed length of the perpendicular to the line from the origin, and ϕ is the angle this perpendicular makes with the x-axis. Note we restrict $0 \le \phi \le \pi$, so $-\infty <$

$d < \infty$. A process of lines $(d_i, \phi_i : -\infty < i < \infty)$ is a Poisson line process of intensity τ if

1. $\cdots < d_{-2} < d_{-1} < d_0 < d_1 < \cdots$ is a Poisson point process of rate τ on the line;

2. the angles ϕ_i are i.i.d. uniform on $[0, \pi)$, independent of (d_i).

An important property of this process is that its distribution is invariant under translations and rotations of the plane. This process cuts the plane into polygons of random area A; we need the result

$$EA = \frac{\pi}{\tau^2}. \qquad \text{(H2a)}$$

H3 A clump size calculation. We now return to the calculation of clump size for $\mathcal{S}$ in Example H1. Condition on $\underset{\sim}{0} \in \mathcal{S}$. Then the disc $D(\underset{\sim}{0}, r)$ contains no particles. Let X_i be the positions of the particles; then a point x near $\underset{\sim}{0}$ is in $\tilde{\mathcal{C}}$ iff $|x - X_i| > r$ for all i. So

$$\tilde{\mathcal{C}} \text{ is the complement of the union } \bigcup_i D(X_i, r).$$

We can approximate $\tilde{\mathcal{C}}$ by the polygon $\tilde{\mathcal{A}}$ obtained by "straightening the edges" of $\tilde{\mathcal{C}}$ in the following way: the edges of $\tilde{\mathcal{A}}$ are the lines l_i which are perpendicular to the line $(\underset{\sim}{0}, X_i)$ and tangent to the disc $D(X_i, r)$. I assert that near $\underset{\sim}{0}$

the lines l_i are approximately a Poisson line process of intensity $\tau = \pi r\theta$. (H3a)

To see this, note first that the distances $r < |X_1| < |X_2| < \cdots$ form a non-homogeneous Poisson point process with intensity $\rho(x) = 2\pi\theta x$ on (r, ∞). Thus for the lines $l_i = (d_i, \phi_i)$, the unsigned distances $|d_i| = |X_i| - r$ are a Poisson point process of rate $2\pi(d + r)\theta$, that is to say approximately $2\pi r\theta$ for d near 0. Finally, taking signed distances halves the rate, giving (H3a).

Thus the clump $\tilde{\mathcal{C}}$ is approximately distributed as the polygon $\tilde{\mathcal{A}}$ containing $\underset{\sim}{0}$ in a Poisson line process of intensity $\tau = \pi r\theta$. So the clumps $\mathcal{C}$ must be like the polygons $\mathcal{A}$, and so their mean areas EC, EA are approximately equal. So (H2a) gives (H1c).

Finally, let us reconsider the clumps $\mathcal{C}_K$ for $\mathcal{S}_K$ in Example H1 associated with the random variables M_K. Suppose K is large. Then the distances $r > |X_1| > |X_2| > |X_3| > \cdots > |X_K|$ form approximately a Poisson process of rate $2\pi x \cdot K/(\pi r^2)$. We can then argue as above that $\tilde{\mathcal{C}}_K$ is like the polygon containing $\underset{\sim}{0}$ in a Poisson line process of intensity $\tau = \pi r K/(\pi r^2) = K/r$. Using (H2a) as above, we get $EC_K \approx r^2(\pi/K^2)$, as stated below (H1.1).

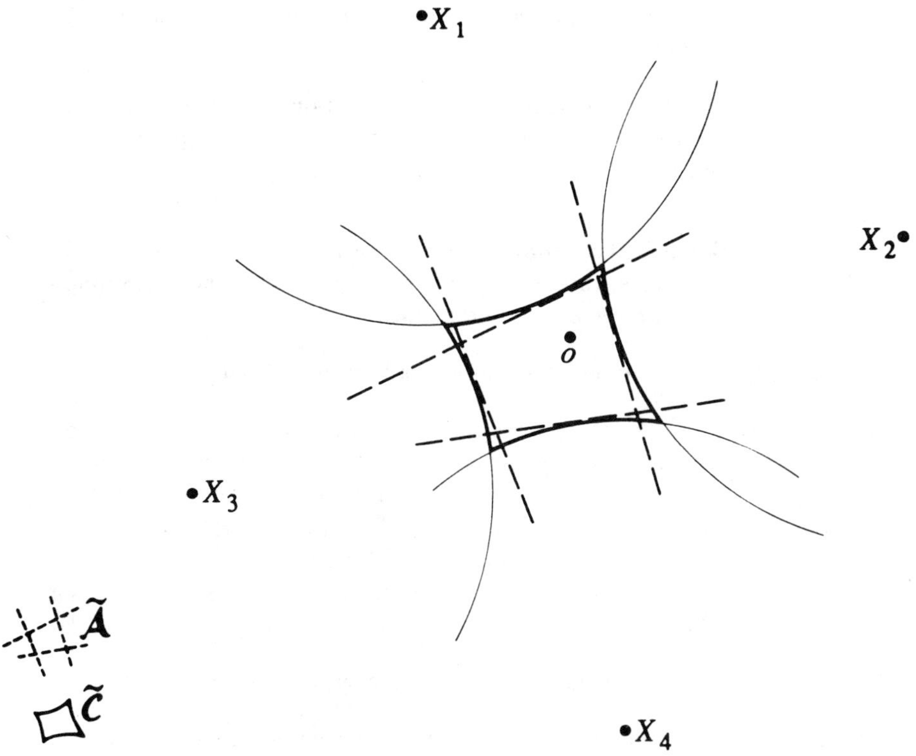

FIGURE H3a.

Remarks Many subsequent examples are variants of the above example, so let us give here some remarks common to all.

1. Problems with a fixed number θ of points can be approximated by problems with a Poisson(θ) number of points, so one may as well use a Poisson model from the start. Then our problems become equivalent to problems about (exact) mosaic processes. Specifically, let $\mathcal{M}$ be the mosaic where centers have rate θ and the constituent random sets $\mathcal{D}$ are discs of radius r. Then $\boldsymbol{P}(L < r)$ in (H1a) is, up to boundary effects, just $\boldsymbol{P}(\mathcal{M}$ covers the unit square). And, if in the definition of "mosaic" we count number of overlaps, then $\boldsymbol{P}(M_K \leq r)$ in (H1b) is just $\boldsymbol{P}(\mathcal{M}$ has K overlaps somewhere in the unit square).

   Hall (1988) gives a detailed treatment of coverage problems for mosaics.

2. In theoretical treatments, boundary effects can be avoided by working on the torus. In any case they are asymptotically negligible; but

(especially in high dimensions) boundary effects may play a large role in non-asymptotic probabilities.

3. Analogous to Section C24, our approximations in (H1d,H1f) are intended for the range of r where the right side is increasing (resp. decreasing) in r.

H4 Example: Empty squares in random scatter. As before, consider a Poisson process, rate θ, of particles in the unit square U. Suppose θ is large. For small s, we shall approximate

$$q(\theta, s) = \boldsymbol{P}(\text{some square of side } s \text{ is empty of particles}).$$

Equivalently, $q(\theta, s) = \boldsymbol{P}(\mathcal{M} \text{ does not cover } U)$, where $\mathcal{M}$ is the random mosaic whose centers have intensity θ and whose constituent random sets are squares of side s. (Here "squares" have sides parallel to the sides of U.) Each such square of side s can be labeled by its center x; note $x \in \widehat{U} = [s/2, 1-s/2] \times [s/2, 1-s/2]$. We apply the heuristic to the random set $\mathcal{S}$ of centers x of empty squares. Since the process of particles is the Poisson process of rate θ,

$$p = \boldsymbol{P}(\text{square with center } x \text{ is empty}) = \exp(-\theta s^2).$$

Condition on the square with center x_0 being empty, and consider the clump $\widetilde{\mathcal{C}}$. Let X_1, X_2, X_3, X_4 be the distances from each side of this square to the nearest particle (see diagram).

Ignoring diagonally-placed nearby particles, we see

$$\widetilde{\mathcal{C}} \text{ is the rectangle } [x_0 - X_1, x_0 + X_3] \times [x_0 - X_4, x_0 + X_2],$$

and so

$$\widetilde{C} = (X_1 + X_3)(X_2 + X_4).$$

Now the X_i are approximately independent, exponential(θs), because

$$\begin{aligned} \boldsymbol{P}(X_i > x) &= \boldsymbol{P}(\text{a certain rectangle of sides } s, x \text{ is empty}) \\ &\approx \exp(-\theta s x). \end{aligned}$$

Thus we get an explicit estimate for the complete distribution of $\widetilde{C}$, so by (A6a) we can compute the distribution of C:

$$C \stackrel{\mathcal{D}}{=} X_1 X_2 \tag{H4a}$$

(this is analogous to the 1-dimensional case (Section A21)). In particular,

$$EC = (\theta s)^{-2}.$$

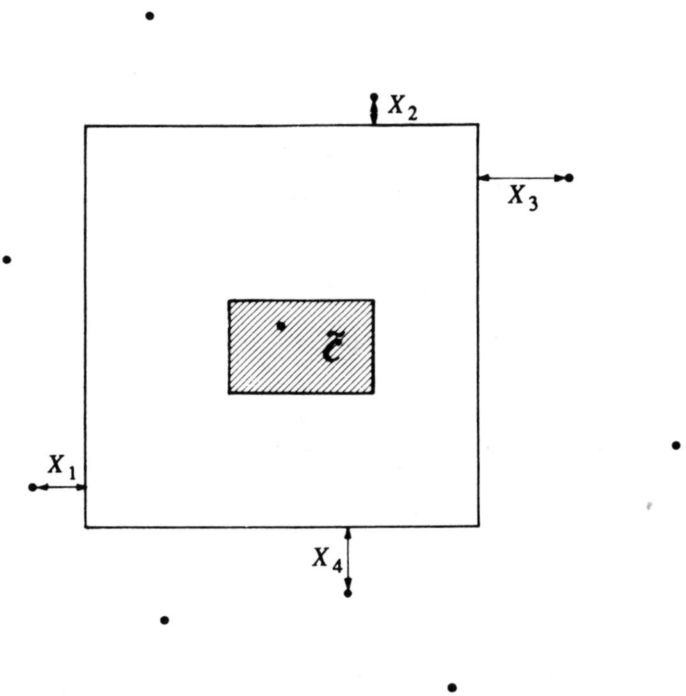

FIGURE H4a.

Getting the clump rate λ from the fundamental identity, we have

$$\begin{aligned} 1 - q(\theta, s) &= \boldsymbol{P}(\mathcal{S} \cap U \text{ empty}) \\ &\approx \exp(-\lambda(1-s)^2) \\ &\approx \exp(-(1-s)^2(\theta s)^2 \exp(-\theta s^2)). \end{aligned} \qquad \text{(H4b)}$$

One reason for treating this example is to add to our collection of explicit clump size distributions. So in the "mosaic" interpretation of the example (remark 1 above), we could use (A4f) to write out an explicit compound Poisson approximation for area$(\mathcal{M}^C \cap U)$. Another reason is to illustrate a technical issue about Poissonization. In developing (H4a) we thought of θ as large and fixed; the quantity at (H4b) is then a reasonable approximation for the distribution function $\boldsymbol{P}(L < s)$ of L = side length of largest empty square. An alternative viewpoint is to consider s as small and fixed, and regard (H4b) as an approximation for the distribution function $\boldsymbol{P}(N \leq \theta)$ of N = number of particles needed until there is no empty square of side s. Again the approximation will be reasonable for the middle of the distribution of N. But the asymptotic form

$$\boldsymbol{P}(N > \theta) \sim (1-s)^2 s^2 \theta^2 \exp(-s^2\theta) \quad \text{as } \theta \to \infty; \qquad s \text{ fixed} \qquad \text{(H4c)}$$

suggested by (H4b) is wrong. Instead of $a(n) = \boldsymbol{P}(N > n) = \boldsymbol{P}$(some empty square amongst n particles), the argument for (H4b) and (H4c) involved treating the particle process as a Poisson process, which really puts a random Poisson(θ) number of particles in U. Thus what we really have at (H4c) is an estimate for the Poisson mixture

$$a^*(\theta) = \sum_n a(n) e^{-\theta} \frac{\theta^n}{n!}.$$

To estimate $a(n)$ we have to "unmix"; doing so (see E2d) gives

$$\boldsymbol{P}(N > n) \sim s^2(1+s)^{-2} n^2 (1-s^2)^n \quad \text{as } n \to \infty; \qquad s \text{ fixed.} \tag{H4d}$$

Remark: There is nothing very special about discs or squares, as far as the use of the heuristic is concerned. What matters is that the class of geometric shapes under study has a finite-dimensional parametrization; we shall illustrate by doing the case of rectangles. Similar problems for Gaussian random fields are treated in Chapter J. Note that we cannot treat a large class like the class of convex sets by this method, since this class doesn't have a finite-dimensional parametrization.

H5 Example: Empty rectangles in random scatter. As before, throw θ particles onto the unit square U. Fix small $A > 0$, and consider the chance that there is no rectangle of area A, with sides parallel to the sides of U, which is empty of particles. We label the rectangles by (x, y, s), where (x, y) is the position of the center and s the length; then the width s' is determined by $s's = A$. To fit inside U requires the constraints

$$A \le s \le 1; \qquad \frac{1}{2}s \le x \le 1 - \frac{1}{2}s; \qquad \frac{1}{2}s' \le y \le 1 - \frac{1}{2}s'. \tag{H5a}$$

Let $I \subset \boldsymbol{R}^3$ be the set of (x, y, s) satisfying constraints (H5a), and let μ be Lebesgue measure (i.e. "volume") on I. Let $\mathcal{S} \subset I$ be the random set of empty rectangles, and apply the heuristic.

$$p = \boldsymbol{P}(\text{rectangle } (x, y, s) \text{ is empty}) \approx \exp(-\theta A).$$

Now condition on a particular rectangle (x_0, y_0, s_0) being in $\mathcal{S}$. As in the last example, let X_1, X_2, X_3, X_4 be the distances from the sides of the rectangle to the nearest particles. I assert that the volume $\widetilde{C} = \mu(\widetilde{\mathcal{C}})$ of the clump $\widetilde{\mathcal{C}}$ of rectangles containing (x_0, y_0, s_0) is approximately

$$\widetilde{C} = \frac{A}{6s_0^2} \cdot \left((X_1 + X_3) + \frac{s_0}{s_0'}(X_2 + X_4) \right)^3. \tag{H5b}$$

To see this, fix s near s_0. Then $(x, y, s) \in \widetilde{\mathcal{C}}$ iff (x, y) is in the rectangle $[x_0 - X_1 - \frac{1}{2}(s_0 - s), x_0 + X_3 + \frac{1}{2}(s_0 - s)] \times [y_0 - X_4 - \frac{1}{2}(s_0' - s'), y_0 + X_2 +$

$\frac{1}{2}(s_0' - s')]$. Since $ss' = s_0 s_0' = A$, we find $(s_0' - s_0) \approx -s_0'/s_0 \cdot (s_0 - s)$, and so

$$\begin{aligned} &\text{area}\{\,(x,y) : (x,y,s) \in \widetilde{C}\,\} \\ &\approx \ (X_1 + X_3 + (s_0 - s))(X_2 + X_4 - \frac{s_0'}{s_0} \cdot (s_0 - s)) \\ &= \ f(s) \qquad \text{say.} \end{aligned}$$

Since $\widetilde{C} = \int f(s)\,ds$, where we integrate over the interval around s_0 in which $f(s) > 0$, we get (H5b) by calculus.

Now as in the last example, the X_i are approximately independent exponential random variables, with parameters $\theta s_0'(i = 1,3)$ and $\theta s_0 (i = 2,4)$. So (H5b) can be rewritten as

$$\widetilde{C} = \frac{s_0}{6A^2\theta^3} \cdot (Y_1 + Y_2 + Y_3 + Y_4)^3;$$

where the Y_i are independent exponential(1). We can now calculate

$$EC = \left(E\left(1/\widetilde{C}\right)\right)^{-1} = \frac{s_0}{A^2\theta^3},$$

using the standard Gamma distribution of $\sum Y_i$. Note that this mean clump size depends on the point (x_0, y_0, s_0), so we are in the non-stationary setting for the heuristic. The fundamental identity gives the non-stationary clump rate

$$\lambda(x,y,s) = \frac{p(s)}{EC(s)} = A^2\theta^3 \exp(-\theta A) s^{-1}.$$

So finally

$$\begin{aligned} &\boldsymbol{P}(\text{no empty rectangle of area } A) \\ &= \ \boldsymbol{P}(S \cap I \text{ empty}) \\ &\approx \ \exp(-\int_I \lambda(x,y,s)\,dx\,dy\,ds). \\ &\approx \ \exp\left(-A^2\theta^3 \exp(-\theta A) \int_A^1 s^{-1}(1-s)\left(1 - \frac{A}{s}\right) ds\right) \qquad \text{(H5c)} \end{aligned}$$

by (H5a). For small A the integral is approximately $\log(1/A) - 2$.

H6 Example: Overlapping random squares. Now consider the analogue of (H1b) for squares instead of discs. That is, fix θ large and s small and throw squares of side s onto the unit square U, their centers forming a Poisson process of rate θ. For $K \geq 2$ we shall estimate

$$q = \boldsymbol{P}(\text{some point of } U \text{ is covered by at least } K \text{ squares}). \qquad \text{(H6a)}$$

We apply the heuristic to $\mathcal{S} = \{x : x \text{ covered by } K \text{ squares}\}$. So

$$\begin{aligned} p &= \boldsymbol{P}(\text{Poisson}(\theta s^2) = K) \\ &= \frac{(\theta s^2)^K}{K!} \cdot \exp(-\theta s^2). \end{aligned} \tag{H6b}$$

Now fix $\underset{\sim}{x}$, condition on $\underset{\sim}{x} \in \mathcal{S}$ and consider the clump $\widetilde{\mathcal{C}}$ containing $\underset{\sim}{x}$. Analogously to the argument in Example H1, there are K points (X_i, Y_i), the centers of the covering squares, which are uniform on the square centered at $\underset{\sim}{x}$. Let $X_{(i)}$ be the order statistics of $X_i - \frac{1}{2}s$, and similarly for $Y_{(i)}$. Ignoring other nearby squares, $\widetilde{\mathcal{C}}$ is the rectangle

$$\widetilde{\mathcal{C}} = [x - (s - X_{(K)}), x + X_{(1)}] \times [y - (s - Y_{(K)}), y + Y_{(1)}] \quad \text{where } \underset{\sim}{x} = (x, y)$$

So

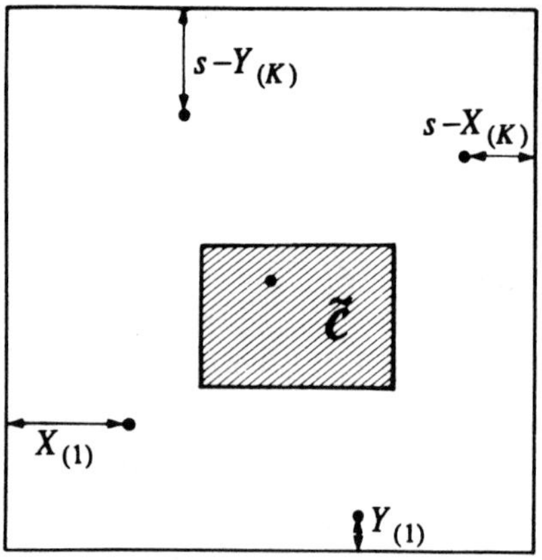

FIGURE H6a.

$$\begin{aligned} \widetilde{C} &= \text{area}(\widetilde{\mathcal{C}}) = (X_{(1)} + s - X_{(K)})(Y_{(1)} + s - Y_{(K)}) \\ &\stackrel{\mathcal{D}}{=} X_{(2)} Y_{(2)} \end{aligned}$$

and it is easy to calculate

$$EC = \text{harmonic mean}(\widetilde{C}) = \left(\frac{s}{K-1}\right)^2. \tag{H6c}$$

The fundamental identity gives

$$\lambda = \frac{p}{EC} = \theta^K s^{2K-2} \frac{(K-1)^2}{K!} \cdot \exp(-\theta s^2). \tag{H6d}$$

So the q in (H6a) satisfies

$$q = 1 - \boldsymbol{P}(\mathcal{S} \cap U \text{ empty}) \approx 1 - \exp(-\lambda). \tag{H6e}$$

H7 Example: Covering K times. Throw discs of radius r, or squares of side s, onto the unit square with their centers forming a Poisson(θ) process; what is the probability

$$q = \boldsymbol{P}\begin{pmatrix}\text{each point of the unit disc is cov-}\\ \text{ered by at least } K \text{ discs [squares]}\end{pmatrix}? \tag{H7a}$$

Examples H1a and H4 treat the cases $K = 1$; the general case is similar. Write $A = \pi r^2$ or s^2 for the area of the discs [squares]. Apply the heuristic to $\mathcal{S} = \{\, x : x \text{ covered } K-1 \text{ times} \,\}$. Then

$$p = \boldsymbol{P}(x \in \mathcal{S}) = \boldsymbol{P}(\text{Poisson}(\theta A) = K-1) = e^{-\theta A}\frac{(\theta A)^{K-1}}{(K-1)!}$$

and the heuristic says

$$q \approx \exp(-\lambda) \approx \exp\left(-\frac{p}{EC}\right). \tag{H7b}$$

But the arguments for EC are exactly the same as for the $K = 0$ case, so

$$EC = \pi^{-1}r^{-2}\theta^{-2} \qquad \text{[discs] by (H1c)} \tag{H7c}$$

$$EC = s^{-2}\theta^{-2} \qquad \text{[squares] by (H4a).} \tag{H7d}$$

Substituting into (H7b) gives an explicit approximation for q.

H8 Example: Several types of particle. As another variant of Example H1b, suppose 3 types of particles are thrown onto the unit square according to Poisson processes of rate θ_1, θ_2, θ_3. Consider

> M =radius of smallest circle containing at least one particle of each type.

Fix r and apply the heuristic to $\mathcal{S} = \{\, x : D(x,r) \text{ contains at least one particle of each type} \,\}$. Then

$$p = \boldsymbol{P}(x \in \mathcal{S}) = \prod_i (1 - \exp(-\theta_i \pi r^2)) \approx (\pi r^2)^3 \theta, \quad \text{where } \theta = \prod_{i=1}^{3} \theta_i.$$

The argument for clump size is exactly as in Example H1b:

$$EC = r^2 c_3, \quad \text{where } c_3 \text{ is the constant at (H1b), conjectured to be } \frac{\pi}{9}.$$

So the heuristic says

$$\boldsymbol{P}(M > r) = \boldsymbol{P}(\mathcal{S} \cap U \text{ empty}) \approx \exp\left(-\frac{p}{EC}\right)$$

$$\approx \exp\left(-\frac{\pi^3 r^4 \theta}{c_3}\right). \tag{H8b}$$

H9 Example: Non-uniform distributions. Consider Example H1 again, but now suppose the points are put down non-uniformly, according to some smooth density function $g(x)$ in U. We can then use the non-stationary form of the heuristic. To study the M_K of (H1b),

$$\boldsymbol{P}(M_K > r) \approx \exp\left(-\int_U \lambda(x)\,dx\right), \tag{H9a}$$

where the clump rate $\lambda(x)$ is the clump rate λ_K of Example H1 with θ replaced by $\lambda g(x)$: that is

$$\lambda(x) = r^{-2} c_K^{-1} \frac{(\theta g(x)\pi r^2)^K}{K!} \cdot \exp(-\theta g(x)\pi r^2).$$

Suppose g attains its maximum at x_0. A little calculus shows that, for θ large, the $g(x)$ in the "exp" term above can be approximated by $g(x_0)$, and so

$$\int_U \lambda(x)\,dx \approx r^{-2} c_K^{-1} \frac{(\theta\pi r^2)^K}{K!} \cdot \exp(-\theta g(x_0)\pi r^2) \cdot \int_U g^K(x)\,dx. \tag{H9b}$$

Substituting into (H9a) gives our approximation for M_K. For the random variable L of (H1a),

$$\boldsymbol{P}(L < r) \approx \exp\left(-\int_U \lambda(x)\,dx\right) \tag{H9c}$$

where the clump rate $\lambda(x)$ is the clump rate λ below (H1c) with θ replaced by $\theta g(x)$: that is

$$\lambda(x) = \pi\theta^2 g^2(x) r^2 \exp(-\theta g(x)\pi r^2)$$

Suppose g attains its minimum at $x^* \in \text{interior}(U)$. Then for large θ the integral is dominated by the contribution from $x \approx x^*$. Write

$$\Delta = \text{determinant}\left(\frac{\partial g(x_1, x_2)}{\partial x_i\,\partial x_j}\right)_{x=x^*}.$$

Then (see (I1d): we use such approximations extensively in Chapter I)

$$\begin{aligned}
\int_U \lambda(x)\,dx &\approx \pi\theta^2 g^2(x^*) r^2 \int_U \exp(-\theta g(x)\pi r^2)\,dx \\
&\approx \pi\theta^2 g^2(x^*) r^2 2\pi(\theta\pi r^2)^{-1} \Delta^{-\frac{1}{2}} \exp(-\theta g(x^*)\pi r^2) \\
&\approx 2\pi\theta g^2(x^*) \Delta^{-\frac{1}{2}} \exp(-\theta g(x^*)\pi r^2).
\end{aligned} \tag{H9d}$$

Substituting into (H9c) gives our explicit approximation for L.

We can formulate discrete versions of our examples — but these are often rather trite, as the next example shows.

H10 Example: Monochrome squares on colored lattice. Take the $N \times N$ square lattice (N large) and color each site red, say, with probability q, independently for each site. Given s, we can ask; what is the chance that there is some $s \times s$ square of red sited within the lattice? To approximate this, label a $s \times s$ square by its lowest-labeled corner (i_1, i_2); this gives an index set $I = \{1, \ldots, N-s+1\} \times \{1, \ldots, N-s+1\}$. Let $\mathcal{S}$ be the random subset of I representing the red $s \times s$ squares. Then

$$p = p(i \in \mathcal{S}) = q^{s^2}.$$

Moreover if s is not small we have clump size $C \approx 1$. For suppose the square labeled (i_1, i_2) is red; in order for an adjacent square, e.g. the one labeled (i_1+1, i_2), to be red we need some $s \times 1$ strip to be red, and this has chance q^s, which will be small unless s is small. So $\boldsymbol{P}(C > 1)$ is small, and we can take $EC \approx 1$. (To be more accurate, copy the argument of Example H4 to conclude $EC \approx (1-q^s)^{-2}$.) So the fundamental identity gives $\lambda \approx q^{s^2}$, and so

$$\begin{aligned} \boldsymbol{P}(\text{no red } s \times s \text{ square}) &\approx \boldsymbol{P}(\mathcal{S} \cap I \text{ empty}) \\ &\approx \exp(-\lambda |I|) \\ &\approx \exp(-(N-s+1)^2 q^{s^2}). \qquad \text{(H10a)} \end{aligned}$$

H11 Example: Caps and great circles. All our examples so far have been more or less direct variations on Example H1. Here we present an example with a slightly different flavor. Let S be the surface of the unit sphere; so area$(S) = 4\pi$. On S throw down small circular caps of radius r, with centers following a Poisson process of rate θ per unit surface area. Consider

$$q(\theta, r) = \boldsymbol{P}(\text{every great circle on } S \text{ intersects some cap}).$$

Equivalently, throw down particles on S according to a Poisson(θ) process; for each great circle γ let D_γ be the distance from γ to the nearest particle; and let $M = \max_\gamma D_\gamma$; then

$$q(\theta, r) = \boldsymbol{P}(M < r).$$

We shall argue that for large θ,

$$q(\theta, r) \approx \exp(-32\theta^2 \exp(-4\theta r \pi)). \qquad \text{(H11a)}$$

Fix a great circle γ. For r small, consider the strip of points on the surface within r of γ; this strip has area $\approx (2\pi)(2r)$, and so

$$p \equiv \boldsymbol{P}(\gamma \text{ intersects no cap}) \approx \exp(-4\pi r \theta). \qquad \text{(H11b)}$$

Now label the points of γ as $[0, 2\pi]$. Condition on γ intersecting no cap, and for each cap near γ let η be the point on the cap which is nearest to γ. Then the η's form approximately a Poisson point process of rate θ per unit area. Let γ' be another great circle such that the maximum distance between γ' and γ is b, small. Then (draw a picture!)

$$\begin{aligned}
&\boldsymbol{P}(\gamma' \text{ intersects no cap} \mid \gamma \text{ intersects no cap}) \\
&\quad\approx \boldsymbol{P}(\text{none of the points } \eta \text{ lie between } \gamma \text{ and } \gamma') \\
&\quad\approx \exp(-\theta \cdot \text{ area between } \gamma \text{ and } \gamma') \\
&\quad\approx \exp\left(-\theta \int_0^{2\pi} |b\cos(t)|\, dt\right) \\
&\quad\approx \exp(-4b\theta). \qquad\qquad (\text{H11c})
\end{aligned}$$

Next, a great circle γ may be parameterized by its "north pole" x_γ. Apply the heuristic to $\mathcal{S} = \{x_\gamma : \gamma \text{ intersects no cap}\}$. Then $p = \boldsymbol{P}(x \in \mathcal{S})$ is given by (H11b). Fix x_γ, condition on $x_\gamma \in \mathcal{S}$ and let $\widetilde{\mathcal{C}}$ be the clump containing x_γ. As in Example H4, for θ large $\widetilde{\mathcal{C}}$ is like the corresponding random polygon in a Poisson line process, whose intensity (by (H11c)) is $\tau = 4\theta$. So by (H2a),

$$EC = \frac{\pi}{(4\theta)^2}.$$

The fundamental identity gives the clump rate

$$\lambda = p/EC = 16\theta^2\pi^{-1}\exp(-4\pi\theta r).$$

Then

$$q(\theta, r) = \boldsymbol{P}(S \cap S \text{ empty}) \approx \exp(-\lambda(2\pi))$$

keeping in mind that diametrically opposite x's describe the same γ; this gives (H11a).

One-dimensional versions of these examples are much easier, and indeed in some cases have tractable exact solutions (e.g. for the probability of randomly-placed arcs of constant length covering a circle). Our type of heuristic asymptotics makes many one-dimensional results rather easy, as the next example shows.

H12 Example: Covering the line with intervals of random length. Let $\mathcal{M}$ be a high-intensity mosaic process on $\boldsymbol{R}^1$, where the constituent random sets $\mathcal{B}$ are intervals of random length B, and whose left endpoints form a Poisson process of rate θ. Let $\mathcal{S} = \mathcal{M}^c$, the uncovered part of the line. Given $x \in \mathcal{S}$, let $R_x > x$ be the next point of $\mathcal{M}$; then $R_x - x$ has exactly exponential(θ) distribution. Thus the intervals of $\mathcal{S}$ have exactly exponential(θ) distribution; in fact, it is easy to see that successive interval lengths of $\mathcal{S}$ and $\mathcal{M}$ form an alternating renewal process.

Applying the heuristic to $\mathcal{S}$:

$$\begin{aligned} p &= \boldsymbol{P}(x \in \mathcal{S}) = \exp(-\theta EB) \\ EC &= \frac{1}{\theta} \\ \lambda &= \frac{p}{EC} = \theta \exp(-\theta EB) \end{aligned}$$

and so

$$\begin{aligned} \boldsymbol{P}(\mathcal{M} \text{ covers } [0,L]) &= \boldsymbol{P}(\mathcal{S} \cap [0,L] \text{ empty}) \\ &\approx \exp(-\lambda L) \\ &\approx \exp(-L\theta \exp(-\theta EB)). \end{aligned} \quad \text{(H12a)}$$

An essentially equivalent problem is to throw a large fixed number θ of arcs of i.i.d. lengths (B_i) onto the circle of unit circumference; this corresponds asymptotically, by Poissonization, to the setting above with $L = 1$, so

$$\boldsymbol{P}(\text{circle covered}) \approx \exp(-\theta \exp(-\theta EB)). \quad \text{(H12b)}$$

In this setting one can ask further questions: for instance, what is the length D of the longest uncovered interval? To answer, recall that intervals of $\mathcal{S}$ occur at rate λ, and intervals have exponential(θ) length, so that intervals of length $\geq x$ occur at rate $\lambda e^{-\theta x}$. So

$$\begin{aligned} \boldsymbol{P}(D < x) &\approx \exp(-\lambda e^{-\theta x}) \\ &\approx \exp(-\theta \exp(-\theta(EB + x))), \quad x > 0. \end{aligned} \quad \text{(H12c)}$$

Similarly, we get a compound Poisson approximation for the total length of uncovered intervals.

Returning to the "mosaic" description and the line $\boldsymbol{R}^1$, consider now the case where the constituent random sets $\mathcal{B}$ are not single intervals but instead are collections of disjoint intervals — say N intervals of total length B, for random N and B. Then the precise alternating renewal property of $\mathcal{S}$ and $\mathcal{M}$ is lost. However, we can calculate exactly

$$\begin{aligned} \psi &\equiv \text{mean rate of intervals of } \mathcal{S} \\ &= \boldsymbol{P}(0 \in \mathcal{S}) \times \lim_{\delta \downarrow 0} \delta^{-1} \boldsymbol{P}(\delta \in \mathcal{M} \mid 0 \in \mathcal{S}) \\ &= \exp(-\theta EB) \times \theta EN. \end{aligned} \quad \text{(H12d)}$$

Provided $\mathcal{S}$ consists of isolated small intervals, we can identify ψ with the clump rate λ of $\mathcal{S}$, as at Section A9 and obtain

$$\boldsymbol{P}(\mathcal{M} \text{ covers } [0,L]) \approx \exp(-L\psi). \quad \text{(H12e)}$$

This should be asymptotically correct, under weak conditions on $\mathcal{B}$.

H13 Example: Clusters in 1-dimensional Poisson processes. Now consider a Poisson process of events, rate ρ say. Fix a length L and an integer K such that

$$p = \boldsymbol{P}(Z \geq K) \text{ is small,} \qquad \text{for } Z \stackrel{\mathcal{D}}{=} \text{Poisson}(\rho L). \tag{H13a}$$

We shall use the heuristic to estimate the waiting time T until the first interval of length L which contains K events. Let $\mathcal{S}$ be the random set of right endpoints of intervals of length L which contain $\geq K$ events. Then p is as at (H13a). We could estimate EC as in the 2-dimensional case by conditioning on an interval containing K events and considering the clump of intervals which contain these K events, ignoring possible nearby events — this would give $EC \approx K/L$ — but we can do better in 1 dimension by using the quasi-Markov estimate (Section D42). In the notation of Section D42,

$$\begin{aligned}
&\boldsymbol{P}([0,L] \text{ contains } K \text{ events, } [\delta, L+\delta] \text{ contains } K-1 \text{ events)} \\
&\quad\approx \boldsymbol{P}([0,\delta] \text{ contains 1 event, } [\delta, L+\delta] \text{ contains } K-1 \text{ events)} \\
&\quad\approx \delta\rho\boldsymbol{P}(Z = K-1) \qquad \text{for } Z \text{ as in (H13a).}
\end{aligned}$$

and so $\psi = \rho\boldsymbol{P}(Z = K-1)$. Now condition on t_0 being the right end of some component interval in a clump $\mathcal{C}$; then there must be 1 event at $t_0 - L$ and $K-1$ events distributed uniformly through $[t_0 - L, t_0]$. Consider the process

$$Y_u = \#\text{ events in } [t_0 - L + u, t_0 + u].$$

Then $Y_0 = K-1$ and we can approximate Y_u by the continuous-time random walk $\widehat{Y}_u$ with transition rates

$$y \to y+1 \text{ rate } \rho; \qquad y \to y-1 \text{ rate } \frac{K-1}{L}.$$

Then

$$\begin{aligned}
EC^+ &\approx E_{K-1}(\text{sojourn time of } \widehat{Y} \text{ in } [K,\infty)) \\
&\approx \rho\left(\frac{K-1}{L} - \rho\right)^{-2} \qquad \text{using (B2i,iv).}
\end{aligned}$$

Estimating λ via (D42d), we conclude T is approximately exponential with mean

$$ET \approx \lambda^{-1} \approx \rho\left(\frac{K-1}{L} - \rho\right)^{-2} (\boldsymbol{P}(Z \geq K))^{-1} + \frac{1}{\rho^{-1}(\boldsymbol{P}(Z = K-1))}. \tag{H13b}$$

COMMENTARY

H14 General references. We have already mentioned Solomon (1978) as a good introduction to stochastic geometry. A more comprehensive account is in Stoyan et al. (1987). The recent book of Hall (1988) gives a rigorous account of random mosaics, including most of our examples which can be formulated as coverage problems. The bibliography of Naus (1979) gives references to older papers with geometric flavors.

H15 Empty regions in random scatter. In the equivalent formulation as coverage problems, our examples H1a, H4, H7 are treated in Hall (1988) Chapter 3 and in papers of Hall (1985a) and Janson (1986). Bounds related to our tail estimate (H4d) are given by Isaac (1987). Hall (1985b) gives a rigorous account of the "vacancy" part of Example H9 (non-uniform distribution of scatter).

Our example H5 cannot be formulated as a standard coverage problem — so I don't know any rigorous treatment, though formalizing our heuristic cannot be difficult. A more elegant treatment would put the natural non-uniform measure μ on I to make a stationary problem. More interesting is

H15.1 Thesis project. What is the size of the largest empty convex set, for a Poisson scatter of particles in the unit square?

H16 Overlaps. The "overlap" problems H1b, H6, H8 do not seem to have been treated explicitly in the literature (note they are different from the much harder "connected components" problems of (H20) below). Rigorous limit theorems can be deduced from the Poisson limit theorem for U-statistics of Silverman and Brown (1978), but this does not identify constants explicitly.

For instance, the constant c_K in Conjecture H1.1 is given by

$$c_K^{-1}\pi^K = \int_{R^2}\cdots\int_{R^2} 1_{(0,x_1,\ldots,x_{K-1} \text{ all in some disc of radius } 1)}\, dx_1 \ldots dx_{K-1}$$

H17 Discrete problems. Problems extending Example H10 are discussed in Nemetz and Kusolitsch (1982) Darling and Waterman (1985; 1986). In our example we have supposed q is not near 1; as $q \to 1$ the problem approaches the continuous Example H4.

H18 1-dimensional coverage. Hall (1988) Chapter 2 gives a general account of this topic.

Stein's method (Section A18) provides a sophisticated method of proving

Poisson-type limits and also getting explicit non-asymptotic bounds. It would be an interesting project to formalize our assertion (H12e) about coverage by mosaics with disconnected constituent sets, using Stein's method. The method can also be used to get explicit bounds in certain 2-dimensional coverage problems: see Aldous (1988c).

H19 Clustering problems in 1-dimension. Clustering problems like Example H13 have been studied extensively. Naus (1982) gives accurate (but complicated) approximations and references; Gates and Westcott (1985) prove some asymptotics. Our estimate (H13b) isn't too accurate. For instance, with $\rho = 1$, $L = 1$, $K = 5$ it gives $ET \approx 95$, whereas in fact $ET \approx 81$.

Samuel-Cahn (1983) discusses this problem when the Poisson process is replaced by a renewal process.

H20 Connected components in moderate-intensity mosaics. This topic, often called "clumping", is much harder: even the fundamental question of when a random mosaic contains an infinite connected component is not well understood. Roach (1968) gave an early discussion of these topics; Hall (1988) Chapter 4 contains an up-to-date treatment.

H21 Covering with discs of decreasing size. Consider a mosaic process of discs on the plane, where for $r < 1$ the rate of centers of discs of radius $(r, r + dr)$ is $(c/r)\,dr$, for a constant c. Let $\mathcal{S}$ be the uncovered region. It is easy to see that $\boldsymbol{P}(x \in \mathcal{S}) = 0$ for each fixed point x, and so $\mathcal{S}$ has Lebesgue measure 0. However, for small c it turns out that $\mathcal{S}$ is not empty; instead, $\mathcal{S}$ is a "fractal" set. See Kahane (1985). The same happens in the 1-dimensional case, covering the line with intervals of decreasing size — see Shepp (1972a; 1972b). This is an example where our heuristic is not applicable.

I Multi-Dimensional Diffusions

I1 Background. Applied to multi-dimensional diffusions, the heuristic helps with theoretical questions such as the distribution of extreme values, and applied questions such as the rate of escape from potential wells. Sections I1–I7 contain some basic definitions and facts we'll need in the examples.

In Chapter D we defined a 1-dimensional diffusion X_t as a continuous-path Markov process such that

$$E(\Delta X_t \mid X_t = x) \approx \mu(x)\Delta t; \qquad \mathrm{var}(\Delta X_t \mid X_t = x) \approx \sigma^2(x)\Delta t \qquad \text{(I1a)}$$

for specified smooth functions $\mu: \boldsymbol{R} \to \boldsymbol{R}$ and $\sigma: \boldsymbol{R} \to \boldsymbol{R}$. A small conditional increment ΔX_t is approximately Normally distributed, so (I1a) can be rewritten in the notation of stochastic differential equations as

$$dX_t = \mu(X_t)\,dt + \sigma(X_t)\,dB_t \qquad \text{(I1b)}$$

where B_t is Brownian motion. This is a more convenient form for discussing the multidimensional case.

As a preliminary, recall some facts about multidimensional Normal distributions. Let $Z = (Z_1, \ldots, Z_d)$ be d-dimensional standard Normal (i.e. with independent $N(0,1)$ components). The general Normal distribution Y can be written as

$$Y = \mu + AZ$$

for some matrix A and some vector μ. And $EY = \mu$, and the matrix $\Sigma = \mathrm{cov}(Y)$ whose entries are the covariances $\mathrm{cov}(Y_i, Y_j)$ is the matrix AA^T. The distribution of Y is called the Normal(μ, Σ) distribution. The covariance matrix Σ is symmetric and non-negative definite. In the non-degenerate case where Σ is positive definite, the Normal$(0, \Sigma)$ distribution has density

$$f(x) = (2\pi)^{-\frac{d}{2}}|\Sigma|^{-\frac{1}{2}} \exp(-\frac{1}{2}x^T\Sigma^{-1}x) \qquad \text{(I1c)}$$

where $|\Sigma| = \det(\Sigma)$. Note that implicit in (I1c) is the useful integration formula

$$\int_{R^d} \exp(-\frac{1}{2}x^T\Sigma^{-1}x)\,dx = (2\pi)^{\frac{d}{2}}|\Sigma|^{\frac{1}{2}}. \qquad \text{(I1d)}$$

Now let $B_t = (B_t^1, \ldots, B_t^d)$ be d-dimensional Brownian motion (i.e. with independent components distributed as 1-dimensional standard Brownian motion). Let $\mu: \boldsymbol{R}^d \to \boldsymbol{R}^d$ and $\sigma: \boldsymbol{R}^d \to \{d \times d \text{ matrices}\}$ be smooth functions. Then the stochastic differential equation

$$dX_t = \mu(X_t)\,dt + \sigma(X_t)\,dB_t \tag{I1e}$$

defines a d-dimensional diffusion, which we think of intuitively as follows: given $X_t = x$, the small increment ΔX_t is approximately $\mu(x)\Delta t + \sigma(x)\Delta B_t$, that is to say the d-dimensional Normal$(\mu(x)\Delta t, \sigma(x)\sigma^T(x)\Delta t)$ distribution. Analogously to (I1a), we can also specify this diffusion as the continuous-path Markov process such that

$$\begin{aligned} E(\Delta X_t \mid X_t = x) &\approx \mu(x)\Delta t \\ \operatorname{cov}(\Delta X_t^i, \Delta X_t^j \mid X_t = x) &\approx \sigma(x)\sigma^T(x)\Big|_{i,j} \Delta t. \end{aligned} \tag{I1f}$$

In practical examples the functions μ, σ are given and we are interested in doing probability calculations with the corresponding diffusion X_t. There are general equations, given below, for hitting probabilities, stationary distributions and mean hitting times. These differential equations are just the intuitively obvious continuous analogues of the corresponding difference equations for Markov chains (B1a,B1b).

Define operators L, L^* acting on smooth functions $f: \boldsymbol{R}^d \to \boldsymbol{R}$ by

$$\begin{aligned} Lf &= \sum_i \mu_i(x)\frac{\partial f}{\partial x_i} + \frac{1}{2}\sum_i\sum_j (\sigma(x)\sigma^T(x))_{i,j}\frac{\partial^2 f}{\partial x_i \partial x_j} \\ L^* f &= -\sum_i \frac{\partial}{\partial x_i}(\mu_i(x)f(x)) \\ &\quad + \frac{1}{2}\sum_i\sum_j \frac{\partial^2}{\partial x_i \partial x_j}\left((\sigma(x)\sigma^T(x))_{i,j} f(x)\right) \end{aligned} \tag{I1g}$$

Let A, B be nice subsets of $\boldsymbol{R}^d$. Then

$$\begin{aligned} &f(x) \equiv \boldsymbol{P}_x(X \text{ hits } A \text{ before } B) \text{ is the solution of} \\ &Lf = 0 \text{ on } (A \cup B)^C; \qquad f = 1 \text{ on } A, \quad f = 0 \text{ on } B. \end{aligned} \tag{I1h}$$

$$\begin{aligned} &\text{If } h(x) \equiv E_x T_A < \infty \text{ (where as usual } T_A \text{ is the hitting time of } X \text{ on } A\text{), then } h \text{ is the solution of} \\ &Lh = -1 \text{ on } A^C; \qquad h = 0 \text{ on } A \end{aligned} \tag{I1i}$$

The diffusion is positive-recurrent iff the equations

$$L^*\pi = 0; \qquad \pi(x) > 0, \quad \int_{\boldsymbol{R}^d} \pi(x)\,dx = 1 \tag{I1j}$$

have a solution, in which case the solution π is the stationary density.

(Positive-recurrence is the property that the mean hitting time on any *open* set is finite. In $d \geq 2$ dimensions, nice diffusions do not hit prespecified single points.)

I2 The heuristic. . The equations above and their derivations are essentially the same in the d-dimensional case as in the 1-dimensional case. The difference is that in the 1-dimensional case these equations have explicit solutions in terms of μ and σ (Section D3), whereas in higher dimensions there is no general explicit solution, so that approximate methods become appealing. Our strategy for using the heuristic to estimate the hitting time T_A to a rarely visited subset A is exactly the same as in the Markov chain case (Section B2); we need only estimate the stationary distribution and the local behavior of the process around the boundary of A.

We start by listing some special cases where the stationary distribution can be found explicitly.

I3 Potential function. Suppose that there is a function $H: \boldsymbol{R}^d \to \boldsymbol{R}$, thought of as a potential function, such that

$$\begin{aligned} \mu(x) &= -\nabla H \qquad (\nabla H = (\frac{\partial H}{\partial x_1}, \ldots, \frac{\partial H}{\partial x_d})) \\ \sigma(x) &= \sigma_0 I, \qquad \text{for a scalar constant } \sigma_0. \end{aligned}$$

Then there is a stationary distribution

$$\pi(x) = c \exp\left(\frac{-2H(x)}{\sigma_0^2}\right)$$

where c is a normalizing constant, provided $H(x) \to \infty$ as $|x| \to \infty$ fast enough that $\int \exp(-2H(x)/\sigma_0^2)\,dx < \infty$. This is analogous to the 1-dimensional case (Example D7).

I4 Reversible diffusions. If the equations

$$\mu_i(x)\pi(x) = \frac{1}{2}\sum_j \frac{\partial}{\partial x_j}((\sigma(x)\sigma^T(x))_{i,j}\pi(x)); \qquad 1 \leq i \leq d$$

have a solution $\pi(x) > 0$ with $\int \pi(x)\,dx = 1$, then π is a stationary density and the stationary diffusion is time-reversible. This is the analogue of the detailed balance equations (Section B28) in the discrete-time case. In 1 dimension (but not in $d \geq 2$ dimensions) all stationary diffusions have this property. It is easy to check that the potential function case (Section I3) gives rise to a reversible diffusion.

I5 Ornstein-Uhlenbeck processes. The general Ornstein-Uhlenbeck process has

$$\mu(x) = -Ax, \qquad \sigma(x) = \sigma,$$

where A and σ are matrices, not dependent on x. This process is *stable* if A is positive-definite, and then there is a stationary distribution π which is Normal$(0, \Sigma)$, where Σ is the symmetric positive-definite matrix satisfying

$$A\Sigma + \Sigma A^T = \sigma\sigma^T. \qquad \text{(I5a)}$$

There are various special cases. Where A is symmetric we have

$$\Sigma = \frac{1}{2}A^{-1}\sigma\sigma^T.$$

If also $\sigma = \sigma_0 I$ for scalar σ_0, then $\Sigma = \frac{1}{2}\sigma_0^2 A^{-1}$ and we are in the setting of Section I3 with potential function $H(x) = \frac{1}{2}x^T Ax$.

I6 Brownian motion on surface of sphere. Let $U(d,r)$ be the ball of center 0, radius r in d dimensions. There is a natural definition of Brownian motion on the surface S_{d-1} of $U(d,r)$. The stationary distribution is uniform, by symmetry. It is useful to record the elementary formulas

$$\text{“volume” of } U(d,r) = \frac{\pi^{d/2}r^d}{(d/2)!} = v_d r^d, \text{ say} \qquad \text{(I6a)}$$

$$\text{“surface area” of } U(d,r) = \frac{2\pi^{d/2}r^{d-1}}{(\frac{1}{2}d-1)!} = s_d r^{d-1}, \text{ say.} \qquad \text{(I6b)}$$

I7 Local approximations. The final ingredient of our heuristic is the local approximation of processes. Locally, a diffusion behaves like general Brownian motion with constant drift and variance: say X_t with

$$\mu(x) = \mu, \qquad \sigma(x) = \sigma,$$

Given a hyperplane $H = \{x : q \cdot x = c\}$ where q is a unit vector, the distance from X_t to H is given by $D_t = q \cdot X_t - c$, and it is clear that

D_t is 1-dimensional Brownian motion with drift $q \cdot \mu$ and variance $q^T\sigma\sigma^T q$. (I7a)

Motion relative to a fixed point is a little more complicated. Let us just consider the case $X_t = \sigma_0 B_t$, where B_t is standard d-dimensional ($d \geq 3$)

Brownian motion and σ_0 is scalar. Let T_r be the first exit time from the ball $U(d,r)$, and let V_r be the total sojourn time within the ball. Then

$$E(T_r \mid X_0 = 0) = \sigma_0^{-2} r^2 d^{-1}. \quad \text{(I7b)}$$
$$E(V_r \mid X_0 = 0) = \sigma_0^{-2} r^2 (d-2)^{-1} \quad \text{(I7c)}$$

from which follows, by subtraction, that

$$E(V_r \mid |X_0| = r) = 2\sigma_0^{-2} r^2 d^{-1} (d-2)^{-1}. \quad \text{(I7d)}$$

See Section I22 for the derivations. When we add a drift term, $X_t = \sigma_0 B_t + \mu t$ say, the exact formulas replacing (I7b–I7d) are complicated but as $r \to 0$ formulas (I7b–I7d) are asymptotically correct, since the drift has negligible effect in time $O(r^2)$.

We are now ready to start the examples.

I8 Example: Hitting times to small balls. Consider a diffusion in $d \geq 3$ dimensions with $\sigma(x) = \sigma_0 I$ and with stationary density π. Fix $r > 0$ small, and fix $x_0 \in \boldsymbol{R}^d$. Let B be the ball, center x_0, radius r, and let T_B be the first hitting time on B. Let $\mathcal{S}$ be the random set of times t such that $X_t \in B$. The "local transience" property of $d \geq 3$ dimensional Brownian motion implies that the heuristic is applicable to $\mathcal{S}$. The mean clump size EC is the mean local sojourn time of X in B, starting from the boundary of B, so by (I7d)

$$EC \approx 2\sigma_0^{-2} r^2 d^{-1} (d-2)^{-1}.$$

And $p = \boldsymbol{P}(X_t \in B) \approx \pi(x_0)\,\text{volume}(B) = \pi(x_0) v_d r^d$ for v_d as in (I6a). Using the fundamental identity, we conclude

T_b is approximately exponential,

$$\text{rate} = \frac{p}{EC} = \frac{1}{2} d(d-2) v_d \sigma_0^2 r^{d-2} \pi(x_0). \quad \text{(I8a)}$$

I9 Example: Near misses of moving particles. Imagine particles moving independently in $d \geq 3$ dimensions, such that at any fixed time their spatial positions form a Poisson process of rate ρ per unit volume. We shall consider two models for the motion of individual particles.

Model 1 Each particle moves in a straight line with constant, random velocity $V = (V_1, \ldots, V_d)$.

Model 2 Each particle performs Brownian motion, with zero drift and σ^2 variance.

Take a large region A, of volume $|A|$, and a long time interval $[0, t_0]$. We shall study M = minimum distance between any two particles, in region A and time interval $[0, t_0]$.

Fix small $x > 0$. At any fixed time, there are about $\rho|A|$ particles in A, so

$$\begin{aligned} p &\equiv \boldsymbol{P}(\text{some two particles in } A \text{ are within } x \text{ of each other}) \\ &\approx \binom{\rho|A|}{2} \frac{v_d x^d}{|A|} \\ &\approx \frac{1}{2} v_d \rho^2 |A| x^d. \end{aligned} \tag{I9a}$$

We apply this heuristic to the random set $\mathcal{S}$ of times t that there exist two particles within distance x. The clump sizes are estimated as follows.

Model 2 The distance between two particular particles behaves as Brownian motion with variance $2\sigma^2$, so as in Example I8 we use (I7d) to get

$$EC = d^{-1}(d-2)^{-1}\sigma^{-2}x^2.$$

Model 1 Fix t and condition on $t \in \mathcal{S}$. Then there are two particles within x; call their positions Z_1, Z_2 and their velocities V^1, V^2. The Poisson model for motion of particles implies that $Y \equiv Z_1 - Z_2$ is uniform on the ball of radius x, and that V^1 and V^2 are copies of V, independent of each other and of Y. So the instantaneous rate of change of distance between the particles is distributed as

$$W \equiv (V^2 - V^1) \cdot \xi, \quad \text{where } \xi \text{ is a uniform unit vector.} \tag{I9b}$$

Now in the notation of Section A9, the ergodic-exit technique,

$$\begin{aligned} f_{C^+}(0) &\approx \delta^{-1}\boldsymbol{P}\left(\begin{matrix}\text{distance between two} \\ \text{particles at time } \delta \text{ is} > x\end{matrix}\middle| \begin{matrix}\text{distance at} \\ \text{time 0 is} < x\end{matrix}\right) \quad \text{as } \delta \to 0 \\ &\approx \delta^{-1}\boldsymbol{P}(|Y| + \delta W > x) \\ &\approx f_{|Y|}(x) EW^+ && \text{(I9c)} \\ &\approx v_d^{-1} s_d x^{-1} EW^+. && \text{(I9d)} \end{aligned}$$

(The argument for (I9c) is essentially the argument for Rice's formula)

In either model, the heuristic now gives

$$\boldsymbol{P}(M > x) \approx \exp(-\lambda t_0) \tag{I9e}$$

where the clump rate λ, calculated from the fundamental identity $\lambda = p/EC$ in Model 2 and from $\lambda = pf^+(0)$ in Model 1, works out as

$$\lambda = \frac{1}{2} s_d \rho^2 |A| EW^+ x^{d-1} \qquad \text{[Model 1]} \tag{I9f}$$

$$\lambda = \frac{1}{2} v_d d(d-2)\rho^2\sigma^2 |A| x^{d-2} \qquad \text{[Model 2]} \tag{I9g}$$

I10 Example: A simple aggregation-disaggregation model. Now imagine particles in 3 dimensions with average density ρ per unit volume. Suppose particles can exist either individually or in linked pairs. Suppose individual particles which come within a small distance r of each other will form a linked pair. Suppose a linked pair splits into individuals (which do not quickly recombine) at rate α. Suppose individual particles perform Brownian motion, variance σ^2. We shall calculate the equilibrium densities ρ_1 of individual particles, and ρ_2 of pairs. Obviously,

$$\rho_1 + 2\rho_2 = \rho. \tag{I10a}$$

But also

$$\alpha\rho_2 = \lambda\rho_1^2 \tag{I10b}$$

where the left side is the disaggregation rate (per unit time per unit volume) and the right side is the aggregation rate. But the calculation (I9g) shows this aggregation rate has

$$\lambda = 2\pi\sigma^2 r. \tag{I10c}$$

Solving the equations gives

$$\rho_1 = \theta^{-1}((1+2\theta\rho)^{\frac{1}{2}} - 1), \quad \text{where } \theta = \frac{8\pi\sigma^2 r}{\alpha}. \tag{I10d}$$

I11 Example: Extremes for diffusions controlled by potentials. Consider the setting of Section I3, where $\mu(x) = -\nabla H$ and $\sigma(x) = \sigma_0 I$. Suppose H is a smooth convex function attaining its minimum at 0 with $H(0) = 0$. Let T_R be the first exit time from the ball B_R with center 0 and radius R, where R is sufficiently large that $\pi(B_R^c)$ is small. Then we can apply the heuristic to the random set of times t such that $X_t \in B_R^c$ (or in a thin "shell" around B_R) to see that T_R will in general have approximately exponential distribution; and in simple cases we can estimate the mean ET_R. There are two qualitatively different situations.

I11.1 Case 1: radially symmetric potentials. In the case $H(x) = h(|x|)$ with radial symmetry, the radial component $|X_t|$ is a 1-dimensional diffusion, and we can apply our 1-dimensional estimates. Specifically, $|X_t|$ has drift $\mu(r) = -h'(r) + \frac{1}{2}(d-1)\sigma_0^2 r^{-1}$ and variance σ_0^2, and from (D4e) and a few lines of calculus we get

$$ET_R \approx R^{1-d}(h'(R))^{-1} K \exp\left(\frac{2H(R)}{\sigma_0^2}\right), \tag{I11a}$$

where

$$K = \int_0^\infty r^{d-1} \exp\left(\frac{-2H(r)}{\sigma_0^2}\right) dr.$$

Of course, we could also implement the heuristic directly with $|X_t|$, and obtain (I11a) by considering clumps of time spend in a shell around B_R.

We can apply this result to the simple Ornstein-Uhlenbeck process , in which $\sigma(x) = \sigma I$ and $\mu(x) = -ax$ for positive scalars a, σ. For here $H(x) = \frac{1}{2}a\sum x_i^2 = \frac{1}{2}ar^2$, and evaluating (I11a) gives

$$ET_R \approx \frac{1}{2}(\frac{1}{2}d-1)!\,a^{-1}\left(\frac{a}{\sigma^2}\right)^{-\frac{1}{2}d} R^{-d}\exp\left(\frac{aR^2}{\sigma^2}\right). \tag{I11b}$$

I11.2 Case 2: non-symmetric potentials. An opposite case occurs when H attains its minimum, over the spherical surface ∂B_R, at a unique point z_0. Since the stationary density (Section I3) decreases exponentially fast as H increases, it seems reasonable to suppose that exits from B_R will likely occur near z_0, and then approximate T_R by T_F, the first hitting time on the hyperplane F tangent to B_R at z_0. Let $q = z_0/|z_0|$ be the unit normal vector at z_0, and let π_F be the density of $q\cdot X$ at $q\cdot z_0$, where X has the stationary distribution π. At z_0 the drift $-\nabla H$ is directed radially inward (since H is minimized on ∂B_R at z_0), and so $q\cdot X_t$ behaves like 1-dimensional Brownian motion with drift $-|\nabla H(z_0)|$ when X_t is near z_0. Thus if we consider clumps

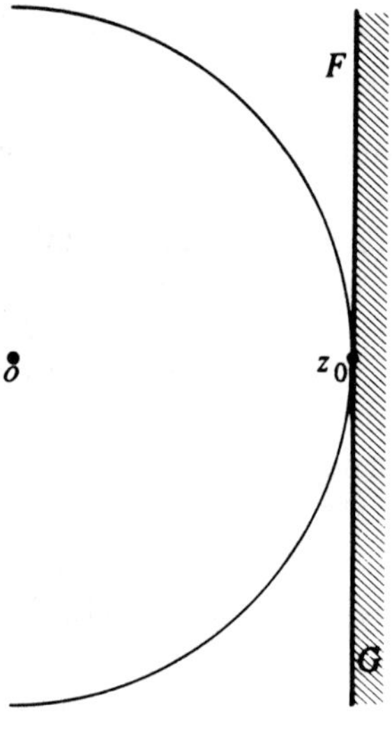

FIGURE I11a.

of time spent in a slice $G = \{\, x : q\cdot z_0 \le q\cdot x \le q\cdot z_0 + \delta \,\}$, we have mean clump size $EC = \delta/|\nabla H(z_0)|$ by Section C5, and $p = \pi(G) = \pi_F\delta$. The fundamental identity gives $\lambda = p/EC$ and hence

$$ET_R \approx \lambda^{-1} \approx \{\pi_F|\nabla H(z_0)|\}^{-1}. \tag{I11c}$$

To make this more explicit we introduce approximations for the stationary distribution π. If $f\colon \boldsymbol{R}^d \to \boldsymbol{R}$ is smooth and has a minimum at x_0, then we can write

$$f(x) \approx f(x_0) + \frac{1}{2}(x-x_0)^T Q(x-x_0); \qquad Q = \frac{\partial^2 f}{\partial x_i \partial x_j}(x_0)$$

for x near x_0. Using the integration formula (I1d) we can get

$$\int_{R^d} \exp(-af(x))\,dx \approx \left(\frac{2\pi}{a}\right)^{\frac{1}{2}d} |Q|^{-\frac{1}{2}} \exp(-af(x_0)) \tag{I11d}$$

for smooth convex f. In particular, we can approximate the normalization constant c in Section I3 to get

$$\pi(x) \approx (\sigma_0^2\pi)^{-\frac{1}{2}d}|Q|^{\frac{1}{2}} \exp(-2\sigma_0^{-2}H(x)) \tag{I11e}$$

when $H(0) = 0$ is the minimum of H, and $Q = \frac{\partial^2 H}{\partial x_i \partial x_j}(0)$.

To evaluate (I11c), take coordinates so that $z_0 = (R, 0, 0, \dots)$. By (I11e),

$$\pi_F \approx (\sigma_0^2\pi)^{-\frac{1}{2}d}|Q|^{\frac{1}{2}} \exp(-2\sigma_0^{-2}H(z_0)) \int_F \exp(-2\sigma_0^{-2}(H(x)-H(z_0))\,dx.$$

But we can estimate the integral by using (I11d) for the $(d-1)$-dimensional hyperplane F, since H is minimized on F at z_0. The integral is approximately

$$(\sigma_0^2\pi)^{(d-1)/2}|Q_1|^{-1/2},$$

where Q_1 is the matrix

$$\frac{\partial^2 H}{\partial x_i \partial x_j}(z_0), \qquad i,j \geq 2.$$

Finally, $\nabla H(z_0) = -\frac{\partial H}{\partial x_1}(z_0)$ and (I11c) gives

$$ET_R \approx \sigma_0 \frac{\pi^{\frac{1}{2}}|Q_1|^{\frac{1}{2}}|Q|^{-\frac{1}{2}}}{-\frac{\partial H}{\partial x_1}(z_0)} \exp\left(\frac{2H(z_0)}{\sigma_0^2}\right). \tag{I11f}$$

The simplest concrete example is the Ornstein-Uhlenbeck process (Section I5) in which we take $\sigma = \sigma_0 I$ and

$$A = \begin{pmatrix} \rho_1 & & & \\ & \rho_2 & & \\ & & \ddots & \\ & & & \rho_2 \end{pmatrix}; \qquad \rho_1 < \rho_2 < \cdots.$$

This corresponds to the potential function $H(x) = \frac{1}{2}\sum \rho_i x_i^2$. Here H has two minima on ∂B_R, at $\pm z_0 = \pm(R, 0, 0, \dots)$, and so the mean exit time is half that of (I11f):

$$ET_R \approx \frac{1}{2}\sigma_0\pi^{\frac{1}{2}}\left(\prod_{i\geq 2}\rho_i \Big/ \prod_{i\geq 1}\rho_i\right)^{\frac{1}{2}} \rho_1^{-1}R^{-1}\exp\left(\frac{\rho_1 R^2}{\sigma_0^2}\right) \tag{I11g}$$

which of course just reduces to the 1-dimensional result. This shows that our method is rather crude, only picking out the extremes of the process in the dominant direction.

I12 Example: Escape from potential wells. We continue in the same general setting: a diffusion X_t with drift $\mu(x) = -\nabla H(x)$ and covariance $\sigma(x) = \sigma_0 I$. Consider now the case where H has two local minima, at z_0 and z_2 say, with a saddle point z_1 between. For simplicity, we consider the 2-dimensional case. The question is: starting in the well near z_0, what is the time T until X escapes over the saddle into the other well? The heuristic will show that T has approximately exponential distribution, and estimate its mean.

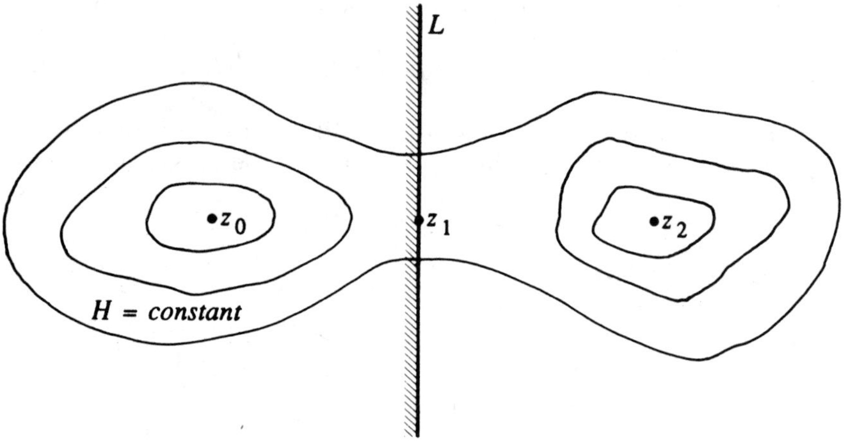

FIGURE I12a.

Take $z_0 = (0,0)$ and $z_1 = (z_1, 0)$ and suppose

$$\frac{\partial^2 H}{\partial x_i \partial x_j}(z_0) = \begin{pmatrix} a_1 & 0 \\ 0 & a_2 \end{pmatrix} \qquad \frac{\partial^2 H}{\partial x_i \partial x_j}(z_1) = \begin{pmatrix} -b_1 & 0 \\ 0 & b_2 \end{pmatrix}$$

Let T_L be the time to hit the line $L = \{(z_1, y) : -\infty < y < \infty\}$. The diffusion will hit L near z_1 and, by symmetry, be equally likely to fall into either well; so $ET \approx 2ET_L$. To estimate ET_L starting from z_0, we may take L to be a reflecting barrier. Apply the heuristic to the random set $\mathcal{S}$ of times t that X_t is in a strip, width δ, bordering L. Near z_1 the x_1-component of X behaves like the unstable 1-dimensional Ornstein-Uhlenbeck process, so by (D24a) the mean clump size is

$$EC \approx \delta \pi^{\frac{1}{2}} b_1^{-\frac{1}{2}} \sigma_0^{-1}.$$

And

$$\begin{aligned} p &= \pi(\text{strip}) \\ &= \delta \int_{-\infty}^{\infty} \pi(z_1, y)\, dy \quad \text{where } \pi(x) = K^{-1} \exp(-2\sigma_0^{-2} H(x)) \\ &\approx \delta K^{-1} \exp(-2\sigma_0^{-2} H(z_1, 0)) \pi^{\frac{1}{2}} \sigma_0 b_2^{-\frac{1}{2}} \end{aligned}$$

using integral approximation (I11d); using this approximation again,

$$K \approx \pi \sigma_0^2 (a_1 a_2)^{-\frac{1}{2}} \exp(-2\sigma_0^{-2} H(z_0)).$$

The heuristic says $ET_L \approx EC/p$, so putting it all together

$$ET \approx 2\pi \sqrt{\frac{b_2}{b_1 a_1 a_2}} \cdot \exp(2\sigma_0^{-2}(H(z_1) - H(z_0))). \tag{I12a}$$

I13 Physical diffusions: Kramers' equation. Our discussion so far of diffusions controlled by potentials is physically unrealistic because, recalling Newton's laws of motion, a potential really acts to cause a change in *velocity* rather than in position. We now describe a more realistic model, which has been much studied by mathematical physicists. Let $H(x)$ be a potential function; we take the 1-dimensional case for simplicity. The position X_t and velocity V_t of a particle moving under the influence of the potential H, of friction (or viscosity), and of random perturbations of velocity, can be modeled as

$$\begin{aligned} dX_t &= V_t\, dt \\ dV_t &= -\alpha H'(X_t)\, dt - \beta V_t\, dt + \eta\, dB_t. \end{aligned}$$

Here α, β, η are constants and B_t is (mathematical) Brownian motion. The pair (X_t, V_t) form a (mathematical) 2-dimensional diffusion, albeit a "degenerate" one. By rescaling space and time, we can reduce the equations above to a canonical form

$$\begin{aligned} dX_t &= V_t\, dt \\ dV_t &= -H'(X_t)\, dt - \gamma V_t\, dt + \sqrt{2\gamma}\, dB_t \end{aligned} \tag{I13a}$$

where γ is a dimensionless constant: this is *Kramers' equation.* It is remarkable that (X_t, V_t) has a simple stationary distribution

$$\pi(x, v) = K \exp(-H(x) - \frac{1}{2}v^2) \tag{I13b}$$

in which position and velocity are independent; velocity is Normally distributed; and position is distributed in the same way as in our earlier models. To understand the role of γ, consider the extreme cases.

(i) As $\gamma \to \infty$, the speeded-up processes $X_{\gamma t}$ converge to the diffusion with $\mu(x) = -H'(x)$ and $\sigma^2(x) = 2$. This gives a sense in which our earlier diffusion models do indeed approximate physically sensible processes.

(ii) As $\gamma \to 0$, the motion approximates that of deterministic frictionless motion under a potential.

Note that in the case $H(x) = \frac{1}{2}ax^2$ the deterministic motion is the "simple harmonic oscillator".

We now repeat the two previous examples in this context.

I14 Example: Extreme values. Let H be smooth convex, with its minimum at x_0, and consider the time T_b until X_t first exceeds a large value b. We give separate arguments for large γ and for small γ.

I14.1 Large γ. Define

$$Z_t = X_t + \gamma^{-1} V_t.$$

Then equations (I13a) yield

$$dZ_t = -\gamma^{-1} H'(X_t)\, dt + \left(\frac{2}{\gamma}\right)^{\frac{1}{2}} dB_t.$$

In particular, for X around b we have

$$dZ_t \approx -\gamma^{-1} H'(b) + \left(\frac{2}{\gamma}\right)^{\frac{1}{2}} dB_t. \tag{I14a}$$

We claim that the first hitting time (on b) of X_t can be approximated by the first hitting time of Z_t. At the exact first time that $Z = b$, we will have $X < b$ and $V > 0$, but by (I14a) Z_t is changing slowly (γ is large) and so at the next time that $V = 0$ we will have $Z \approx b$ and hence $X \approx b$.

Now by (I14a) and (D4a) the clump rate for hits of Z on b is

$$\lambda_b = f_Z(b) \cdot \gamma^{-1} H'(b).$$

From the joint density (I13b) we calculate

$$f_Z(b) \approx K \exp(-H(b) + \frac{1}{2}\gamma^{-2}(H'(b))^2).$$

Thus the hitting time T_b has approximately exponential distribution with mean

$$ET_b \approx \lambda_b^{-1} \approx \gamma K^{-1} (H'(b))^{-1} \exp(H(b) - \frac{1}{2}\gamma^{-2}(H'(b))^2). \tag{I14b}$$

We may further approximate K by using the Gaussian approximation around x_0, as at Example D7, to get $K \approx (H''(x_0)/2\pi)^{1/2}\exp(H(x_0))$ and hence

$$ET_b \approx \gamma\left(\frac{2\pi}{H''(x_0)}\right)^{\frac{1}{2}}(H'(b))^{-1}\exp(H(x_0)-H(b))\exp(-\frac{1}{2}\gamma^{-2}(H'(b))^2). \tag{I14c}$$

Note that as $\gamma \to \infty$, our estimate for ET_b/γ tends to the estimate (D7b) for the mathematical diffusion, as suggested by (I13i).

I14.2 Small γ. Consider first the deterministic case $\gamma = 0$. Starting at $x_1 > x_0$ with velocity 0, the particle moves to $\widehat{x}_1 < x_0$ such that $H(\widehat{x}_1) = H(x_1)$ and then returns to x_1: call this a x_1-cycle. "Energy"

$$E_t \equiv H(X_t) + \frac{1}{2}V_t^2 \tag{I14d}$$

is conserved, so we can calculate

$$\begin{aligned} D(x_1) \quad &\equiv \quad \text{duration of } x_1\text{-cycle} \\ &= 2\int_{\widehat{x}_1}^{x_1}(\text{velocity at } x)^{-1}\,dx \\ &= 2\int_{\widehat{x}_1}^{x_1}(2(H(x_1)-H(x))^{-\frac{1}{2}}\,dx \qquad \text{(I14e)} \\ I(x_1) \quad &\equiv \quad \int V^2(t)\,dt \text{ over a } x_1\text{-cycle} \qquad \text{(I14f)} \\ &= 2\int_{\widehat{x}_1}^{x_1}(2(H(x_1)-H(x))^{\frac{1}{2}}\,dx \qquad \text{(I14g)} \end{aligned}$$

In the stochastic ($\gamma > 0$) case we calculate, from (I13a),

$$dE_t = \gamma(1-V_t^2)\,dt + (2\gamma)^{\frac{1}{2}}V_t\,dB_t \tag{I14h}$$

Let $\widehat{X}_1, \widehat{X}_2, \widehat{X}_3, \ldots$ be the right-most extremes of successive cycles. Integrating (I14h) over a cycle gives

$$H(\widehat{X}_{n+1}) \approx H(\widehat{X}_n) + \gamma(D(\widehat{X}_n) - I(\widehat{X}_n)) + \text{Normal}(0, 4\gamma I(\widehat{X}_n)). \tag{I14i}$$

So for $x \approx b$, the process $H(\widehat{X})$ over many cycles can be approximated by Brownian motion with some variance and with drift $-\gamma(I(b)-D(b))$ per cycle, that is $-\gamma(I(b)/D(b)-1)$ per unit time. Now the clump rate for hits of X on b is $1/D(b)$ times the rate for hits of $H(\widehat{X})$ on $H(b)$, so by the Brownian approximation for $H(\widehat{X})$ and (D2b)

$$\lambda_b \quad = \quad f_{H(X)}(H(b))\cdot\gamma\left(\frac{I(b)}{D(b)}-1\right)$$

$$= \quad (H'(b) + H'(\widehat{b}))^{-1} f_X(b) \cdot \gamma \left(\frac{I(b)}{D(b)} - 1 \right), \quad \text{where } H(\widehat{b}) = H(b)$$

$$= \quad \gamma K_X \left(\frac{I(b)}{D(b)} - 1 \right) (H'(b) + H'(\widehat{b}))^{-1} \exp(-H(b)). \tag{I14j}$$

and as before K_X can be estimated in terms of x_0 to give

$$\lambda_b \quad \approx \quad \left(\frac{H''(x_0)}{2\pi} \right)^{\frac{1}{2}} \gamma \left(\frac{I(b)}{D(b)} - 1 \right) \tag{I14k}$$

$$\times (H'(b) + H'(\widehat{b}))^{-1} \exp(H(x_0) - H(b)). \tag{I14l}$$

In the "simple harmonic oscillator" case $H(x) = \frac{1}{2}x^2$, we have

$$D(b) = 2\pi, \qquad I(b) = \pi b^2$$

and the expression (I14k) becomes

$$\lambda_b \approx \frac{1}{4} \gamma b \phi(b), \qquad b \text{ large,} \tag{I14m}$$

where ϕ as usual is the standard Normal density function.

I15 Example: Escape from potential well. Now consider the case of a double-welled potential H such that the stationary distribution is mostly concentrated near the bottoms of the two wells. *Kramers' problem* asks: for the process (Section I13) started in the well near x_0, what is the mean time ET to escape into the other well? We shall consider the case where γ is not small: we want to ignore the possibility that, after passing over the hump at x_1 moving right (say), the particle will follow approximately the deterministic trajectory to around z and then return and re-cross the hump. There is no loss of generality in assuming that H is symmetric about x_1.

Let

$$a = -H''(x_1); \qquad b = H''(x_0). \tag{I15a}$$

To study the motion near the hump x_1, set

$$Z_t = X_t - x_1 + \rho V_t$$

where ρ is a constant to be specified later. Since $H'(x) \approx -a(x - x_1)$ for x near x_1, we have for X_t near x_1

$$dZ_t \quad = \quad dX_t - \rho\, dV_t = V_t\, dt + \rho\, dV_t$$

$$dV_t \quad \approx \quad a(Z_t - \rho V_t) - \gamma V_t\, dt + \sqrt{2\gamma}\, dB_t$$

using the basic equations (I13a). Rearranging,

$$dZ_t \quad \approx \quad \rho a Z_t\, dt + \rho\sqrt{2\gamma}\, dB_t + (1 - a\rho^2 - \gamma\rho) V_t\, dt.$$

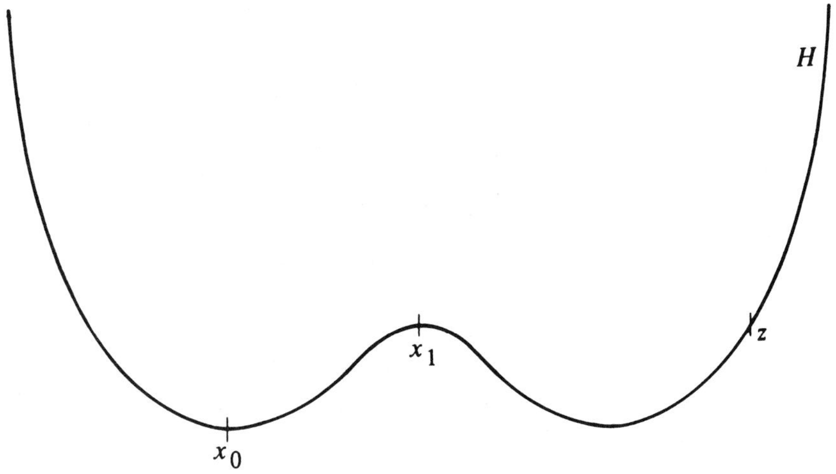

FIGURE I15a.

Choosing $\rho > 0$ to make the last term vanish

$$a\rho^2 + \gamma\rho - 1 = 0 \tag{I15b}$$

we see that Z_t approximates a 1-dimensional unstable Ornstein-Uhlenbeck process. Essentially, what's happening is that a particle starting with (x, v) for x near x_1 would, in the absence of random perturbations, fall into one or other well according to the sign of $z = x + \rho v$. Thus $ET \approx 2E\widehat{T}$, where $\widehat{T}$ is the time until Z first reaches 0. Applying the heuristic to the clumps of time that $Z_t \in [0, \delta]$ we get

$$E\widehat{T} \approx \frac{EC}{\delta f_Z(0)} \tag{I15c}$$

where f_Z is the stationary density of Z and the mean clump size is, by (D24a),

$$EC = \frac{1}{2}\pi^{\frac{1}{2}}(a\rho)^{-\frac{1}{2}}(\rho\sqrt{2\gamma})^{-1}\delta.$$

To estimate $f_Z(0)$, first note that from (I11d) the normalizing constant K for the stationary distribution (I13b) has

$$K^{-1} \approx 2(2\pi)^{\frac{1}{2}}\left(\frac{\pi}{b}\right)^{\frac{1}{2}}\exp(-H(x_0))$$

where the first 2 arises from symmetry about x_1. So for x near x_1,

$$\pi(x, v) \approx A\exp(\frac{1}{2}a(x - x_1)^2 - \frac{1}{2}v^2)$$

for $A = K^{-1}\exp(H(x_1))$. Since $Z = X - x_1 + \rho V$,

$$\begin{aligned} f_Z(0) &= \int_{-\infty}^{\infty} \pi\left(x_1 + u, \frac{-u}{\rho}\right) \rho^{-1}\, du \\ &\approx A\rho^{-1} \int \exp(\frac{1}{2}au^2 - \frac{1}{2}u^2\rho^{-2})\, du \\ &= A\rho^{-1}(2\pi)^{\frac{1}{2}}(-a + \rho^{-2})^{\frac{1}{2}} \\ &= A(2\pi)^{\frac{1}{2}}\gamma^{-\frac{1}{2}}\rho^{-\frac{1}{2}} \qquad \text{using the definition (I15b) of } \rho. \end{aligned}$$

Putting it all together,

$$ET \approx 2\pi(ab)^{-\frac{1}{2}}\rho^{-1}\exp(H(x_1) - H(x_0)) \tag{I15d}$$

where ρ^{-1} is, solving (I15b),

$$\rho^{-1} = \frac{1}{2}\gamma + \sqrt{\frac{1}{2}\gamma + a}.$$

This is the usual solution to Kramers' problem. As $\gamma \to \infty$ it agrees with the 1-dimensional diffusion solution (D25b), after rescaling time.

As mentioned before, for $\gamma \approx 0$ the behavior of X_t changes, approximating a deterministic oscillation: this case can be handled by the method of Example I14.2.

We now turn to rather different examples.

I16 Example: Lower boundaries for transient Brownian motion. The usual LIL and integral test (D15) extend easily to d-dimensional Brownian motion B_t to give

$$\limsup_{t\to\infty} \frac{|B_t|}{(2t\log\log t)^{\frac{1}{2}}} = 1 \quad \text{a.s.} \tag{I16a}$$

$$B_t \leq t^{\frac{1}{2}}\psi(t) \text{ ultimately iff } \int^{\infty} t^{-1}\psi^d(t)\exp(-\frac{1}{2}\psi^2(t))\, dt < \infty \tag{I16b}$$

Now for $d \geq 3$ Brownian motion is transient, so we can consider lower boundaries too. First we consider the d-dimensional Ornstein-Uhlenbeck process X_t (whose components are independent 1-dimensional standard Ornstein-Uhlenbeck processes). Let $b(t) \downarrow 0$ smoothly, and apply the heuristic to $\mathcal{S} = \{\, t : |X_t| \leq b(t)\,\}$. So

$$p(t) \equiv \boldsymbol{P}(|X_t| \leq b(t)) \sim (2\pi)^{-\frac{1}{2}d} v_d b^d(t).$$

We estimate the clump size by approximating the boundary around t by the level at $b(t)$; then (I7d) gives

$$EC(t) \approx 2b^2(t)d^{-1}(d-2)^{-1}.$$

So the clump rate is

$$\lambda(t) = \frac{p(t)}{EC(t)} = a_d b^{d-2}(t) \qquad \text{for a certain constant } a_d.$$

So as at (D15)

$$X_t \geq b(t) \text{ ultimately} \qquad \text{iff} \int^{\infty} \lambda(t)\,dt < \infty$$

$$\text{iff} \int^{\infty} b^{d-2}(t)\,dt < \infty. \qquad \text{(I16c)}$$

Now given d-dimensional Brownian motion B_t, then $X_t \equiv e^{-t}B(e^{2t})$ is the d-dimensional Ornstein-Uhlenbeck process, and (I16c) translates to

$$|B_t| \geq t^{\frac{1}{2}} b(t) \text{ ultimately iff} \int^{\infty} t^{-1} b^{d-2}(t)\,dt < \infty. \qquad \text{(I16d)}$$

In particular,

$$|B_t| \geq t^{\frac{1}{2}} \log^{-\alpha}(t) \text{ ultimately iff } \alpha > \frac{1}{d-2}.$$

I17 Example: Brownian motion on surface of sphere. For a quite different type of problem, consider Brownian motion on the surface S_{d-1} of the ball of radius R in $d \geq 4$ dimensions. As mentioned at Section I6, the stationary distribution is uniform. Let B be a "cap" of radius r (r small) on the surface S_{d-1}, and let T_B be the first hitting time on B. We shall use the heuristic to show that T_B is approximately exponentially distributed with mean

$$ET_B \approx \pi^{\frac{1}{2}} \frac{((d-5)/2)!}{((d-2)/2)!} \cdot R^{d-1} r^{3-d} = \bar{t}(r) \qquad \text{say.} \qquad \text{(I17a)}$$

For consider the clumps of time spent in B. Around B the process behaves like $(d-1)$ dimensional Brownian motion, and this local transience makes the heuristic applicable. By (I7d)

$$EC \approx 2(d-1)^{-1}(d-3)^{-1} r^2.$$

And

$$p = \pi(B) \approx \frac{v_{d-1} r^{d-1}}{s_d R^{d-1}} = \frac{1}{2}\pi^{-\frac{1}{2}} \frac{((d-2)/2)!}{((d-1)/2)!} \cdot r^{d-1} R^{1-d}$$

where v_d and s_d are the volume and surface area (Section I6) of the d-dimensional unit ball. Then the heuristic estimate $ET_B \approx EC/p$ gives (I17a).

We can also treat the continuous analogue of the coupon-collector's problem of Chapter F. Let V_r be the time taken until the path has passed within distance r of every point on the surface S_{d-1}. We shall use the heuristic to obtain

$$V_r \sim (d-1)\bar{t}(r)\log(1/r) \quad \text{as } r \to 0. \tag{I17b}$$

for $\bar{t}(r)$ as at (I17a). The upper estimate is easy. Fix $\epsilon > 0$. Choose a set A of points in S_{d-1} such that every point in S_{d-1} is within ϵr of some point of A; we can do this with $|A| = O(r^{1-d})$. Then V_r is bounded above by the time V^* taken to pass within distance $(1-\epsilon)r$ from each point of A. Then using (I17a),

$$\boldsymbol{P}(V^* > t) \lesssim |A| \exp\left(\frac{-t}{\bar{t}((1-\epsilon)r)}\right).$$

So as $r \to 0$, V^* is bounded by

$$v = \bar{t}((1-\epsilon)r)\,(d-1)\log(1/r)\cdot(1+\eta); \qquad \eta > 0 \text{ fixed.}$$

Since ϵ and η are arbitrary, this gives the upper bound in (I17b).

For the lower bound we use a continuous analog of the argument at the end of Section F12. Fix $\epsilon > 0$ and let

$$t = (1-\epsilon)(d-1)\bar{t}(r)\log(1/r).$$

We shall apply the heuristic to the random set $\mathcal{S}$ of points $x \in S_{d-1}$ such that the path up to time t has not passed within distance r of t. Here

$$p = \boldsymbol{P}(x \in \mathcal{S}) \quad \approx \quad \exp\left(\frac{-t}{\bar{t}(r)}\right) \qquad \text{by (I17a)}$$

We shall argue that the clump rate $\widehat{\lambda}$ for $\mathcal{S}$ satisfies

$$\widehat{\lambda} \equiv \frac{p}{EC} \to \infty \quad \text{as } r \to 0$$

implying

$$\boldsymbol{P}(V_r < t) = \boldsymbol{P}(\mathcal{S} \text{ empty}) \to 0 \quad \text{as } r \to 0,$$

giving the lower bound in (I17b). Thus what we must prove is

$$EC = o(p) \equiv o(\exp(-t/\bar{t}(r))) \quad \text{as } r \to 0. \tag{I17c}$$

Now fix $x_0 \in \mathcal{S}$. Then, as at Section A6,

$$\begin{aligned} EC \leq E\widetilde{C} &= \int_{S_{d-1}} \boldsymbol{P}(x \in \mathcal{S} \mid x_0 \in \mathcal{S})\,dx \\ &\leq \int_0^\infty s_{d-1} y^{d-2} q(y)\,dy \end{aligned} \tag{I17d}$$

where $q(y) = \boldsymbol{P}(x \in \mathcal{S} \mid x_0 \in \mathcal{S})$ when $|x - x_0| = y$. We shall show

$$q(y) \approx \exp(-cty/r\bar{t}(r)), \qquad y \text{ small} \tag{I17e}$$

where c depends only on the dimension d, and then it is straightforward to verify (I17c). To establish (I17e), fix x with $|x - x_0| = y$ for small y. Let D be the union of the caps B with radii r centered at x_0 and at x. Then $\mathrm{vol}(D) \approx \mathrm{vol}(B)(1 + cy/r)$ for some c depending only on d. Use the heuristic to compare the first hitting time T_D with the first hitting time T_B: the clump sizes are essentially the same, so we get $ET_B/ET_D \approx 1 + cy/r$. Then

$$\frac{\boldsymbol{P}(T_D > t)}{\boldsymbol{P}(T_B > t)} \approx \exp(-\frac{t}{ET_D} + \frac{t}{ET_B}) \approx \exp(-\frac{cyt}{rET_B}).$$

and this is (I17e).

Remark: This was fairly easy because we only needed a crude upper bound on EC above. Finding the exact asymptotics of EC needed for the "convergence in distribution" improvement of (I17b), seems harder: see Section I27.

I18 Rice's formula for conditionally locally Brownian processes. The basic type of problem discussed in this chapter concerns the time for a stationary d-dimensional diffusion X_t to exit a region $A \subset \boldsymbol{R}^d$. Such problems can usually be viewed in another way, as the time for a 1-dimensional process $Y_t = h(X_t)$ to hit a level b. Of course, one typically loses the Markov property in passing to 1 dimension. Instead, one gets the following property. There is a stationary process (Y_t, μ_t, σ_t) such that

given (Y_t, μ_t, σ_t), we have $Y_{t+\delta} \approx Y_t + \delta\mu_t + \sigma_t B_\delta$ for small $\delta > 0$. (I18a)

Write $f(y, \mu, \sigma)$ for the marginal density of (Y_t, μ_t, σ_t). Consider excursions of Y_t above level b. If these excursions are brief and rare, and if μ and σ do not change much during an excursion, then the clump rate λ_b for $\{t : Y_t \geq b\}$ is

$$\lambda_b \approx E(\mu^- \mid Y = b) f_Y(b) \tag{I18b}$$

where f_Y is the marginal density of Y_t and $\mu^- = \max(0, -\mu)$. Note the similarity to Rice's formula (C12.1) for smooth processes, even though we are dealing with locally Brownian processes. To argue (I18b), let $\lambda_{\mu,\sigma}\, d\mu\, d\sigma$ be the rate of clumps which start with $\mu_t \in (\mu, \mu + d\mu)$, $\sigma_t \in (\sigma, \sigma + d\sigma)$. Then as at Section D2,

$$\lambda_{\mu,\sigma} = -\mu f(b, \mu, \sigma), \qquad \mu < 0.$$

Since $\lambda_b \approx \iint \lambda_{\mu,\sigma}\, d\mu\, d\sigma$, we get (I18b).

Of course, one could state this result directly in terms of the original process X, to get an expression involving the density of X at each point of the boundary of A and the drift rate in the inward normal direction.

I19 Example: Rough $\mathcal{X}^2$ processes. For $1 \leq i \leq n$ let $X_i(t)$ be independent stationary Ornstein-Uhlenbeck processes, covariance

$$R_i(t) = \exp(-\theta_i|t|).$$

Let $Y(t) = \sum_{i=1}^n a_i X_i^2(t)$, and consider extremes of Y_t. In s.d.e. notation,

$$\begin{aligned} dX_i &= -\theta_i X_i\, dt + (2\theta_i)^{\frac{1}{2}}\, dB_t \\ dX_i^2 &= -2\theta_i(X_i^2 - 1)\, dt + \text{ terms in } dB_t \end{aligned}$$

and so

$$dY_t = -2\sum a_i\theta_i(X_i^2 - 1)\, dt + \text{ terms in } dB_t.$$

Thus the extremal behavior of Y_t is given by

$$\boldsymbol{P}(\sup_{0 \leq s \leq t} Y_s \leq b) \approx \exp(-\lambda_b t), \qquad b \text{ large},$$

where the clump rate λ_b is, using (I18b),

$$\lambda_b = f_Y(b)\sum_{i=1}^n 2a_i\theta_i E(X_i^2 - 1 \mid \sum a_i X_i^2 = b) \tag{I19a}$$

where the X_i have standard Normal distribution.

This can be made explicit in the special case $a_i \equiv 1$, where Y has $\mathcal{X}^2$ distribution and (I19a) becomes

$$\lambda_b = f_Y(b) \cdot 2\left(\frac{b}{n} - 1\right)\sum \theta_i.$$

COMMENTARY

I20 General references. I don't know any good introductory account of multi-dimensional diffusions — would someone like to write a short monograph in the spirit of Karlin and Taylor's (1982) account of the 1-dimensional case? Theoretical works such as Dynkin (1962), Stroock and Varadhan (1979), are concerned with questions of existence and uniqueness, of justifications of the basic equations (Section I1), and these issues are somewhat removed from the business of doing probability calculations. An applied mathematician's treatment is given by Schuss (1980). Gardiner (1983) gives the physicist's approach.

I21 Calculation of stationary distributions. Durrett (1985) gives a nice discussion of reversibility. Gardiner (1983) discusses Ornstein-Uhlenbeck processes.

Another setting where stationary distributions can be found explicitly concerns Brownian motion with particular types of reflecting boundary. Such processes occur in the heavy traffic limit of queueing systems — Harrison and Williams (1987).

I22 Radial part of Brownian motion. For $X_t = \sigma_0 B_t$, the radial part $|X_t|$ is the 1-dimensional diffusion with

$$\mu(x) = \frac{1}{2}(d-1)x^{-1} \qquad \sigma(x) = \sigma_0,$$

called the *Bessel(d) process* (Karlin and Taylor (1982)). Then the formulas in Section I7 can be derived from 1-dimensional formulas. Alternatively, (I7b) holds by optional stopping of the martingale $|X_t|^2 - d\sigma_0^2 t$, and (I7c) by direct calculus.

When $X_t = \sigma B_t$ for a general matrix σ, there must be some explicit formulas corresponding to those in Section I7, but I don't know what!

I23 Hitting small balls. Berman (1983b) Section 6 treats a special case of Example I8; Baxendale (1984) gives the associated integral test. Clifford et al. (1987) discuss some "near miss" models in the spirit of Example I9 and give references to the chemistry literature.

I24 Potential wells. Gardiner (1983) and Schuss (1980) give textbook accounts of our examples. Matkowsky et al. (1982; 1984) describe recent work on Kramers' problem. The simple form of the stationary distribution (I13b) in Kramers' problem arises from physical reversibility: see Gardiner (1983) p. 155.

One can attempt to combine the "large γ" and "small γ" arguments: this is done in Matkowsky et al. (1984).

I25 Formalizations of exit problems. Our formulations of exit problems for diffusions don't readily lend themselves to formalizations as limit theorems. There is an alternative set-up, known as Ventcel-Friedlin theory, where one considers a diffusion $X^\epsilon(t)$ with drift $\mu(x)$ and variance $\epsilon\sigma(x)$, and there is a fixed region $A \subset \boldsymbol{R}^d$. Let T_ϵ be the exit time from A for X^ϵ; then one can study the asymptotics of T_ϵ as $\epsilon \to 0$. This is part of large deviation theory — see Varadhan (1984) for the big picture. Day (1987) is a good survey of rigorous results concerning exit times and places. Day (1983) gives the exponential limit law for exit times in this setting.

It is important to realize that the Ventcel-Friedlin set-up is not always natural. Often a 1-parameter family X^ϵ arises from , say, a 3-parameter physical problem by rescaling, as in Kramers' equation (I13a) for X^γ. Though it is mathematically natural to take limits in ϵ, the real issue is understanding the effects of changes in the original physical parameters: such changes affect not only ϵ but also the potential function H and the boundary level b.

I26 Boundary layer expansions. Schuss (1980) develops an analytic technique, describable as "singular perturbations" or "boundary layer expansions", which is in essence similar to our heuristic but presented in quite a different way: the approximations are done inside a differential equations setting instead of directly in the probabilistic setting. In principle this is more general than our heuristic: some of the mathematical physics examples such as the "phase-locked loops" of Schuss (1980) Chapter 9 genuinely require such analytic techniques. On the other hand many of the examples, particularly those in queueing models (Knessl et al. (1985; 1986b; 1986a)) can be done more simply and directly via our heuristic.

I27 Brownian motion on surface of d-sphere. Matthews (1988a) gives a rigorous treatment of Example I17. Our argument suggests the clump size at (I17a) should satisfy

$$EC \sim a_d(r\log(1/r))^{d-1} \quad \text{as } r \to 0$$

for some (unknown) constant a_d. If so, the heuristic yields the convergence in distribution result

$$\begin{aligned} &\boldsymbol{P}(V_r \le \bar{t}(r)((d-1)\log(1/r) - (d-1)\log\log(1/r) + w)) \\ &\qquad \to \exp(-s_d R^{d-1} a_d^{-1} e^{-w}) \quad \text{as } r \to 0. \qquad \text{(I27a)} \end{aligned}$$

Finding a_d explicitly, and justifying (I27a), looks hard.

I28 Rough $\mathcal{X}^2$ processes. More complicated variations of Example I19 involving infinite sums arise in some applications — see Walsh (1981) — but good estimates are unknown.

I29 Smooth stationary processes. We can also consider exit problems for stationary non-Markov processes $X(t)$ in $\boldsymbol{R}^d$. Suppose the process has smooth paths, so that $V(t) = \frac{d}{dt}X(t)$ exists, and let $A \subset \boldsymbol{R}^d$ be a large region such that $\boldsymbol{P}(X(t) \not\in A)$ is small. Suppose A has a smooth boundary ∂A. Then we can in principle estimate exit times as in the 1-dimensional case (Chapter C), using Rice's upcrossing formula. That is, the exit time T_A will have approximately exponential distribution with rate λ_A given by

$$\lambda_A = \int_{\partial A} \rho(x)\,dx \qquad \text{(I29a)}$$

where $\rho(x)$ is the "outcrossing rate" at $x \in \partial A$ defined by

$$\rho(x)|dB|\,dt = \boldsymbol{P}(X \text{ crosses from } A \text{ to } A^C \text{ through } dB \text{ during } [t, t+dt])$$

where $x \in dB \subset \partial A$ and $|dB|$ is the area of the boundary patch dB. Now Rice's formula (Section C12) gives

$$\rho(x) = E(\langle \eta_x, v \rangle^+ \mid X = x) f_X(x) \qquad \text{(I29b)}$$

where η_x is the unit vector normal to ∂A at x, and where $\langle\, ,\, \rangle$ denotes dot product.

As a special case, suppose $X(t) = (X_1(t); 1 \leq i \leq d)$ is Gaussian, mean zero, with independent components such that $EX_i(0)X_i(t) \sim 1 - \frac{1}{2}\theta_i t^2$ as $t \to 0$. Then $V_i(t)$ has Normal$(0, \theta_i)$ distribution, and $V(t)$ is independent of $X(t)$, so (I29b) becomes

$$\rho(x) = (2\pi)^{-\frac{1}{2}} < \eta_x, \theta > f_X(x) \quad \text{where } \theta = (\theta_i). \qquad \text{(I29c)}$$

Even in this case, explicitly evaluating the integral in (I29a) is hard except in the simplest cases. For instance, one may be interested in extremes or boundary crossing for $h(X(t))$, where $h\colon \boldsymbol{R}^d \to \boldsymbol{R}$ is given, and this involves evaluation (I29a) for $\partial A = \{\, x : h(x) = h_0 \,\}$. Lindgren (1984a) gives a nice treatment of this problem.

J Random Fields

In this chapter we look at the topics of chapters C and D — extrema and boundary crossings – for d-parameter processes instead of 1-parameter processes. A *random field* $X(t)$ or X_t is just a real-valued process parametrized by $t = (t_1, \ldots, t_d)$ in $\boldsymbol{R}^d$ or some subset of $\boldsymbol{R}^d$. Since this concept may be less familiar to the reader than earlier types of random process, let us start by mentioning several contexts where random fields arise.

J1 Spatial processes. Think of t as a point in physical space ($d = 2$ or 3, say) and think of $X(t)$ as the value of some physical quantity at point t (pressure, temperature, etc.). These give the most natural examples of random fields.

J2 In analysis of 1-parameter processes. For ordinary Brownian motion $(B_t : 0 \le t < \infty)$ one may be interested in quantities like

$$\sup_{\substack{0 \le t_1 < t_2 \le T \\ t_2 - t_1 \ge \delta}} \frac{B_{t_2} - B_{t_1}}{(t_2 - t_1)^{\frac{1}{2}}}. \tag{J2a}$$

It is useful to think of this as the supremum of a random field $X(t_1, t_2)$ over a certain region: such topics are treated in Chapter K.

J3 Gaussian fields and white noise. As in the 1-parameter setting, a d-parameter Gaussian process is determined by its mean function (which we assume to be zero unless otherwise stated) and its covariance function. But many natural Gaussian processes can be constructed more explicitly from white noise, as explained below. Let μ be a positive non-atomic measure on $\boldsymbol{R}^d$. Associated with μ is the *μ-white noise* process $(W(A) : A \subset \boldsymbol{R}^d, \quad \mu(A) < \infty)$ specified by

$$W(A) \stackrel{\mathcal{D}}{=} \text{Normal}(0, \mu(A)) \tag{J3a}$$

$$\text{for disjoint } (A_i) \text{ the } W(A_i) \text{ are independent} \tag{J3b}$$

$$W(A \cup B) + W(A \cap B) = W(A) + W(B) \quad \text{a.s.} \tag{J3c}$$

For $\mu =$ Lebesgue measure ("volume") on $\boldsymbol{R}^d$, this is *white noise.* If μ is a probability measure then we can also define *$\mu-$Brownian sheet*

$$Z(A) = W(A) - \mu(A)W(\boldsymbol{R}^d) \qquad A \subset \boldsymbol{R}^d. \tag{J3d}$$

For $\mu =$ Lebesgue measure on $[0,1]^d$ this is *Brownian sheet.* W and Z are set-indexed Gaussian processes with mean zero and covariances

$$EW(A)W(B) = \mu(A \cap B) \tag{J3e}$$

$$EZ(A)Z(B) = \mu(A \cap B) - \mu(A)\mu(B). \tag{J3f}$$

It is also useful to note

$$EZ(A)Z(B) = \frac{1}{4} - \frac{1}{2}\mu(A \Delta B) \qquad \text{if } \mu(A) = \frac{1}{2}. \tag{J3g}$$

These set-indexed processes can be regarded as point-indexed processes by restricting attention to a family $\mathcal{A}$ of subsets A with finite-dimensional parametrization. Let $(A_t : t \in \boldsymbol{R}^{\widehat{d}})$ be a family of subsets of $\boldsymbol{R}^d$, for example the family of discs in $\boldsymbol{R}^2$ (where say (t_1, t_2, t_3) indicates the disc centered at (t_1, t_2) with radius t_3). Then $X(t) \equiv W(A_t)$ or $X(t) \equiv Z(A_t)$ defines a $\widehat{d}$ parameter mean-zero Gaussian random field, whose covariance is given by (J3e,J3f).

J4 Analogues of the Kolmogorov-Smirnov test. The distribution of the supremum of a Gaussian random field occurs naturally in connection with Kolmogorov-Smirnov type tests. Let (ξ_i) be i.i.d. with distribution μ and let μ^N be the empirical distribution of the first N observations:

$$\mu^N(\omega, A) = N^{-1} \sum_{i=1}^{N} 1_{(\xi_i \in A)}.$$

Consider the normalized empirical distribution

$$Z^N(A) = N^{\frac{1}{2}}(\mu^N(A) - \mu(A)). \tag{J4a}$$

The central limit theorem says that, as $N \to \infty$ for fixed $A \subset \boldsymbol{R}^d$,

$$Z^N(A) \xrightarrow{\mathcal{D}} \text{Normal}(0, \mu(A)(1 - \mu(A))) \stackrel{\mathcal{D}}{=} Z(A), \tag{J4b}$$

where Z is the μ-Brownian sheet (J3d). Now consider a family $(A_t : t \in \boldsymbol{R}^{\widehat{d}})$ of subsets of R^d. Under mild conditions (J4b) extends to

$$(Z^N(A_t) : t \in \boldsymbol{R}^{\widehat{d}}) \xrightarrow{\mathcal{D}} (Z(A_t) : t \in \boldsymbol{R}^{\widehat{d}}) \tag{J4c}$$

in the sense of weak convergence of processes; in particular

$$\sup_t Z^N(A_t) \xrightarrow{\mathcal{D}} \sup_t Z(A_t). \tag{J4d}$$

For an i.i.d. sequence of observations, a natural statistical test of the hypothesis that the distribution is μ is to form the normalized empirical distribution Z^N and compare the observed value of $\sup_t Z^N(A_t)$, for a suitable family (A_t), with its theoretical ("null") distribution. And (J4d) says that for large N this null distribution can be approximated by the supremum of a Gaussian random field.

J5 The heuristic. The basic setting for our heuristic analysis is where $X(t)$ is a d-dimensional stationary random field without long-range dependence, and where we are interested in the supremum

$$M_A = \sup_{t \in A} X(t) \tag{J5a}$$

for some nice subset (cube or sphere, usually) A of $\boldsymbol{R}^d$. So we fix a high level b and consider the random set $\mathcal{S}_b = \{t : X(t) \geq b\}$. Suppose this resembles a mosaic process with some clump rate λ_b, clump shape $\mathcal{C}_b$ and clump volume C_b; then

$$\boldsymbol{P}(M_A \leq b) \approx \exp(-\lambda_b |A|); \qquad |A| = \text{volume}(A) \tag{J5b}$$

$$\lambda_b = \frac{\boldsymbol{P}(X(t) \geq b)}{EC_b}. \tag{J5c}$$

This is completely analogous to the 1-parameter case (Section C4). The practical difficulty is that two of the most useful techniques for calculating EC_b and hence λ_b are purely 1-parameter: the "renewal-sojourn" method (Section A8) and the "ergodic-exit" method (Section A9). We do still have the "harmonic mean" method (Section A6) and the "conditioning on semi-local maxima" method (Section A7), but it is hard to get explicit constants that way. Otherwise, known results exploit special tricks.

J6 Discrete processes. In studying the maximum M_A of a discrete process $(X(t); t \in \boldsymbol{Z}^d)$, some of the ideas in the 1-parameter case extend unchanged. For instance, the "approximate independence of tail values" condition (C7a)

$$\boldsymbol{P}(X(t) \geq b \mid X(0) \geq b) \to 0 \quad \text{as } b \to \infty; \quad t \neq 0 \tag{J6a}$$

is essentially enough to ensure that M_A behaves as if the $X(t)$ were independent. And arguments for the behavior of moving average processes (Section C5) extend fairly easily; for here the maximum is mostly due to a single large value of the underlying i.i.d. process. Such processes are rather uninteresting and will not be pursued.

J7 Example: Smooth Gaussian fields. This is the analogue of Section C23. Let $X(t)$ be stationary Gaussian with $X(t) \stackrel{\mathcal{D}}{=} \text{Normal}(0,1)$ and with correlation function

$$R(t) \equiv EX(0)X(t). \tag{J7a}$$

Suppose $R(t)$ has the form

$$R(t) = 1 - \frac{1}{2}t^T\Lambda t + O(|t|^2) \quad \text{as } t \to 0, \tag{J7b}$$

in which case Λ is the positive-definite matrix

$$\Lambda_{ij} = -\frac{\partial^2 R(t)}{\partial t_i \partial t_j}. \tag{J7c}$$

Let $|\Lambda| = \text{determinant}(\Lambda)$ and let ϕ be the Normal(0,1) density. We shall show that for b large the clump rate is

$$\lambda_b = (2\pi)^{-\frac{1}{2}d}|\Lambda|^{\frac{1}{2}}b^{d-1}\phi(b) \tag{J7d}$$

and then (J5b) gives the approximation for maxima M_A.

We argue (J7d) by the "conditioning on semi-local maxima" method. The argument at (C26e), applied to clumps $\{\, t : X_t \geq b \,\}$ rather than slices $\{\, t : X_t \in (y, y+dy) \,\}$, gives

$$\boldsymbol{P}(X(t) \geq b) = \int_b^\infty L(x)m(x,b)\,dx \tag{J7e}$$

where $L(x)\,dx$ is the rate of local maxima of heights in $[x, x+dx]$, and where $m(x,b) = E\,\text{volume}\{\, t : X_t \geq b \,\}$ in a clump around a local maximum of height x. The key fact is that, around a high level x, the process $X(t)$ is almost deterministic;

$$\text{given } X(0) = x \text{ and } \partial X(t)/\partial t_i = v_i,$$
$$X(t) \approx x + v \cdot t - \frac{1}{2}x(t^T\Lambda t) \qquad \text{for } t \text{ small.} \tag{J7f}$$

This follows from the corresponding 1-parameter result (C25a) by considering sections. So in particular, if $X(0) = x$ is a local maximum then $X(t) \approx x - \frac{1}{2}x(t^T\Lambda t)$ and so

$$\begin{aligned} m(x,b) &\approx \text{volume}\{\, t : t^T\Lambda t \leq (x-b)/(\tfrac{1}{2}x) \,\}; \qquad x > b \\ &= v_d|\Lambda|^{-\frac{1}{2}}\left(\frac{2(x-b)}{x}\right)^{\frac{1}{2}d} \end{aligned}$$

where v_d is the volume of the unit sphere in d dimensions:

$$v_d = \frac{2\pi^{\frac{1}{2}d}}{d\Gamma(\frac{1}{2}d)}. \tag{J7g}$$

Substituting into (J7e) and writing $x = b + u$,

$$\boldsymbol{P}(X(t) \geq b) \approx v_d|\Lambda|^{-\frac{1}{2}}2^{\frac{1}{2}d}\int_0^\infty \left(\frac{u}{x}\right)^{\frac{1}{2}d} L(b+u)\,du.$$

We want to solve for L. Anticipating a solution of the form $L(x) = g(x)\phi(x)$ for g varying slowly relative to ϕ, we have $L(b+u) \approx g(b)\phi(b+u) \approx g(b)\phi(b)e^{-bu}$ and $\boldsymbol{P}(X(t) \geq b) \approx \phi(b)/b$, giving

$$b^{-1}\phi(b) \approx v_d|\Lambda|^{-\frac{1}{2}}\left(\frac{2}{b}\right)^{\frac{1}{2}d} g(b)\phi(b)\int_0^\infty u^{\frac{1}{2}d}e^{-bu}\,du.$$

This reduces to

$$g(b) \approx (2\pi)^{-\frac{1}{2}d}|\Lambda|^{\frac{1}{2}}b^d.$$

So $L(x) \approx (2\pi)^{-d/2}|\Lambda|^{1/2}x^d\phi(x)$. But the clump rate λ_b satisfies

$$\lambda_b = \int_b^\infty L(x)\,dx,$$

giving (J7d) as the first term.

This technique avoids considering the mean clump size EC_b, but it can now be deduced from (J5c):

$$EC_b = (2\pi)^{\frac{1}{2}d}b^{-d}|\Lambda|^{-\frac{1}{2}}. \tag{J7h}$$

In the case where Λ is diagonal, so that small increments of $X(t)$ in orthogonal directions are uncorrelated, we see from (J7h) that EC_b is just the product of the clump sizes of the 1-parameter marginal processes.

Some final remarks will be useful later. For a local maximum Y of height at least b, the "overshoot" $\xi = Y - b$ will satisfy $\boldsymbol{P}(\xi > x) = \lambda_{b+x}/\lambda_b \approx e^{-bx}$ for large b. So

$$\xi \stackrel{\mathcal{D}}{\approx} \text{exponential}(b) \qquad \text{for } b \text{ large.} \tag{J7i}$$

This agrees with the 1-parameter case. We can now calculate

$$\frac{EC_b^2}{(EC_b)^2} \approx \binom{d}{d/2} \qquad \text{for } b \text{ large.} \tag{J7j}$$

For conditional on a local maximum having height $b+x$, we have the clump size $C_b \approx ax^{d/2}$ for a independent of x. So

$$\begin{aligned} EC_b &= a\int_0^\infty be^{-bx}x^{\frac{1}{2}d}\,dx = ab^{\frac{1}{2}d}(d/2)! \\ EC_b^2 &= a^2\int_0^\infty be^{-bx}x^d\,dx = a^2b^dd! \end{aligned}$$

J8 Example: 2-dimensional shot noise. Given $0 < \rho < \infty$ and a distribution μ on $\boldsymbol{R}^+$, let (T_i) be the points of a Poisson process in $\boldsymbol{R}^2$ of rate ρ and associate i.i.d. (μ) variables (ξ_i) with each T_i. Let $h : \boldsymbol{R}^+ \to \boldsymbol{R}^+$ be decreasing with $h(0) = 1$. Under mild conditions we can define the stationary random field

$$X(t) = \sum \xi_i h(|t - T_i|); \qquad t \in \boldsymbol{R}^2. \tag{J8a}$$

This is partly analogous to Examples C13, C15. One can write down an expression for the transform of the marginal distribution $\widehat{X} = X(0)$. Let us consider the special case

$$\xi_i \text{ has exponential}(\alpha) \text{ distribution.} \tag{J8b}$$

The Poisson property implies that the distribution $X(T)$ at a typical point (T, ξ) of the Poisson process is

$$X(T) \stackrel{\mathcal{D}}{=} \widehat{X} + \xi; \quad \text{where } (\widehat{X}, \xi) \text{ are independent.} \tag{J8c}$$

Thus $X(T)$ has density

$$\begin{aligned} f_{X(T)}(x) &= \int_0^x f_{\widehat{X}}(y) \alpha e^{-\alpha(x-y)}\, dy \\ &\approx A\alpha e^{-\alpha x} \qquad \text{for large } x, \text{ where } A = Ee^{\alpha \widehat{X}}. \end{aligned} \tag{J8d}$$

Let $L(x)\, dx$ be the rate of local maxima of $X(t)$ of heights in $[x, x + dx]$. At high levels, local maxima of X should occur only at points T of the Poisson process, so from (J8d)

$$L(x) = A\alpha\rho e^{-\alpha x} \qquad \text{for large } x.$$

The heuristic now says that the rate λ_b of clumps of $\{\, t : X_t \geq b \,\}$ satisfies

$$\lambda_b = \int_b^\infty L(x)\, dx \approx A\rho e^{-\alpha b}.$$

As usual, (J5b) gives the heuristic approximation for maxima M_A of X.

J9 Uncorrelated orthogonal increments Gaussian processes. Let $X(t)$, $t \in \boldsymbol{R}^d$ be a stationary Gaussian random field. For fixed large b let λ_b and EC_b be the rate and mean volume of clumps of X above b. Write $(z_1, \ldots, z_d)$ for the orthogonal unit vectors in $\boldsymbol{R}^d$ and write $X_i(s) = X(sz_i)$, $s \in \boldsymbol{R}$, for the marginal processes. Let λ_b^i, EC_b^i be clump rate and size for $X_i(s)$. Suppose x has the *uncorrelated orthogonal increments* property

$$(X_1(s), X_2(s), \ldots, X_d(s)) \text{ become uncorrelated as } s \to 0. \tag{J9a}$$

It turns out that this implies a product rule for mean clump sizes:

$$EC_b = \prod_{i=1}^{d} EC_b^i. \tag{J9b}$$

Then the fundamental identity yields

$$\lambda_b = \frac{P(X(0) > b)}{\prod_{i=1}^{d} EC_b^i} = \frac{\prod_{i=1}^{d} \lambda_b^i}{P^{d-1}(X(0) \geq b)}; \tag{J9c}$$

everything reduces to 1-parameter problems.

This product rule is a "folk theorem" for which there seems no general known proof (Section J32) or even a good general heuristic argument. We have already remarked, below (J7h), that it holds in the smooth case; let us now see that it holds in the following fundamental example.

J10 Example: Product Ornstein-Uhlenbeck processes. Consider a 2-parameter stationary Gaussian random field $X(t)$ with covariance of the form

$$R(t) \equiv EX(0)X(t) \approx 1 - \mu_1|t_1| - \mu_2|t_2| \quad \text{as } t \to 0. \tag{J10a}$$

An explicit example is

$X(t) = W(A_t)$; where W is 2-parameter white noise (Section J3) and A_t is the unit square with lower-left corner t. (J10b)

We want to consider the shape of the clumps $\mathcal{C}_b$ where $X \geq b$. Fix t_0 and suppose $X(t_0) = x > b$. Consider the increments processes $\widehat{X}_i(s) = X(t_0 + sz_i) - X(t_0)$, s small. From the white noise representation in (J10b), or by calculation in (J10a), we see that $\widehat{X}_1(s)$ and $\widehat{X}_2(s)$ are almost independent for s small. The clump $\mathcal{C}_b$ is essentially the set $\{\, t_0 + (s_1, s_2) : \widehat{X}_1(s_1) + \widehat{X}_2(s_2) \geq b - x \,\}$. As in Example J7, $\mathcal{C}_b$ is not anything simple like a product set. But this description of $\mathcal{C}_b$ suggests looking at sums of independent 1-parameter processes, which we now do.

Let $Y_1(t_1)$, $Y_2(t_2)$ be stationary independent Gaussian 1-parameter processes such that

$$R_i(t) \equiv EY_i(0)Y_i(t) \approx 1 - 2\mu_i|t| \quad \text{as } t \to 0. \tag{J10c}$$

Consider

$$X(t_1, t_2) = 2^{-\frac{1}{2}}(Y_1(t_1) + Y_2(t_2)). \tag{J10d}$$

This is of the required form (J10a). Let $M_i = \sup_{0 \leq t \leq 1} Y_i(t)$. By (D10e) the Y_i have clump rates

$$\lambda_b^i = 2\mu_i b\phi(b) \tag{J10e}$$

and hence M_i has density of the form

$$f_{M_i}(x) \sim 2\mu_i x^2 \phi(x) \quad \text{as } x \to \infty. \tag{J10f}$$

Let us record a straightforward calculus result:

If M_1, M_2 are independent with densities $f_{M_i}(x) \sim a_i x^{n_i}\phi(x)$ as $x \to \infty$, then $2^{-1/2}(M_1 + M_2) \equiv M$ has density $f_M(x) \sim a_1 a_2 (x/\sqrt{2})^{n_1+n_2}\phi(x)$ as $x \to \infty$. (J10g)

In the present setting this shows that $M = \sup_{0 \le t_1, t_2 \le 1} X(t_1, t_2)$ has density of the form

$$f_M(x) \sim \mu_1\mu_2 x^4 \phi(x) \quad \text{as } x \to \infty. \tag{J10h}$$

The clump rate λ_b for X is such that $\boldsymbol{P}(M \le b) \approx \exp(-\lambda_b)$, and so (J10h) yields

$$\lambda_b \approx \mu_1\mu_2 b^3 \phi(b). \tag{J10i}$$

Since the extremal behavior of a Gaussian $X(t)$ depends only on the behavior of $R(t)$ near $t = 0$, this conclusion holds for any process of the form (J10a) (without long-range dependence, as usual), not just those of the special form (J10d).

By the fundamental identity, the mean clump sizes for $X(t)$ are

$$EC_b = (\mu_1\mu_2)^{-1} b^{-4}. \tag{J10j}$$

Comparing with (D10f) we see that the product rule (J9b) works in this example.

This example is fundamental; several subsequent examples are extensions in different directions. Let us record the obvious d-parameter version:

$$\text{If } R(t) \approx 1 - \prod_{i=1}^{d} \mu_i |t_i| \quad \text{as } t \to 0 \quad \text{in } \boldsymbol{R}^d, \text{ then } \lambda_b = \Big(\prod_{i=1}^{d} \mu_i\Big) b^{2d-1}\phi(b). \tag{J10k}$$

J11 An artificial example. A slightly artificial example is to take $X(t_1, t_2)$ stationary Gaussian and smooth in one parameter but not in the other, say

$$R(t_1, t_2) \approx 1 - \mu|t_1| - \frac{1}{2}\rho t_2^2 \quad \text{as } t \to 0. \tag{J11a}$$

In this case we can follow the argument above, but with $Y_2(t_2)$ now a smooth 1-parameter Gaussian process, and we conclude

$$\lambda_b = (2\rho)^{-\frac{1}{2}} \pi^{\frac{1}{2}} \mu b^2 \phi(b). \tag{J11b}$$

Again, the product rule (J9b,J9c) gives the correct answer.

The remaining examples in this chapter concern "locally Brownian" fields, motivated by the following discussion.

J12 Maxima of μ-Brownian sheets. For stationary processes $X(t)$, $t \in \boldsymbol{R}^d$, our approximations

$$\boldsymbol{P}(M_A \leq b) \approx \exp(-\lambda_b|A|)$$

correspond to the limit assertion

$$\sup_b |\boldsymbol{P}(M_A \leq b) - \exp(-\lambda_b|A|)| \to 0 \quad \text{as } |A| \to \infty. \tag{J12a}$$

Consider now $M = \sup_t X(t)$, where X arises from a μ-Brownian sheet Z via $X(t) = Z(A_t)$, as in Section J3. Typically $X(t)$ is a non-stationary mean zero Gaussian process. We can still use the non-stationary form of the heuristic. Write $\mathcal{S}_b$ for the random set $\{t : X_t \geq b\}$ for b large; $EC_b(t_0)$ for the mean volume of clumps of $\mathcal{S}$ which occur near t_0; $\lambda_b(t_0)$ for the clump rate at t_0; and $p_b(t_0) = \boldsymbol{P}(X(t_0) \geq b)$. Then the heuristic approximation is

$$\lambda_b(t) = \frac{p_b(t)}{EC_b(t)} \tag{J12b}$$

$$\boldsymbol{P}(M \leq b) \approx \exp(-\int \lambda_b(t)\,dt) \tag{J12c}$$

and so

$$\boldsymbol{P}(M > b) \approx \int \lambda_b(t)\,dt \qquad \text{for large } b. \tag{J12d}$$

As discussed in Section A10, in this setting we cannot hope to make (J12c) into a limit assertion, since we have only one M; on the other hand (J12d) corresponds to the limit assertion

$$\boldsymbol{P}(M > b) \sim \int \lambda_b(t)\,dt \quad \text{as } b \to \infty; \tag{J12e}$$

i.e. the asymptotic behavior of the tail of M. Note that in the statistical application to generalized Kolmogorov-Smirnov tests (Section J4), this tail, or rather the values of b which make $\boldsymbol{P}(M > b) = 0.05$ or 0.01, say, are of natural interest.

Note we distinguish between the 1-sided maximum $M = \sup_t X(t)$ and the 2-sided maximum $M^* = \sup_t |X(t)|$. By symmetry

$$\boldsymbol{P}(M^* > b) \approx 2\boldsymbol{P}(M > b) \approx 2\int \lambda_b(t)\,dt \qquad \text{for large } b. \tag{J12f}$$

J13 1-parameter Brownian bridge. Let μ be a continuous distribution on $\boldsymbol{R}^1$ and let Z be μ-Brownian sheet. Let

$$\begin{aligned} X(t) &= Z(-\infty, t). \\ M &= \sup_t X(t). \end{aligned} \tag{J13a}$$

The natural space transformation which takes μ to the uniform distribution on $(0,1)$ will take X to Brownian bridge; hence M has exactly the same distribution for general μ as for Brownian bridge. At Example D17 we gave a heuristic treatment of M for Brownian bridge; it is convenient to give a more abstract direct treatment of the general case here, to exhibit some calculations which will extend unchanged to the multiparameter case. Here and in subsequent examples we make heavy use of the Normal tail estimates, and Laplace's method of approximating integrals, discussed at Section C21.

For a stationary Gaussian process with covariance of the form $R(s) = f_0 - g_0|s|$ as $s \to 0$, we know from (D10f) that the mean clump sizes are

$$EC_b = f_0^2 g_0^{-1} b^{-2}. \tag{J13b}$$

Next, consider a 1-parameter non-stationary mean-zero Gaussian process $X(t)$ satisfying

$$EX(t)X(t+s) \approx f(t) - g(t)|s| \quad \text{as } s \to 0; \tag{J13c}$$

$$f \text{ achieves its maximum at } 0;\ f(t) \approx \tfrac{1}{4} - \alpha t^2 \text{ as } t \to 0. \tag{J13d}$$

Write $g_0 = g(0)$. Around $t = 0$ the process is approximately stationary and so by (J13b)

$$EC_b(t) \approx 2^{-4} g_0^{-1} b^{-2} \qquad \text{for } t \approx 0. \tag{J13e}$$

Next, for t near 0 consider

$$\begin{aligned} p_b(t) &= \boldsymbol{P}(X(t) \geq b) \\ &= \overline{\Phi}(bf^{-\frac{1}{2}}(t)) \\ &\approx (2b)^{-1}\phi(2b(1+2\alpha t^2)) \quad \text{using } \begin{array}{l} f^{-\frac{1}{2}}(t) \approx 2(1+2\alpha t^2) \\ \overline{\Phi}(x) \approx \dfrac{\phi(x)}{x} \end{array} \\ &\approx 2b^{-1}\phi(2b)\exp(-8b^2\alpha t^2). \end{aligned} \tag{J13f}$$

Next,

$$\begin{aligned} \lambda_b(t) &= \frac{p_b(t)}{EC_b(t)} \\ &\approx 8g_0 b\phi(2b)\exp(-8b^2\alpha t^2) \end{aligned}$$

using (J13f) and (J13e); since $p(t)$ decreases rapidly as $|t|$ increases there is no harm in replacing $EC(t)$ by its value at 0. Thus

$$\int \lambda_b(t)\, dt \approx 2\frac{g_0}{\sqrt{\alpha}} e^{-2b^2}. \tag{J13g}$$

In the setting (J13a), there is some t_0 such that $\mu(-\infty, t_0) = \frac{1}{2}$, and by translating we may assume $t_0 = 0$. Suppose μ has a density a at 0. Then for t, s near 0,

$$EX(t+s)X(t) \approx (\frac{1}{2} + at) - (\frac{1}{2} + at)(\frac{1}{2} + at + as) \approx \frac{1}{4} - a^2t^2 - \frac{1}{2}a|s|.$$

Thus we are in the setting of (J13c,J13d) with $\alpha = a^2$, $g_0 = \frac{1}{2}a$, and so

$$\begin{aligned} \boldsymbol{P}(M > b) &\approx \int \lambda_b(t)\,dt && \text{by (J12d)} \\ &\approx \exp(-2b^2) && \text{by (J13g).} \end{aligned} \qquad \text{(J13h)}$$

This is our heuristic tail estimate for M; as remarked in Example D17, it is mere luck that it happens to give the exact non-asymptotic result in this one example.

J14 Example: Stationary × Brownian bridge processes. Consider a 2-parameter mean-zero Gaussian process $X(t_1, t_2)$ which is "stationary in t_1 but Brownian-bridge-like in t_2". More precisely, suppose the covariance is of the form

$$\begin{aligned} &EX(t_1, t_2)X(t_1 + s_1, t_2 + s_2) \\ &\quad\approx f(t_2) - g_1(t_2)|s_1| - g_2(t_2)|s_2| \quad \text{as } |s| \to 0; \end{aligned} \qquad \text{(J14a)}$$

f has its maximum at $t_2 = 0$ and $f(t_2) \approx \frac{1}{4} - \alpha t_2^2$ as $|t_2| \to 0$. (J14b)

Write

$$M_T = \sup_{\substack{0 \le t_1 \le T \\ t_2}} X(t_1, t_2);$$

we are supposing X is defined for t_2 in some interval around 0, whose exact length is unimportant. We shall show

$$\boldsymbol{P}(M_T > b) \approx 32 g_1(0) g_2(0) \alpha^{-\frac{1}{2}} T b^2 \exp(-2b^2) \qquad \text{for large } b. \qquad \text{(J14c)}$$

Concrete examples are in the following sections. To argue (J14c), in the notation of Section J12

$$\begin{aligned} \boldsymbol{P}(M_T > b) &\approx \int \frac{p_b(t)}{EC_b(t)}\,dt \\ &\approx \iint \frac{p_b(0, t_2)}{EC_b(0,0)}\,dt_1\,dt_2 \end{aligned} \qquad \text{(J14d)}$$

because $p_b(t_1, t_2)$ does not depend on t_1; and also because $EC_b(t_1, t_2) = EC_b(0, t_2)$ and this may be approximated by $EC(0,0)$ because $p_b(0, t_2)$

decreases rapidly as t_2 increases away from 0. Around $(0,0)$, the process X behaves like the stationary field with covariance

$$R(s_1, s_2) \approx \frac{1}{4} - g_1(0)|s_1| - g_2(0)|s_2| \quad \text{as } |s| \to 0.$$

At Sections J9,J10 we saw the product rule for mean clump sizes held in this setting; so $EC_b(0,0) = EC_b^1 EC_b^2$, where EC_b^i are the mean clump sizes for the i-parameter processes with covariances

$$R^i(s_i) \approx \frac{1}{4} - g_i(0)|s_i| \quad \text{as } |s_i| \to 0.$$

By (J14d),

$$\begin{aligned} \boldsymbol{P}(M_T > b) &\approx \frac{T}{EC_b^1} \int \frac{p_b(0, t_2)}{EC_b^2}\, dt_2 \\ &\approx \frac{T}{EC_b^1} 2g_2(0)\alpha^{-\frac{1}{2}} \exp(-2b^2) \end{aligned}$$

since this is the same integral evaluated at (J13g). From (J13b) we find $EC_b^1 = (16g_1(0))^{-1}b^{-2}$, and we obtain (J14c).

J15 Example: Range of Brownian bridge. Let $B^0(t)$ be Brownian bridge on $[0,1]$ and let

$$X(t_1, t_2) = B^0(t_2) - B^0(t_1), \qquad M^* = \sup_{0 \le t_1 \le t_2} |X(t_1, t_2)|.$$

So

$$M^* = \sup_t B^0(t) - \inf_t B^0(t). \tag{J15a}$$

Also, if Z is μ-Brownian sheet for any continuous 1-parameter μ, then

$$M^* \stackrel{\mathcal{D}}{=} \sup_{t_1 < t_2} |Z(t_1, t_2)|. \tag{J15b}$$

We shall argue

$$\boldsymbol{P}(M^* > b) \approx 8b^2 \exp(-2b^2) \qquad \text{for } b \text{ large}, \tag{J15c}$$

which agrees with the exact asymptotics (Section J30.4). Note that, for

$$M = \sup_{t_1 < t_2} Z(t_1, t_2) = \sup_{t_1 < t_2} (B^0(t_2) - B^0(t_1))$$

we can deduce from (J12f) that,

$$\boldsymbol{P}(M > b) \approx 4b^2 \exp(-2b^2) \qquad \text{for } b \text{ large}. \tag{J15d}$$

Write $\widehat{X}(t_1,t_2) = X(t_1, t_1 \oplus \frac{1}{2} \oplus t_2)$, where $\oplus$ is addition modulo 1 and $0 \le t_1 < 1$, $-\frac{1}{2} \le t_2 < \frac{1}{2}$. Then $M^* = \sup_{t_1,t_2} \widehat{X}(t_1,t_2)$. And $\widehat{X}$ has covariance

$$E\widehat{X}(t_1,t_2)\widehat{X}(t_1+s_1,t_2+s_2) = \frac{1}{4} - t_2^2 - (\frac{1}{2} - t_2)|s_1| - (\frac{1}{2} + t_2)|s_2|$$

after a little algebra. Thus (J14a) holds with $\alpha = 1$, $g_1(0) = g_2(0) = \frac{1}{2}$ and then (J14c) gives (J15c).

J16 Example: Multidimensional Kolmogorov-Smirnov. Let μ be a continuous distribution on $\boldsymbol{R}^2$ and let Z be the μ-Brownian sheet (Section J3). For $t \in \boldsymbol{R}^2$ let $A_t = (-\infty, t_1) \times (-\infty, t_2)$ and let $X(t) = Z(A_t)$,

$$M = \sup_t X(t) = \sup_t Z(A_t). \qquad \text{(J16a)}$$

Let $L = \{\, t : \mu(A_t) = \frac{1}{2} \,\}$ and $L^+ = \{\, t : \mu(A_t) > \frac{1}{2} \,\}$. We shall argue

$$\boldsymbol{P}(M > b) \approx 8(\frac{1}{2} - \mu(L^+))b^2 \exp(-2b^2) \qquad \text{for large } b. \qquad \text{(J16b)}$$

In particular, consider the case where μ is uniform on the square $[0,1] \times [0,1]$. Here $L = \{\, (t, 1/(2t)); \frac{1}{2} < t < 1 \,\}$ and then

$$\mu(L^+) = \int_{\frac{1}{2}}^{1} \left(1 - \frac{1}{2t}\right) dt = \frac{1}{2}(1 - \log 2),$$

so the constant in (J16b) becomes $4 \log 2$. The same holds for any product measure with continuous marginals, by a scale change. Our argument supposes some smoothness conditions for μ, but the result seems true without them. Consider the two examples of the uniform distributions on the downward diagonal $D_1 = \{\, (t_1,t_2) : t_1 + t_2 = 1 \,\}$ and on the upward diagonal $D_2 = \{\, (t_1,t_2) : t_1 = t_2 \,\}$ of the unit square. For the downward diagonal D_1 we have $\mu(L^+) = 0$ and so the constant in (J16b) is 4. But here M is exactly the same as M in Example J15, the maximal increment of 1-parameter Brownian bridge over all intervals, and our result here agrees with the previous result (J15b). For D_2 we have $\mu(L^+) = \frac{1}{2}$ and the constant in (J16b) is 0; and this is correct because in this case M is just the ordinary maximum of Brownian bridge (Section J13) whose distribution is given by (J13h).

To argue (J16b), let m_1, m_2 be the medians of the marginal distribution of μ. Under smoothness assumptions on μ, L is a curve as shown in the diagram. For $t_1 > m_1$ define $L_2(t_1)$ by: $(t_1, L_2(t_1)) \in L$. Define $L_1(t_2)$ similarly.

Fix b large. In the notation of Section J12,

$$\boldsymbol{P}(M > b) = \iint \frac{p_b(t_1,t_2)}{EC_b(t_1,t_2)}\, dt_1\, dt_2. \qquad \text{(J16c)}$$

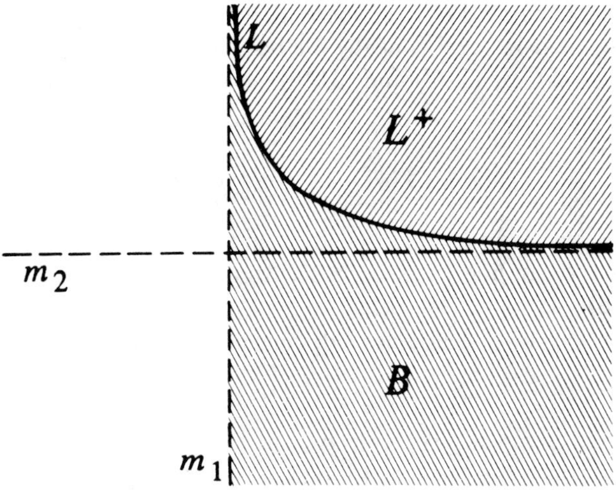

FIGURE J16a.

For convenience we drop subscripts b. Since $p_b(t) = \boldsymbol{P}(X(t) > b)$ and $X(t)$ has variance $\mu(A_t)(1 - \mu(A_t))$, the integrand becomes small as t moves away from L, and we may approximate $EC(t_1, t_2)$ by $EC(t_1, L_2(t_1))$. Fix $t_1 > m_1$ and let $\widehat{t_1} = (t_1, L_2(t_1)) \in L$. Then

$$EX(\widehat{t_1})X(\widehat{t_1} + s) = \frac{1}{4} - \frac{1}{2}F_1(\widehat{t_1})|s_1| - \frac{1}{2}F_2(\widehat{t_1})|s_2| + O(|s|) \quad \text{as } |s| \to 0, \tag{J16d}$$

where $F_i(t) = \frac{\partial}{\partial t_i}\mu(A_t)$. Around $\widehat{t_1}$ the random field $X(t)$ behaves like the stationary random field with covariance of the form (J16d). So as in Sections J9,J10 the mean clump size is the product

$$EC(\widehat{t_1}) = EC_1(\widehat{t_1})EC_2(\widehat{t_1}) \tag{J16e}$$

of mean clump sizes for the marginal processes $X_i(u)$, $u \in \boldsymbol{R}$, which are (locally) stationary Gaussian with covariances

$$EX_i(u)X_i(u + s) = \frac{1}{4} - \frac{1}{2}F_i(\widehat{t_1})|s| \quad \text{as } |s| \to 0. \tag{J16f}$$

We can now rewrite (J16c) as

$$\boldsymbol{P}(M > b) = \int \frac{1}{EC_1(\widehat{t_1})} \left(\int \frac{p(t_1, t_2)}{EC_2(t_1)}\, dt_2 \right) dt_1. \tag{J16g}$$

But for each fixed $t_1 > m_1$ the inner integral is the integral evaluated at (J13g,J13h), the process $t_2 \to X(t_1, t_2)$ being of the form considered in Section J13. Thus the inner integral is approximately $\exp(-2b^2)$. For $t_1 < m_1$ the inner integral is negligible because the line of integration does not meet L. Next, by (J16f) and (J13b) we see $EC_1(\widehat{t}_1) \approx (8F_1(\widehat{t}_1))^{-1}b^{-2}$. So (J16g) becomes

$$\boldsymbol{P}(M > b) \approx 8b^2 \exp(-2b^2) \int_{m_1}^{\infty} F_1(\widehat{t}_1)\, dt_1. \tag{J16h}$$

But this integral is just $\mu(B)$, where B is the region below L and to the right of the median line $\{\, t : t_1 = m_1 \,\}$. And $\mu(B) = \frac{1}{2} - \mu(L^+)$, giving the result (J16b).

The same argument works in d dimensions, where $A_t = \prod_{i=1}^{d}(-\infty, t_i)$. Here we find

$$\boldsymbol{P}(M > b) \approx K_\mu (b^2)^{d-1} \exp(-2b^2) \tag{J16i}$$

for K_μ defined as follows. Write $F(t) = \mu(A_t)$ for the distribution function, and $F_i(t) = \partial F(t)/\partial t_i$. Let S_0 be the set of $s = (s_1, \ldots, s_{d-1})$ such that $F(s_1, \ldots, s_{d-1}, \infty) > \frac{1}{2}$; for $s \in S_0$ define $\widehat{s} \in \boldsymbol{R}^d$ by $\widehat{s} = (s_1, \ldots, s_{d-1}, s_d)$ where $F(\widehat{s}) = \frac{1}{2}$. Then

$$K_\mu = 8^{d-1} \int_{S_0} F_1(\widehat{s}) F_2(\widehat{s}) \cdots F_{d-1}(\widehat{s})\, ds. \tag{J16j}$$

The argument uses smoothness assumptions; it is not clear whether K_μ simplifies. For a product distribution μ in $d = 3$ dimensions, we get $K_\mu = 8 \log^2(2)$. In $d \geq 3$ it is not clear which μ maximizes K_μ.

J17 Example: Rectangle-indexed sheets. As in the previous example let Z be the μ-Brownian sheet associated with a distribution μ on $\boldsymbol{R}^2$. Instead of considering semi-infinite rectangles $(-\infty, t_1) \times (-\infty, t_2)$ it is rather more natural to consider the family $\mathcal{A} = (A_t; t \in I_0)$ of all finite rectangles $[s_1, s_2] \times [t_1, t_2]$, where $s_1 < s_2$ and $t_1 < t_2$, and where t denotes a 4-tuple (s_1, s_2, t_1, t_2). Let $X(t) = Z(A_t)$ and

$$M = \sup_{A \in \mathcal{A}} Z(A) = \sup_t X(t). \tag{J17a}$$

The argument in Example J16 goes through, under smoothness assumptions on μ, to show

$$\boldsymbol{P}(M > b) \approx K_\mu b^6 \exp(-2b^2) \qquad \text{for large } b, \tag{J17b}$$

where K_μ is defined as follows. Let $F(t)$ be the distribution function of μ and let $F_i = \partial F/\partial t_i$. Let I_1 be the set of 3-tuples (s_1, s_2, t_1) such that for some $t_2 = t_2(s_1, s_2, t_1)$ the rectangle $(s_1, s_2) \times (t_1, t_2)$ has μ-measure equal to $\frac{1}{2}$. Then

$$\begin{aligned} K_\mu \;=\; 2^9 \iiint & (F_1(s_1,t_2) - F_1(s_1,t_1))\,(F_2(s_2,t_1) - F_2(s_1,t_1)) \\ & \times (F_1(s_2,t_2) - F_1(s_2,t_1))\, ds_1\, ds_2\, dt_1 \end{aligned} \tag{J17c}$$

where $t_2 = t_2(s_1, s_2, t_1)$. For general μ this expression does not seem to simplify; for a product measure we find

$$K_\mu = 16(3 - 4\log 2). \tag{J17d}$$

J18 Isotropic Gaussian processes. The explicit results obtained so far rely on the product rule (Section J9) for Gaussian processes whose increments in orthogonal directions are uncorrelated. We now consider a different but natural class, the isotropic processes.

It is convenient to do this in some generality. Let $d \geq 1$ be dimension, let $0 < \alpha \leq 2$ and $0 < a < \infty$ be parameters, and consider a stationary mean-zero Gaussian random field $X(t)$, $t \in \boldsymbol{R}^d$ with covariance of the form

$$R(s) \equiv EX(t)X(t+s) \approx 1 - a|s|^\alpha \quad \text{as } |s| \to 0. \tag{J18a}$$

We shall see later that for such a process the clump rate is

$$\lambda_b = K_{d,\alpha} a^{d/\alpha} b^{2d/\alpha - 1} \phi(b), \qquad \text{for } b \text{ large,} \tag{J18b}$$

where $0 < K_{d,\alpha} < \infty$ depends only on d and α, and as usual this implies the approximation for the maximum M_A of X over a large set $A \subset \boldsymbol{R}^d$:

$$\boldsymbol{P}(M_A \leq b) \approx \exp(-\lambda_b |A|).$$

From (J7d), with $\Lambda_{ij} = 2^{1/2} a 1_{(i=j)}$, and from (D10g), we get

$$K_{d,2} = \pi^{-\frac{1}{2}d}, \quad d \geq 1; \qquad K_{1,1} = 1. \tag{J18c}$$

These are the only values for which $K_{d,\alpha}$ is known explicitly. There are several non-explicit expressions for $K_{d,\alpha}$, all of which involve the following process.

Given d and α, define $Z(t)$, $t \in \boldsymbol{R}^d$, as follows. $Z(0)$ is arbitrary. Given $Z(0) = z_0$, the process Z is (non-stationary) Gaussian with

$$EZ(t) = z_0 - |t|^\alpha \tag{J18d}$$

$$\text{cov}(Z(s), Z(t)) = |t|^\alpha + |s|^\alpha - |t - s|^\alpha. \tag{J18e}$$

To understand this definition, fix b large and define a rescaled version of X:

$$Y^b(t) = b(X(\gamma t) - b); \qquad \gamma = (ab^2)^{-1/\alpha}. \tag{J18f}$$

Note that $Y^b(0)$ bounded as $b \to \infty$ implies $X(0) - b \to 0$ as $b \to \infty$. By computing means and covariances it is not hard to see

$$\text{dist}(Y^b(t), t \in \boldsymbol{R}^d \mid Y^b(0) = y_0) \xrightarrow{\mathcal{D}} \text{dist}(Z(t), t \in \boldsymbol{R}^d \mid Z(0) = y_0) \quad \text{as } b \to \infty. \tag{J18g}$$

Thus Z approximates a rescaled version of X around height b. This generalizes several previous results.

We can now give the most useful expression for $K_{d,\alpha}$. Give $Z(0)$ the exponential(1) distribution and let

$$\begin{aligned} \widetilde{\mathcal{C}} &= \{ t \in \boldsymbol{R}^d : Z(t) \geq 0 \} \\ \widetilde{C} &= \text{volume}(\widetilde{\mathcal{C}}) \\ K_{d,\alpha} &= E\left(\frac{1}{\widetilde{C}}\right). \end{aligned} \tag{J18h}$$

With this definition of K, the heuristic argument for (J18b) is just the harmonic mean estimate of clump size (Section A6) together with (J18g). In detail, write C_b^X and $\widetilde{C}_b^X$ for the ordinary and the conditioned clump sizes of X above b. Then

$$\begin{aligned} \lambda_b &= \frac{\boldsymbol{P}(X(0) > b)}{EC_b^X} && \text{by the fundamental identity} \\ &= \boldsymbol{P}(X(0) > b) E\left(\frac{1}{\widetilde{C}_b^X}\right) && \text{by the harmonic mean formula} \\ &\approx b^{-1}\phi(b) E\left(\frac{1}{\widetilde{C}_b^X}\right) && \text{by the Normal tail estimate} \\ &\approx b^{-1}\phi(b)(ab^2)^{d/\alpha} E\left(\frac{1}{\widetilde{C}_0^{Y^b}}\right) && \text{by the scaling (J18f),} \end{aligned}$$

where $\widetilde{C}_0^{Y^b}$ is the size of clump of $\{ t : Y^b(t) > 0 \}$ given $Y^b(0) > 0$. Now $\text{dist}(Y^b(0) \mid Y^b(0) > 0) = \text{dist}(b(X(0) - b) \mid X(0) > b) \stackrel{\mathcal{D}}{\to} \text{exponential}(1) \stackrel{\mathcal{D}}{=} \text{dist}(Z(0))$, and so from (J18g)

$$\widetilde{C}_0^{Y^b} \stackrel{\mathcal{D}}{\to} \widetilde{C} = \widetilde{C}_0^Z,$$

completing the heuristic argument for (J18b).

Alternative expression for $K_{d,\alpha}$ are given at Section J37. As mentioned before, exact values are not known except in cases (J18c), so let us consider bounds.

J19 Slepian's inequality. Formalizations of our heuristic approximations as limit theorems lean heavily on the following result (see Leadbetter et al. (1983) §4.2,7.4).

Lemma J19.1 *Let X, Y be Gaussian processes with mean zero and variance one. Suppose there exists $\delta > 0$ such that $EX(t)X(s) \leq EY(t)Y(s)$ for all $|t-s| \leq \delta$. Then*

$$\boldsymbol{P}(\sup_{t \in A} X(t) \geq b) \geq \boldsymbol{P}(\sup_{t \in A} Y(t) \geq b) \quad \textit{for all } b, \textit{ and } A \textit{ with } \text{diam}(A) \leq \delta.$$

Note in particular this formalizes the idea that, for a stationary Gaussian process, the asymptotic behavior of extrema depends only on the behavior of the covariance function at 0. Note also that in our cruder language, the conclusion of the lemma implies

$$\lambda_b^X \geq \lambda_b^Y.$$

As an application, fix d and α, let X be the isotropic field (J18a) with $a = 1$, and let $Y_a(t)$, $t \in \boldsymbol{R}^d$, be the field with covariances

$$EY_a(t)Y_a(t+s) \approx 1 - a\left(\sum_{i=1}^{d} |s_i|^\alpha\right).$$

Assuming the product rule (Section J9) works for general α (we know it for $\alpha = 1$ and 2), applying (J18b) to the marginal processes gives the clump rate for Y_a:

$$\lambda_b^Y = K_{1,\alpha}^d a^{d/\alpha} b^{2d/\alpha - 1} \phi(b). \tag{J19a}$$

But it is clear that

$$EY_1(0)Y_1(t) \leq EX(0)X(t) \leq EY_a(0)Y_a(t) \qquad \text{for small } t\text{; where } a = d^{-\frac{1}{2}}.$$

Then Slepian's inequality, together with (J19a) and (J18b), gives

$$d^{-\frac{1}{2}d/\alpha} K_{1,\alpha}^d \leq K_{d,\alpha} \leq K_{1,\alpha}^d. \tag{J19b}$$

In particular,

$$d^{-\frac{1}{2}d} \leq K_{d,1} \leq 1. \tag{J19c}$$

J20 Bounds from the harmonic mean. We ought to be able to get better bounds from the explicit expression (J18h) for $K_{d,\alpha}$. For $\widetilde{C}$ defined there, we have

$$K_{d,\alpha} = E\left(\frac{1}{\widetilde{C}}\right) \geq \frac{1}{E\widetilde{C}}. \tag{J20a}$$

We shall show

$$E\widetilde{C} = \pi^{\frac{1}{2}d} \frac{(2d/\alpha)!}{(d/\alpha)!(\frac{1}{2}d)!}. \tag{J20b}$$

These give a lower bound for $K_{d,\alpha}$. For $\alpha = 1$ this is rather worse than the lower bound of (J19c), but it has the advantage of being explicit for all α.

Fix d and α, take $Z(t)$ as in Section J18 with $Z(0) \stackrel{\mathcal{D}}{=}$ exponential(1) and let $Z_1(t)$ be the 1-parameter marginal process. Then

$$E\widetilde{C} = \int_{\boldsymbol{R}^d} \boldsymbol{P}(Z(t) \geq 0)\, dt$$

But the integrand is a function of $|t|$ only, so

$$
\begin{aligned}
E\widetilde{C} &= a_d \int_0^\infty t^{d-1} \boldsymbol{P}(Z_1(t) \geq 0)\, dt; \qquad a_d = \frac{2\pi^{\frac{1}{2}d}}{(\frac{1}{2}d-1)!}. \\
&= a_d \int_0^\infty \int_0^\infty t^{d-1} e^{-z} \boldsymbol{P}(Z_1(t) - Z_1(0) \geq -z)\, dt\, dz \\
&= a_d \int_0^\infty \int_0^\infty t^{d-1} e^{-z} \overline{\Phi}((2t^\alpha)^{-\frac{1}{2}}(t^\alpha - z))\, dt\, dz \\
&= \alpha^{-1} a_d \int_0^\infty \int_0^\infty 2s^{-1} s^{2d/\alpha} e^{-z} \overline{\Phi}(2^{-\frac{1}{2}} s^{-1}(s^2 - z))\, ds\, dz \qquad (s = t^{\frac{1}{2}\alpha}) \\
&= \alpha^{-1} a_d I\left(\frac{2d}{\alpha}\right), \quad \text{say.} \qquad \text{(J20c)}
\end{aligned}
$$

We can avoid evaluating the integral by appealing to what we know in the case $\alpha = 2$. For $X(t)$ with covariance $EX(0)X(t) \approx 1 - |t|^2$ as $|t| \to 0$, (J7h) gives $EC_b = \pi^{d/2} b^{-d}$. So in this case, scaling shows the clump sizes C, $\widetilde{C}$ for Y^b (which as in Section J18 approximate the clump sizes for Z) satisfy $EC = \pi^{d/2}$. So

$$
\begin{aligned}
E\widetilde{C} &= \frac{EC^2}{EC} \qquad \text{by (A15b)} \\
&= \pi^{\frac{1}{2}d} \frac{EC^2}{(EC)^2} \\
&= \pi^{\frac{1}{2}d} \binom{d}{d/2} \qquad \text{by (J7j),} \qquad \text{(J20d)}
\end{aligned}
$$

this ratio being unaffected by the scaling which takes X to Y^b. Comparing (J20c) with (J20d) we conclude

$$\frac{1}{2} a_n I(n) = \pi^{\frac{1}{2}n} \binom{n}{n/2}; \qquad n \geq 1 \text{ integer.}$$

Substituting the formula for a_n,

$$I(n) = (\frac{1}{2}n - 1)! \binom{n}{n/2}. \qquad \text{(J20e)}$$

Assuming this holds for non-integer values n, (J20c) gives (J20b).

J21 Example: Hemispherical caps. Here is a simple example where locally isotropic fields arise in the Kolmogorov-Smirnov setting. Let S be the 2-sphere, that is the surface of the 3-dimensional ball of unit radius. Let μ be the uniform distribution on S. Let $Z(A)$ be the μ-Brownian sheet. Let $\mathcal{A} = \{ A_t : t \in S \}$ be the set of hemispherical caps, indexed by their "pole" t, and consider $M = \sup_{\mathcal{A}} Z(A_t)$. Here $\mu(A_t) \equiv \frac{1}{2}$, and

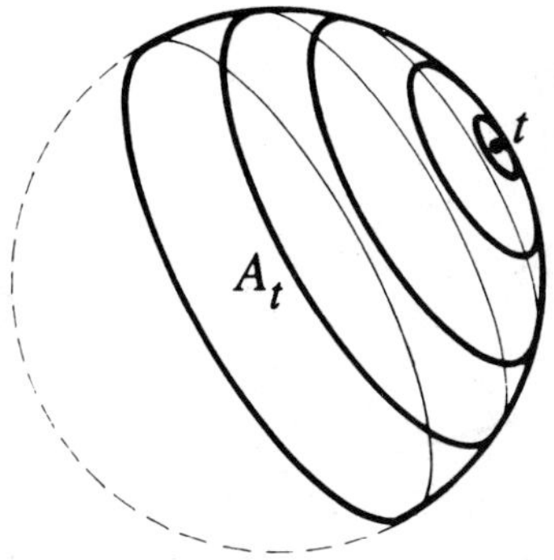

FIGURE J21a.

$$\begin{aligned} EZ(A_t)Z(A_s) &= \frac{1}{4} - \frac{1}{2}\mu(A_t \Delta A_s) \\ &\approx \frac{1}{4} - \frac{1}{2}\pi^{-1}|t-s| \qquad \text{for } |t-s| \text{ small,} \end{aligned}$$

by an easy calculation. Now the scaled form of (J18b) says: if $X(t)$, $t \in \boldsymbol{R}^2$, is stationary with $EX(t)X(s) \sim \frac{1}{4} - a|t-s|$ as $|t-s| \to 0$, then the clump rate is $\lambda_b = 128K_{2,1}a^2b^3\exp(-2b^2)$. In this example we have $a = (2\pi)^{-1}$, giving

$$\begin{aligned} \boldsymbol{P}(M > b) &\approx \lambda_b \,\text{area}(S) && \text{(J21a)} \\ &\approx 128\pi^{-1}K_{2,1}b^3\exp(-2b^2). && \text{(J21b)} \end{aligned}$$

The reader may like to work through the similar cases where the caps have some other fixed size, or have arbitrary size.

J22 Example: Half-plane indexed sheets. For our last example of Kolmogorov-Smirnov type, let μ be a distribution on $\boldsymbol{R}^2$ which is rotationally invariant, so its density is of the form $f(x_1, x_2) = g(r)$, $r^2 = x_1^2 + x_2^2$. Let $Z(A)$ be the μ-Brownian sheet, let $\mathcal{A}$ be the family of all half-spaces and let $M = \sup_{\mathcal{A}} Z(A)$. We shall argue

$$\boldsymbol{P}(M > b) \sim K_\mu b^2 e^{-2b^2} \quad \text{as } b \to \infty, \qquad \text{(J22a)}$$

where K_μ has upper bound 16 and a lower bound depending on μ given at (J22f).

A directed line in the plane can be parametrized as (d, θ), $-\infty < d < \infty$, $0 \le \theta \le 2\pi$, and a half-space $A_{d,\theta}$ can be associated with each directed line, as in the diagram.

Write $X(d,\theta) = Z(A_{d,\theta})$. Then X is stationary in θ and Brownian-bridge like in d. For $d \approx 0$ the process X is approximately stationary. Using the

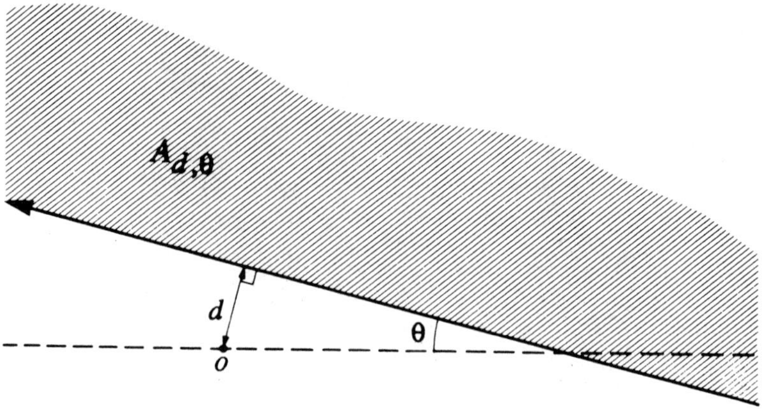

FIGURE J22a.

fact $\text{cov}(Z(A), Z(B)) = \frac{1}{4} - \frac{1}{2}\mu(A \Delta B)$ when $\mu(A) = \frac{1}{2}$, we can calculate the covariance near $d = 0$:

$$EX(0, \theta_0)X(d, \theta_0 + \theta) \approx \frac{1}{4} - \frac{1}{2}\int_{-\infty}^{\infty} |\theta r - d| g(r)\, dr; \qquad \theta,\ d \text{ small.} \quad \text{(J22b)}$$

Write γ for the marginal density at 0:

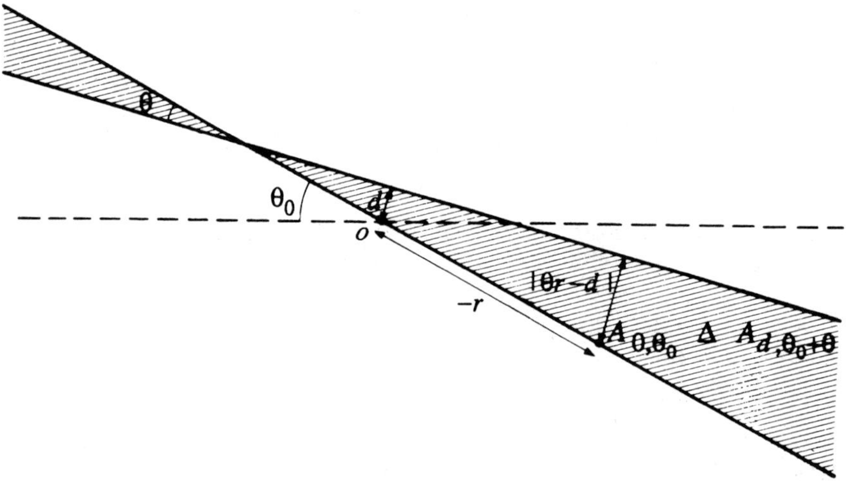

FIGURE J22b.

$$\gamma = \int_{-\infty}^{\infty} f(0, x_2)\, dx_2 = \int_{-\infty}^{\infty} g(r)\, dr.$$

Suppose that the covariance in (J22b) worked out to have the form

$$\frac{1}{4} - a_1|\theta| - a_2|d|; \qquad \theta,\ d \text{ small.} \tag{J22c}$$

Then we would be in the setting of Example J14, with $\alpha = \gamma^2$, and (J14c) would give

$$\boldsymbol{P}(M > b) \sim 32 a_1 a_2 \gamma^{-1} (2\pi) b^2 \exp(-2b^2). \tag{J22d}$$

Though the covariance (J22b) is not of form (J22c), we can upper and lower bound it in this form, and then Slepian's inequality (Section J19) justifies bounding by the corresponding quantities (J22d).

The upper bound is easy. By rotational invariance

$$\int_{-\infty}^{\infty} |x| g(x)\, dx = \frac{1}{\pi},$$

and so

$$\int_{-\infty}^{\infty} |\theta r - d| g(r)\, dr \le \pi^{-1}|\theta| + \gamma|d|.$$

Appealing to (J22d), we get

$$\boldsymbol{P}(M > b) \le 16 b^2 \exp(-2b^2);$$

that is, we get the bound $K_\mu \le 16$ for (J22a).

For the lower bound, we seek a_1, a_2 such that

$$\int |\theta r - d| g(r)\, dr \ge a_1|\theta| + a_2|d| \qquad \text{for all } \theta,\ d$$

and subject to this constraint we wish to maximize $a_1 a_2$. This is routine calculus: put

$$\psi(c) = \int_{-\infty}^{c} g(x)\, dx - \int_{c}^{\infty} g(x)\, dx$$

and define $c^* > 0$ by

$$2c^* \psi(c^*) = \int_{-\infty}^{\infty} |x - c^*| g(x)\, dx. \tag{J22e}$$

Then the maximum product works out to be $c^* \psi^2(c^*)$. Plugging into (J22d) gives the lower bound

$$K_\mu \ge 16\pi\gamma^{-1} c^* \psi^2(c^*). \tag{J22f}$$

In the case where μ is uniform on a centered disc, the lower bound is 4.

J23 The power formula. The arguments used in the examples may be repeated to give the following heuristic rule. Let μ be distribution which is "essentially d_1-dimensional". Let Z be the μ-Brownian sheet. Let $\mathcal{A} = \{ A_t : t \in \boldsymbol{R}^{d_2} \}$ be a family of subsets with d_2-dimensional parametrization. Let d_3 be the dimension of $\{ t : \mu(A_t) = \frac{1}{2} \}$. Then

$$\boldsymbol{P}(\sup Z(A) > b) \sim K_\mu b^\alpha \exp(-2b^2) \quad \text{as } b \to \infty, \tag{J23a}$$

where α is given by *the power formula*

$$\alpha = d_1 + d_3 - 1. \tag{J23b}$$

Note that usually we have $d_3 = d_2 - 1$, except where (as in Example J21) the sets A_t are especially designed to have $\mu(A_t) \equiv \frac{1}{2}$.

J24 Self-normalized Gaussian fields. The Kolmogorov-Smirnov type examples involved Gaussian fields Z such that $\operatorname{var} Z(A) = \mu(A)(1 - \mu(A))$. We now turn to related examples involving fields which are self-normalized to have variance 1, as in the basic example (Example J10) of product Ornstein-Uhlenbeck process. For these examples a slightly different form of the product-rule heuristic is useful. For $t = (t_1, t_2)$ let $Y(t)$ be a stationary Gaussian field with $EY(t) = -b$ and such that, given $Y(0,0) = y \approx 0$,

$$Y(t) \overset{\mathcal{D}}{\approx} y + \sigma B_1(t_1) + \sigma B_2(t_2) - a|t_1| - a|t_2| \qquad \text{for } |t| \text{ small;} \tag{J24a}$$

where B_1, B_2 are independent Brownian motions. For small δ consider the random set $\{t : Y(t) \in [0, \delta]\}$ and write δEC for the mean clump size. Then EC depends only on a and σ. As in Example J10, we can now calculate EC by considering the special case where Y arises as the sum of 2 independent 1-parameter processes, and we find

$$EC = \frac{1}{2}\sigma^2 a^{-3}. \tag{J24b}$$

J25 Example: Self-normalized Brownian motion increments. Let B_t be standard Brownian motion. For large T consider

$$M_T = \sup_{\substack{0 \leq s < t \leq T \\ t - s > 1}} \frac{B_t - B_s}{(t-s)^{\frac{1}{2}}}.$$

We shall argue

$$\boldsymbol{P}(M_T \leq b) \approx \exp(-\frac{1}{4} b^3 \phi(b) T). \tag{J25a}$$

Write $X(s,t) = (B_t - B_2)/(t-s)^{1/2}$. Fix b, and apply the heuristic to the random set $\{ (s,t) : X(s,t) \in [b, b+\delta] \}$. The heuristic says

$$\begin{aligned} \boldsymbol{P}(M_T \leq b) &\approx \exp\left(-\iint \lambda_b(s,t)\, ds\, dt\right) \\ &\approx \exp\left(-\iint \frac{p_b(s,t)}{EC(s,t)}\, ds\, dt\right) \\ &\approx \exp\left(-\phi(b) \iint \frac{1}{EC(s,t)}\, ds\, dt\right) \qquad \text{(J25b)} \end{aligned}$$

where the integral is over $\{0 \leq s < t \leq T, t-s \geq 1\}$. To estimate the clump sizes, fix t_1^*, t_2^* and condition on $X(t_1^*, t_2^*) = b + y$ for small y. Then $B(t_2^*) - B(t_1^*) = b(t_2^* - t_1^*)^{1/2} + y'$ say. Conditionally, $B(t)$ has drift $b/(t_2^* - t_1^*)^{1/2}$ on $[t_1^*, t_2^*]$ and drift 0 on $[0, t_1^*]$ and on $[t_2^*, T]$. We can now compute the conditional drift of $X(t_1^* + t_1, t_2^* + t_2)$ for small, positive and negative, t_1 and t_2.

$$\left.\frac{d}{dt_2} EX(t_1^*, t_2^* + t_2)\right|_{0+} = 0 + \left.\frac{d}{dt_2} \frac{b(t_2^*-t_1^*)^{\frac{1}{2}}}{(t_2^*+t_2-t_1^*)^{\frac{1}{2}}}\right|_0 = -\frac{\frac{1}{2}b}{t_2^*-t_1^*}$$

$$\left.\frac{d}{dt_2} EX(t_1^*, t_2^* + t_2)\right|_{0-} = \frac{b/(t_2^*-t_1^*)^{\frac{1}{2}}}{(t_2^*-t_1^*)^{\frac{1}{2}}} + \left.\frac{d}{dt_2} \frac{b(t_2^*-t_1^*)^{\frac{1}{2}}}{(t_2^*+t_2-t_1^*)^{\frac{1}{2}}}\right|_0 = +\frac{\frac{1}{2}b}{t_2^*-t_1^*}$$

and similarly for d/dt_1. We conclude that the process

$$Y(t_1, t_2) = X(t_1^* + t_1, t_2^* + t_2) - b$$

is of the form (J24a) with $a = \frac{1}{2}b(t_2^* - t_1^*)^{-1}$ and $\sigma = (t_2^* - t_1^*)^{-1/2}$. So (J24b) gives the clump size

$$EC(t_1^*, t_2^*) = 4b^{-3}(t_2^* - t_1^*)^2. \qquad \text{(J25c)}$$

Thus the integral in (J25b) becomes

$$\begin{aligned} \frac{1}{4}b^3 \iint (t-s)^{-2}\, ds\, dt &= \frac{1}{4}b^3 \int_1^T (T-u)u^{-2}\, du \\ &\sim \frac{1}{4}b^3 T \qquad \text{for large } T, \end{aligned}$$

yielding the result in (J25a).

J26 Example: Self-normalized Brownian bridge increments. Let B^0 be standard Brownian bridge. For small δ consider

$$M_\delta = \sup_{\substack{0 \leq s < t \leq 1 \\ \delta \leq t-s \leq 1-\delta}} \frac{B^0(t) - B^0(s)}{g(t-s)}$$

where $g(u) = s.d.(B^0(s+u) - B^0(s)) = \sqrt{u(1-u)}$. We shall argue

$$P(M_\delta \le b) \approx \exp(-\frac{1}{4}b^3\phi(b)\delta^{-1}). \qquad \text{(J26a)}$$

At the heuristic level, this is very similar to the previous example. However, considering the usual representation of B^0 in terms of B, there seems no exact relationship between the M's in the two examples. We write $X(s,t) = (B^0(t) - B^0(s))/g(t-s)$ and apply the heuristic to $\{ (s,t) : X(s,t) \in [b, b+\eta] \}$. As in the previous example,

$$P(M_\delta \le b) \approx \exp\left(-\phi(b) \iint \frac{1}{EC(s,t)}\, ds\, dt\right). \qquad \text{(J26b)}$$

Condition on $X(t_1^*, t_2^*) = b + y$ for small y. We can then calculate

$$\left.\frac{d}{dt_2} EX(t_1^*, t_2^* + t_2)\right|_{0+} = -\frac{\frac{1}{2}b}{g^2(t_2^* - t_1^*)}$$
$$\left.\frac{d}{dt_2} EX(t_1^*, t_2^* + t_2)\right|_{0-} = \frac{\frac{1}{2}b}{g^2(t_2^* - t_1^*)}$$

and similarly for d/dt_1. Then the process

$$Y(t_1, t_2) = X(t_1^* + t_1, t_2^* + t_2) - b$$

is of the form (J24a) with $a = \frac{1}{2}b/g^2(t_2^* - t_1^*)$ and $\sigma = 1/g(t_2^* - t_1^*)$. so (J24b) gives the clump size

$$EC(t_1^*, t_2^*) = 4b^{-3}g^4(t_2^* - t_1^*). \qquad \text{(J26c)}$$

So the integral in (J26b) becomes

$$\frac{1}{4}b^3 \iint \frac{1}{g^4(t-s)}\, ds\, dt = \frac{1}{4}b^3 \int_\delta^1 \frac{1-u}{(u(1-u))^2}\, du$$
$$\sim \frac{1}{4}b^3\delta^{-1} \quad \text{as } \delta \to 0.$$

yielding the result (J26a).

J27 Example: Upturns in Brownian bridge with drift. Let $X(t) = B^0(t) - wt$, $0 \le t \le 1$, where B^0 is standard Brownian bridge. Equivalently, X is Brownian motion conditioned on $X(1) = -w$. Consider $M = \sup_{0 \le s \le t < 1}(X(t) - X(s))$. We shall argue

$$P(M > b) \approx 2(w + 2b)(w + b)\exp(-2b(w + b)); \qquad \text{(J27a)}$$

provided the right side is small and decreasing in $b' > b$. The case $w = 0$ was given in Example J15.

Fix b and w. Consider the random set $\{ (t_1,t_2) : X(t_2) - X(t_1) \in [b, b+\delta] \}$. The heuristic is

$$\begin{aligned} \boldsymbol{P}(M > b) &\approx \iint_{0 \leq t_1 \leq t_2 \leq 1} \lambda(t_1,t_2)\, dt_1\, dt_2 \\ &\approx \iint \frac{p(t_1,t_2)}{EC(t_1,t_2)}\, dt_1\, dt_2. \end{aligned} \tag{J27b}$$

Here

$$\begin{aligned} p(t_1,t_2) &= \delta^{-1}\boldsymbol{P}(X(t_2) - X(t_1) \in [b, b+\delta]) \\ &= (s(1-s))^{-\frac{1}{2}}\phi((ws+b)(s(1-s))^{-\frac{1}{2}}); \qquad s = t_2 - t_1. \end{aligned}$$

Let $u = b/(w+2b)$. Then $p(t_1,t_2)$ is maximized when $s \equiv t_2 - t_1 = u$. Thus we consider clump size $EC(t_1^*, t_2^*)$ for $t_2^* - t_1^* = u$. Given there is a clump near t^* we have $X(t_2^*) - X(t_1^*) = b + y_0$ for y_0 small. Let

$$Y(t_1,t_2) = X(t_2^* + t_2) - X(t_1^* + t_1) - b.$$

Then for small t, Y behaves as at (J24a) for $a = w + 2b$ and $\sigma = 1$. Thus (J24a) gives the clump size $EC(t_1^*, t_2^*) = \frac{1}{2}(w+2b)^{-3}$.

We can now evaluate (J27b), using the fact that the integral is dominated by contributions from $t_2 - t_1 \approx u$.

$$\begin{aligned} \boldsymbol{P}(M > b) &\approx 2(w+2b)^3(u(1-u))^{-\frac{1}{2}} \iint \phi(f(t_2 - t_1))\, dt_1\, dt_2 \\ &\approx 2(w+2b)^3(u(1-u))^{-\frac{1}{2}}(1-u) \int_0^1 \phi(f(s))\, ds \end{aligned}$$

where $f(s) = (ws+b)(s(1-s))^{-1/2}$. Evaluating the integral using (C21e), the result simplifies to (J27a).

J28 Example: 2-parameter LIL. The 2-parameter law of the iterated logarithm can be derived in the same way as the classical 1-parameter case treated in (D15). First consider the stationary Gaussian field $X(t_1,t_2)$ with covariance of the form

$$R(t_1,t_2) = \exp(-|t_1| - |t_2|). \tag{J28a}$$

Let $b(\underset{\sim}{t})$ be such that $b(\underset{\sim}{t}) \to \infty$ slowly as $\underset{\sim}{t} \to \infty$, that is as $\min(t_1,t_2) \to \infty$. We apply the heuristic to the random set $\{ \underset{\sim}{t} : X(\underset{\sim}{t}) \in [b(\underset{\sim}{t}), b(\underset{\sim}{t}) + \delta] \}$. Around a fixed point $\underset{\sim}{t}^*$ we approximate the sloping boundary $b(\underset{\sim}{t})$ by the level boundary $b(\underset{\sim}{t}^*)$, and then (J10i) gives the clump rate $\lambda(\underset{\sim}{t}^*) = b^3(\underset{\sim}{t}^*)\phi(b(\underset{\sim}{t}^*))$. So the heuristic gives

$$\boldsymbol{P}(X(\underset{\sim}{t}) \leq b(\underset{\sim}{t}) \text{ for all } \underset{\sim}{t} \in [s_0, s_1]^2) \approx \exp\left(-\int_{s_0}^{s_1}\int_{s_0}^{s_1} b^3(\underset{\sim}{t})\phi(b(\underset{\sim}{t}))\, d\underset{\sim}{t} \right). \tag{J28b}$$

This translates to the integral test

$$\boldsymbol{P}(\limsup_{\underset{\sim}{t}\to\infty}(X(\underset{\sim}{t})-b(\underset{\sim}{t}))\le 0)=\begin{cases}1 & \text{if } \int_0^\infty\int_0^\infty b^3(\underset{\sim}{t})\phi(b(\underset{\sim}{t}))\,d\underset{\sim}{t}<\infty\\ 0 & \text{if } \int_0^\infty\int_0^\infty b^3(\underset{\sim}{t})\phi(b(\underset{\sim}{t}))\,d\underset{\sim}{t}=\infty\end{cases}.$$

If $b(\underset{\sim}{t})$ is of the form $\widehat{b}(\|\underset{\sim}{t}\|)$ where $\|\underset{\sim}{t}\|$ is the Euclidean norm, then changing to polar coordinates the integral becomes

$$\int_0^\infty \widehat{b}^3(r)\phi(\widehat{b}(r))r\,dr<\infty. \tag{J28c}$$

Putting $\widehat{b}(r)=(c\log r)^{1/2}$, the critical case is $c=4$, and so in particular we have the crude result

$$\limsup\frac{X(t)}{(4\log\|t\|)^{\frac{1}{2}}}=1 \quad \text{a.s.} \tag{J28d}$$

Now let $Y(\underset{\sim}{t})$ be 2-parameter Brownian motion, in other words $Y(t_1,t_2)=W([0,t_1]\times[0,t_2])$ for white noise W. Then as in the 1-parameter case, $X(t_1,t_2)=e^{-(t_1+t_2)}Y(e^{2t_1},e^{2t_2})$ is of form (J28a), and then (J28d) gives

$$\limsup_{\underset{\sim}{t}\to\infty}\frac{Y(\underset{\sim}{t})}{(4t_1t_2\log\log\|\underset{\sim}{t}\|)^{\frac{1}{2}}}=1 \quad \text{a.s.} \tag{J28e}$$

COMMENTARY

J29 General references. The best, and complementary, books on random fields are by Adler (1981), who gives a careful theoretical treatment, and by Vanmarcke (1982),who gives a more informal treatment with many interesting engineering applications.

We have looked only at a rather narrow topic, approximations for maxima M_A when volume(A) gets large, and tail approximations for fixed M_A. In one direction, the study of Gaussian processes with "infinite-dimensional" parameter sets, and the related study of convergence of normalized empirical distributions, has attracted much theoretical attention: see e.g. Gaenssler (1983). In an opposite direction, non-asymptotic upper and lower bounds for tails $\boldsymbol{P}(M_A>x)$ for fixed A have been studied for 2-parameter Brownian motion and other special fields: see Abrahams (1984b) for a survey and bibliography. There is a huge literature on maxima of Gaussian fields, much of it from the Russian school. See Math. Reviews section 60G15 for recent work.

J30 Comments on the examples.

J30.1 Convergence of empirical distributions (Section J4). See Gaenssler (1983) p. 132 for technical conditions on (A_t) under which the convergence (J4d) holds.

J30.2 Smooth Gaussian fields. See Adler (1981) Chapter 6 for a detailed treatment.

J30.3 Shot noise. See e.g. Orsinger and Battaglia (1982) for results and references.

J30.4 Range of Brownian bridge. The M^* in Example J15 has exact distribution

$$\boldsymbol{P}(M^* > x) = \sum_{n=1}^{\infty}(4n^2x^2 - 1)\exp(-2n^2x^2).$$

J30.5 Brownian bridge increments. Examples J26 and J27 are discussed in Siegmund (1988) and Hogan and Siegmund (1986). These papers, and Siegmund (1986), give interesting statistical motivations and more exact second-order asymptotics for this type of question.

J30.6 d-parameter LIL. Sen and Wichura (1984b; 1984a) give results and references.

J31 Rice's formula. It is natural to hope that smooth random fields can be studied via some d-parameter analogue of Rice's formula (C12.1). This leads to some non-trivial integral geometry — see Adler (1981) Chapter 4.

J32 The product rule. This rule (Section J9) is fundamental in the subsequent examples, and deserves more theoretical attention than it has received. The technique in Example J10 of noting that fields with uncorrelated orthogonal increments can be approximated as sums of 1-parameter processes has been noted by several authors: Abrahams and Montgomery (1985), Siegmund (1988). Whether there is any analogue for non-Gaussian fields is an interesting question — nothing seems known.

Adler (1981) p. 164 discusses formal limit theorems in the context of Example J10, locally product Ornstein-Uhlenbeck processes.

J33 Generalized Kolmogorov-Smirnov: suprema of μ-Brownian sheets. Example J16 gave the most direct 2-dimensional analog of the classical 1-dimensional Kolmogorov-Smirnov statistic. From the viewpoint of the heuristic, deriving our formulas for the tail behavior of the distribution is not really hard. However, theoretical progress in this direction has been painfully slow — in only the simplest case (product measure μ) of this basic Example J16 has the leading constant ($4 \log 2$, here) been determined in the literature: see Hogan and Siegmund (1986). Similar remarks apply to the power formula (Section J23). If the subject had been developed by physicists, I'm sure they would have started by writing down this formula, but it is hard to extract from the mathematical literature. Weber (1980) gives bounds applicable to a wide class of set-indexed Gaussian processes, but when specialized to our setting they do not give the correct power. Presumably a differential geometric approach will ultimately lead to general rigorous theorems — the techniques based on metric entropy in e.g. Adler and Samorodnitsky (1987) seem to be an inefficient way of exploiting finite-dimensional structure.

Adler and Brown (1986), Adler and Samorodnitsky (1987) give accounts of the current state of rigorous theory.

From the practical statistical viewpoint, one must worry about the difference between the true empirical distribution and the Gaussian field approximation; no useful theory seems known here. An interesting simulation study of the power of these tests is given in Pyke and Wilbour (1988).

J34 General formalizations of asymptotics. In Section D38 we discussed Berman's formalization of the heuristic for sojourn distributions, involving approximating the distribution $(X(t) \mid X(0) > b)$ by a limiting process. The same method works in the d-parameter case: see Berman (1986b) who treats Gaussian fields and multiparameter stable processes.

An interesting approach to the Poisson point process description (Section C3) in the multiparameter setting is given by Norberg (1987) using semicontinuous processes.

J35 Bounds for the complete distribution function. For maxima of μ-Brownian sheets, our heuristic can only give tail estimates. For the "standard" 2-parameter bridge, i.e. $\mu =$ product measure, some explicit analytic bounds are known: see Cabana and Wchebor (1982).

Special cases of other Gaussian fields are discussed in Adler (1984), Abrahams (1984a), and Orsingher (1987): see Abrahams (1984b) for more references.

J36 Lower bounds via the second moment method. For arbitrary fields X_t, $t \in \boldsymbol{R}^d$, with finite second moments, and arbitrary $A \subset \boldsymbol{R}^d$, we can obtain lower bounds for the tail of $M_A \equiv \sup_{t \in A} X_t$ in an elementary and rigorous way by using the second moment inequality (Section A15). For any density function f,

$$P(M_A \geq b) \geq \frac{\left(\int_A P(X_t > b) f(t)\, dt\right)^2}{\int_A \int_A P(X_s > b, X_t > b) f(s) f(t)\, ds\, dt}. \qquad \text{(J36a)}$$

Note this is the opposite of our experience with, say, the classical LIL, where the upper bounds are easier than the lower bounds.

As at Section A15, the bound (J36a) is typically "off" by a constant as $b \to \infty$. In the context of, say, trying to prove the power formula (Section J23) for Kolmogorov-Smirnov type statistics, this will not matter. Working through the calculus to compute the bound (J36a) in our examples is a natural thesis project, being undertaken in Schaper (1988).

J37 The isotropic constant. As discussed in Section J18, the asymptotic behavior of isotropic Gaussian fields involves the high-level approximating process $Z(t)$ and a constant $K_{d,\alpha}$. The usual expression for this constant, due to Qualls and Watanabe (1973), and Bickel and Rosenblatt (1973), is

$$K_{d,\alpha} = \lim_{T \to \infty} T^{-d} \int_{-\infty}^{0} P(\sup_{t \in [0,T]^d} Z(t) > 0 \mid Z(0) = z) e^{-z}\, dz. \qquad \text{(J37a)}$$

It is easy to derive this heuristically. We use the notation of Section J18. By (J18f,J18g), for b large

$$P(\sup_{t \in [0,1]^d} X(t) \geq b \mid X(0) = b + \frac{b}{z}) \approx P(\sup_{t \in [0, a^{1/\alpha} b^{2/\alpha}]^d} Z(t) > 0 \mid Z(0) = z). \qquad \text{(J37b)}$$

Since $\phi(b + z/b) \approx \phi(b) e^{-z}$,

$$\begin{aligned}
\lambda_b &= P(\sup_{t \in [0,1]^d} X(t) > b) \\
&\approx \int_{-\infty}^{\infty} P(\sup_{t \in [0,1]^d} X(t) > b \mid X(0) = b + \frac{z}{b}) \phi(b) e^{-z} b^{-1}\, dz \\
&\approx b^{-1} \phi(b) \int_{-\infty}^{\infty} P(\sup_{t \in [0, a^{1/\alpha} b^{1/\alpha}]^d} Z(t) > 0 \mid Z(0) = z) e^{-z}\, dz \qquad \text{by (J37b)} \\
&\approx b^{-1} \phi(b) a^{d/\alpha} b^{2d/\alpha} K_{d,\alpha} \qquad \text{using definition (J37a) of } K.
\end{aligned}$$

This is the same formula as was obtained with definition (J18h) of K. Of course, with either definition one needs to argue that $K \in (0, \infty)$. But the "harmonic mean" formula (J18h) is clearly nicer, as it avoids both the limit and the supremum.

K Brownian Motion: Local Distributions

This opaque title means "distributions related to local sample path properties of Brownian motion". I have in mind properties such as Lévy's estimate of the modulus of continuity, the corresponding results on small increments, the paradoxical fact that Brownian motion has local maxima but not points of increase, and self-intersection properties in d dimensions. Although these are "0–1" results, they can be regarded as consequences of stronger "distributional" assertions which can easily be derived via our heuristic. The topics of this section are more theoretical than were previous topics, though many are equivalent to more practical-looking problems on boundary-crossing.

B_t denotes standard 1-dimensional Brownian motion.

K1 Modulus of continuity. For a function $f(t)$, $0 \leq t \leq T$ let

$$w(\delta, T) = \sup_{\substack{0 \leq t_1 < t_2 \leq T \\ t_2 - t_1 \leq \delta}} |f(t_2) - f(t_1)|.$$

Then f is continuous iff $w(\delta, T) \to 0$ as $\delta \to 0$. Let $W(\delta, T)$ be this (random) modulus of continuity applied to Brownian motion, considered as a random function. Path-continuity of Brownian motion is the result that

$$W(\delta, t) \to 0 \quad \text{a.s.} \quad \text{as } \delta \to 0. \tag{K1a}$$

and it is natural to ask about the rate of convergence. By scaling,

$$W(\delta, T) \stackrel{\mathcal{D}}{=} \delta^{\frac{1}{2}} W(1, T/\delta), \tag{K1b}$$

so there is little difference between studying $W(\delta, 1)$ as $\delta \downarrow 0$ and $W(1, T)$ as $T \to \infty$. The basic result is

Theorem K1.1 (Lévy's theorem)

$$\frac{W(\delta, 1)}{(2\delta \log(1/\delta))^{\frac{1}{2}}} \to 1 \quad a.s. \quad as\ \delta \downarrow 0.$$

We shall present the heuristic arguments for two refinements of this, as follows.

K1.2 (The Chung-Erdos-Sirao Test) *Let* $\psi(\delta) \to \infty$ *as* $\delta \downarrow 0$. *Then* $W(\delta, 1) \leq \delta^{1/2}\psi(\delta)$ *for all sufficiently small* δ, *a.s., iff*

$$\int_{0+} u^{-2}\psi^3(u)\phi(\psi(u))\,du < \infty.$$

K1.3 (The Asymptotic Distribution of W.) *As* $\delta \downarrow 0$,

$$\sup_{x \geq 1} |\boldsymbol{P}(W(\delta, 1) \leq \delta^{\frac{1}{2}} x) - \exp(-x^3\phi(x)\delta^{-1})| \to 0.$$

The relationship between these results is closely analogous to that described in Section D15 between the classical LIL, the associated integral test and the last crossing distribution; such a triple of increasingly refined results is associated with many a.s. properties of Brownian motion.

K2 Example: The Chung-Erdos-Sirao test. Set $X(t_1, t_2) = (B(t_2) - B(t_1))/(t_2 - t_1)^{1/2}$. Take ψ as in (K1.2) above and consider the random set

$$\mathcal{S} = \{\, (t_1, t_2) : X(t_1, t_2) \in [\psi(t_2 - t_1), \psi(t_2 - t_1) + \epsilon]\,\}$$

We are interested in whether $\mathcal{S}$ is bounded away from the diagonal $\Delta = \{\, (t,t) : 0 \leq t \leq 1\,\}$. But we have already considered this process X at Example J24, and calculated the clump rate at level b to be

$$\lambda_b(t_1, t_2) = \frac{1}{4} b^3 (t_2 - t_1)^{-2}\phi(b).$$

Around a fixed point (t_1^*, t_2^*) we can approximate the (slowly) sloping boundary $b(t_1, t_2)$ by the level boundary $b(t_1^*, t_2^*)$, and thus take the clump rate of $\mathcal{S}$ to be

$$\lambda(t_1, t_2) = \frac{1}{4}\psi^3(t_2 - t_1)\,(t_2 - t_1)^{-2}\phi(\psi(t_2 - t_1)).$$

Considering $\mathcal{S}$ as a Poisson clump process with this rate λ, the condition for $\mathcal{S}$ to be bounded away from Δ is

$$\lim_{\eta \downarrow 0} \iint_{\substack{0 \leq t_i \leq 1 \\ t_2 > t_1 + \eta}} \lambda(t_1, t_2)\,dt_1\,dt_2 = 0$$

and this is equivalent to the integral test at (K1.2) above.

K3 Example: The asymptotic distribution of W. The method above does not work for estimating the distribution of $W(\delta, T)$ itself. The difficulty is that we want to study the maximum of a random field X over a region A, and the variance of X is maximal on the boundary of A, so that we cannot safely ignore edge effects as in the usual application of the heuristic: it could be that many clumps in A overlap its edge. Instead we use a different form of the heuristic. We shall argue that, for T large,

$$\boldsymbol{P}(W(1,T) \leq b) \approx \exp(-b^3\phi(b)T) \tag{K3a}$$

where as usual (recall Section C24) only the region where $b^3\phi(b)$ is decreasing is relevant. Scaling gives (K1.3).

Write $\underset{\sim}{t} = (t_1, t_2)$, where $t_1 \leq t_2 \leq t_1 + 1$. Fix b large and consider the random set

$$\mathcal{S} = \{\underset{\sim}{t} : |B_{t_2} - B_{t_1}| \geq b\}.$$

So $p(\underset{\sim}{t}) = \boldsymbol{P}(\underset{\sim}{t} \in \mathcal{S}) = \boldsymbol{P}(|B_{t_2} - B_{t_1}| \geq b) = q(t_2 - t_1)$ say, where

$$q(v) = 2\overline{\Phi}\left(\frac{b}{v^{\frac{1}{2}}}\right). \tag{K3b}$$

For each clump $\mathcal{C}$ of $\mathcal{S}$ there is some "interior length"

$$u = \min_{t \in \mathcal{C}}(t_2 - t_1) \leq t_1.$$

Let $\lambda(u)\,du$ be the (1-dimensional) rate at which clumps occur with interior length $\in [u, u + du]$. Then the heuristic says

$$\boldsymbol{P}(W(1,T) \leq b) \approx \exp(-\lambda T), \quad \text{where } \lambda = \int_0^1 \lambda(u)\,du. \tag{K3c}$$

We shall estimate λ using the "marked clumps" variation (Section A20.3) of the fundamental identity:

$$q(v) = \int_0^v \lambda(u) EC_{u,v}\,du \tag{K3d}$$

where $C_{u,v}$ is the area of $\{\underset{\sim}{t} : t_2 - t_1 \leq v, B_{t_2} - B_{t_1} \geq b\}$ in a clump with interior length u. We shall argue later that

$$EC_{u,v} \approx \frac{1}{4}(v-u)^2. \tag{K3e}$$

Then from (K3b) and (K3d),

$$\int_0^v \frac{1}{2}(v-u)^2\lambda(u)\,du = 4\overline{\Phi}\left(\frac{b}{v^{\frac{1}{2}}}\right)$$

and differentiating twice

$$\begin{aligned} \lambda = \int_0^1 \lambda(u)\,du &= 4\frac{d^2}{dv^2}\overline{\Phi}\left(\frac{b}{v^{\frac{1}{2}}}\right)\Bigg|_{v=1} \\ &\approx b^3\phi(b), \qquad \text{to first order.} \end{aligned}$$

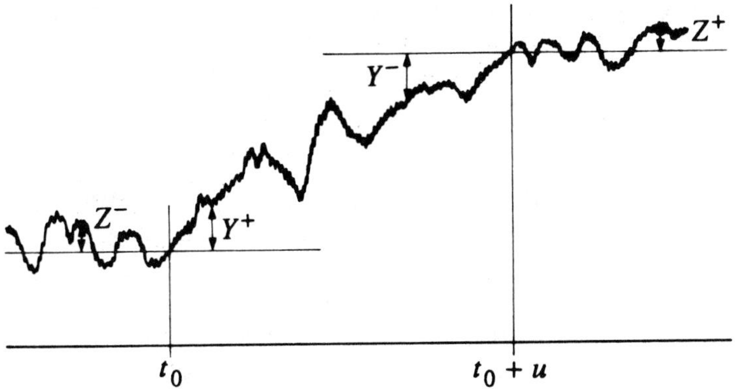

FIGURE K3a.

Thus (K3c) gives (K3a).

To argue (K3e), consider a clump $\mathcal{C}$ with interior length u taken at $(t_0, t_0 + u)$ say, with $B_{t_0+u} - B_{t_0} = b$ say. Consider the qualitative behavior of the incremental processes

$$Y_s^+ = B_{t_0+s} - B_{t_0}; \qquad Z_s^+ = B_{t_0+u+s} - B_{t_0+u}$$

for small $s > 0$. Since u is the interior length, we must have $Y_s^+ \geq 0$ and $Y_s^+ \geq \max_{s'<s} Z_{s'}^+$, and we can regard Y^+, Z^+ as Brownian motions conditioned on these events. The first conditioning makes Y^+ increase rapidly at 0 (as BES(3) — see Section K14), and then the second conditioning will have little effect on Z^+, which will be roughly like standard Brownian motion. Similarly for the left-increments Y^-, Z^-. To estimate $C_{u,v}$, we consider only times of the form $\underset{\sim}{t} = (t_0 - s_1, t_0 + u + s_2)$ in view of the rapid increase of Y^+ and Y^-; then

$$EC_{u,v} \approx \iint_{\substack{s_1, s_2 \geq 0 \\ s_1 + s_2 \leq v-u}} \boldsymbol{P}(Z_{s_1}^+ < Z_{s_2}^-)\, ds_1\, ds_2.$$

But the integrand is $\approx \frac{1}{2}$, since Z^+ and Z^- are roughly like standard Brownian motions, and this gives (K3e).

Remark: Various other results about large increments of Brownian motion can be obtained in essentially the same way — see Section K10. Let us instead look at something slightly different.

K4 Example: Spikes in Brownian motion. For a function $f(t)$ the modulus

$$w^*(\delta, T) = \sup_{\substack{0 \le t_1 < t_2 < t_3 \le T \\ t_3 - t_1 \le \delta}} \min(f(t_2) - f(t_1), f(t_2) - f(t_3))$$

represents the height of the largest "spike" of width $\le \delta$. We shall study the random modulus $W^*(\delta, T)$ for Brownian motion. One would intuitively expect $W^*(\delta, 1) \approx \frac{1}{2} W(\delta, 1)$ by reflection at the top of the spike; but such arguments can be misleading.

Write $\underset{\sim}{t} = (t_1, t_3)$, where $t_1 < t_3 \le t_1 + 1$. Fix b large and consider the random set

$$\mathcal{S} = \{\underset{\sim}{t} : \exists t_2 \in (t_1, t_3) \text{ such that } B_{t_2} - B_{t_1} \ge b \quad \text{and } B_{t_2} - B_{t_3} \ge b\}.$$

Here $p(\underset{\sim}{t}) = \boldsymbol{P}(\underset{\sim}{t} \in \mathcal{S}) = q(t_3 - t_1)$, where

$$\begin{aligned} q(v) &= \boldsymbol{P}(B_v^* \ge b, B_v^* \ge b + B_v) \quad \text{where } B_v^* = \max_{t \le v} B_t \\ &= 2\boldsymbol{P}(B_v^* \ge b, B_v < 0) \quad \text{by a symmetry argument, considering whether } B_v > 0 \text{ or } B_v < 0, \\ &= 2\boldsymbol{P}(B_v > 2b) \quad \text{by the reflection principle} \\ &= 2\overline{\Phi}\left(\frac{2b}{v^{\frac{1}{2}}}\right). \end{aligned} \tag{K4a}$$

Each clump $\mathcal{C}$ of $\mathcal{S}$ has an interior width $u = \min_{t \in \mathcal{C}}(t_3 - t_1)$. As in Example K3 we consider the area $C_{u,v}$ of $\{\underset{\sim}{t} : t_3 - t_1 \le v\} \cap \mathcal{C}$ in a clump $\mathcal{C}$ of interior width u. Consider such a clump. The interior width is taken at $(t_0, t_0 + u)$, say, and so $B_{t_0} = B_{t_0+u}$ and there exists $t_2 \in (t_0, t_0 + u)$ such that $B_{t_2} = \max_{t_0 < t < t_0 + u} B_t = B_{t_0} + b$; also we must have $B_t \ge B_{t_0}$ on $(t_0, t_0 + u)$. The situation here is actually much *simpler* than in Example K3. For a point $\underset{\sim}{t} = (t_1, t_3)$ near $(t_0, t_0 + u)$ can be in $\mathcal{S}$ only if $t_1 \le t_0$ and $t_3 \ge t_0 + u$. Thus the incremental processes

$$Z_s^+ = B_{t_0+u+s} - B_{t_0+u}; \qquad Z_s^- = B_{t_0-s} - B_{t_0}$$

which are *a priori* Brownian motion conditioned so that there is no spike with width $< u$; are in fact almost exactly genuine Brownian motions since the conditioning event is almost certain. So

$$\begin{aligned} EC_{u,v} &\approx \iint_{\substack{s_1, s_2 > 0 \\ s_1 + s_2 \le v - u}} \boldsymbol{P}(Z_{s_1}^+ \le 0, Z_{s_2}^- \le 0)\, ds_1\, ds_2 \\ &\approx \frac{(v-u)^2}{8} \end{aligned} \tag{K4b}$$

since the integrand $\approx \frac{1}{4}$.

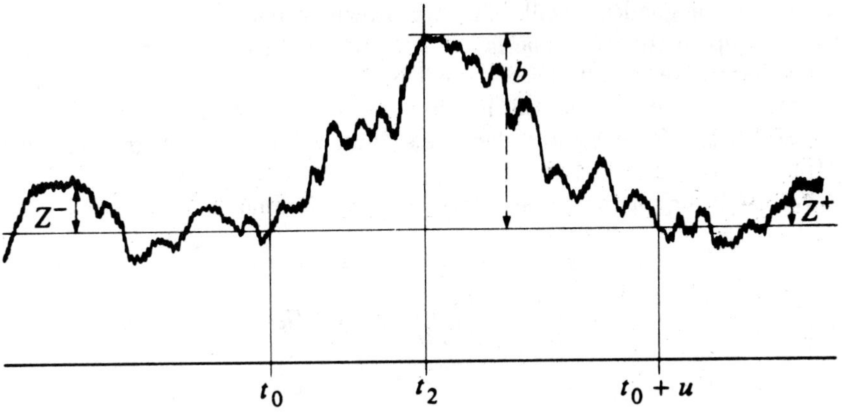

FIGURE K4a.

We now repeat the method of Example K3. The heuristic says

$$P(W^*(1,T) < b) \approx \exp(-\lambda T), \quad \text{where } \lambda = \int_0^1 \lambda(u)\, du$$

and where $\lambda(u)\, du$ is the rate of clumps with interior width $\in [u, u+du]$. The "marked clumps" formula (A20.3)

$$q(v) = \int_0^v \lambda(u) EC_{u,v}\, du$$

together with the estimates (K4a), (K4b), enable us to solve for λ:

$$\lambda = 16b^3\phi(2b) \qquad \text{to first order.}$$

So we obtain the distributional estimate

$$P(W^*(1,T) \le b) \approx \exp(-16b^3\phi(2b)T). \tag{K4c}$$

By scaling we can derive the analogue of Lévy's theorem:

$$\frac{W^*(\delta,1)}{\sqrt{\frac{1}{2}\delta\log(1/\delta)}} \to 1 \quad \text{a.s.} \quad \text{as } \delta \to 0. \tag{K4d}$$

We can also develop the analogue of the Chung-Erdos-Sirao test — exercise! (see Section K11 for the answer).

K5 Example: Small increments. The quantity

$$M(\delta,T) = \inf_{0\le t\le T-\delta}\ \sup_{t\le u\le v\le t+\delta} \frac{1}{2}|B_v - B_u|$$

measures the smallest oscillation of a Brownian path over any interval of length δ up to time T. (The factor $\frac{1}{2}$ is included to facilitate comparison with alternative definitions of "small increments"). As with the modulus of continuity, we can use the heuristic to obtain distributional estimates for $M(1,T)$ as $T \to \infty$, and derive a.s. limit results and integral tests for $M(\delta,1)$ as $\delta \to 0$.

We need to quote some estimates for Brownian motion. Let

$$\begin{aligned} H_b &= \inf\{\, t : |B_t| = b\,\} \\ R_t &= \frac{1}{2}\big|\sup_{u\le t} B_u - \inf_{u\le t} B_t\big|. \end{aligned}$$

Then, for $t/b^2 \to \infty$,

$$\boldsymbol{P}(H_b \ge t) = \boldsymbol{P}(\sup_{u\le t}|B_u| \le b) \sim 4\pi^{-1}\exp\left(-\frac{\pi^2 t}{8b^2}\right) \tag{K5a}$$

$$\boldsymbol{P}(R_t \le b) = \boldsymbol{P}(\sup_{u\le v\le t}|B_u - B_v| \le 2b) \sim 2tb^{-2}\exp\left(-\frac{\pi^2 t}{8b^2}\right) \tag{K5b}$$

These can be derived from the series expansion for the exact distribution of $(\max_{u\le t} B_u, \min_{u\le t} B_u)$, but are more elegantly obtained using the eigenvalue method (M7b). We record two immediate consequences: for t/b^2 large,

(K5c) $H_b - t$,conditional on $\{H_b > t\}$, has approximately exponential distribution with mean $8b^2/\pi^2$.

(K5d) $b - R_t$,conditional on $\{R_t < b\}$, has approximately exponential distribution with mean $4b^3/\pi^2 t$.

Now fix $b > 0$, small. We shall apply the heuristic to the random set $\mathcal{S}$ of times t such that $\sup_{t\le u<v\le t+1}\frac{1}{2}|B_u - B_v| \le b$. From (K5b),

$$p = \boldsymbol{P}(t \in \mathcal{S}) \approx 2b^{-2}\exp\left(-\frac{\pi^2}{8b^2}\right). \tag{K5e}$$

Condition on $t_0 \in \mathcal{S}$, and consider the size $\widetilde{C}$ of the clump $\widetilde{\mathcal{C}}$ containing t_0. Write $\widetilde{C} = \widetilde{C}_+ + \widetilde{C}_-$, where $\widetilde{C}_+$ is the size of $\widetilde{\mathcal{C}} \cap [t_0,\infty]$. Write $(y_1,y_2) = (\min_{t_0\le u\le t_0+1} B_u, \max_{t_0\le u\le t_0+1} B_u)$. Then $y_2 - y_1 = 2b - \delta$, say, where δ will be small relative to b. For small $u > 0$, we have $t_0 + u \in \mathcal{S}$ iff $u \le T^* = \min\{\, t > 0 : B_{t_0+1+t} = y_2 + \delta \text{ or } y_1 - \delta\,\}$, neglecting the small chance that extreme values y_1, y_2 of B on $[t_0, t_0+1]$ are taken in $[t_0, t_0+T^*]$. So $\widetilde{C}_+ \overset{\mathcal{D}}{\approx} T^*$. But since δ is small,

$$T^* \overset{\mathcal{D}}{\approx} \widehat{T} = \min\{\, t > 0 : B_{t_0+1+t} = y_1 \text{ or } y_2\,\}.$$

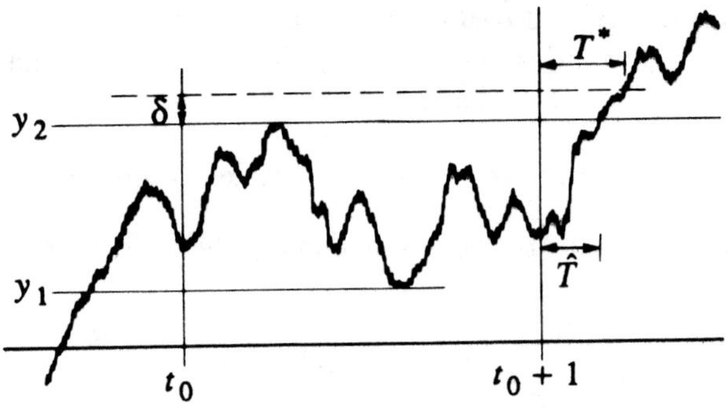

FIGURE K5a.

Now the distribution of B_{t_0+1} is like the distribution of Brownian motion at time 1 when it has been conditioned to stay within $[y_1, y_2]$ during $0 \leq u \leq 1$; hence from (K5c)

$$\widetilde{C}_+ \overset{\mathcal{D}}{\approx} T^* \overset{\mathcal{D}}{\approx} \text{exponential, mean } \frac{8b^2}{\pi^2}. \tag{K5f}$$

Similarly for $\widetilde{C}_-$, and so as at Section A21 we get $EC = 8b^2/\pi^2$. The fundamental identity gives

$$\lambda = \frac{p}{EC} \approx \frac{1}{4}\pi^2 b^{-4} \exp\left(-\frac{\pi^2}{8b^2}\right),$$

and the heuristic says

$$\begin{aligned} \boldsymbol{P}(M(1,T) > b) &= \boldsymbol{P}(\mathcal{S} \cap [0,T] \text{ empty}) \\ &\approx \exp(-\lambda T) \\ &\approx \exp\left(-\frac{1}{4}\pi^2 b^{-4} T \exp\left(\frac{-\pi^2}{8b^2}\right)\right). \end{aligned} \tag{K5g}$$

This is the basic distribution approximation. By scaling

$$\boldsymbol{P}(M(\delta,1) > x) \approx \exp\left(-\frac{1}{4}\pi^2 \delta x^{-4} \exp\left(\frac{-\pi^2 \delta}{8x^2}\right)\right).$$

For fixed $\alpha > 0$ we find

$$\boldsymbol{P}(M(\delta,1) > \alpha\sqrt{\delta/\log(1/\delta)}\,) \to \begin{cases} 0 & \text{as } \delta \to 0 \quad (\alpha > \pi/\sqrt{8}) \\ 1 & \text{as } \delta \to 0 \quad (\alpha < \pi/\sqrt{8}) \end{cases},$$

exponentially fast in each case, so using monotonicity

$$\frac{M(\delta,1)}{\sqrt{\delta/\log(1/\delta)}} \to \frac{\pi}{\sqrt{8}} \quad \text{a.s.} \quad \text{as } \delta \to 0. \tag{K5h}$$

K6 Example: Integral tests for small increments. In the same setting, we now consider the associated "integral test" problem: what functions $\psi(\delta) \to 0$ as $\delta \to 0$ are such that a.s.

$$M(\delta, 1) \geq \delta^{\frac{1}{2}}\psi(\delta) \qquad \text{for all sufficiently small } \delta? \tag{K6a}$$

Write $\underset{\sim}{t} = (t_1, t_2)$ for $0 \leq t_1 < t_2 \leq 1$. and apply the heuristic to the random set $\mathcal{S}$ of $\underset{\sim}{t}$ such that

$$\sup_{t_1 \leq u < v \leq t_2} \frac{1}{2}|B_u - B_v| \leq (t_2 - t_1)^{\frac{1}{2}}\psi(t_2 - t_1).$$

By (K5b),

$$p(\underset{\sim}{t}) = \boldsymbol{P}(\underset{\sim}{t} \in \mathcal{S}) \approx 2\psi^{-2}(t_2 - t_1)\exp\left(\frac{-\pi^2}{8\psi^2(t_2 - t_1)}\right).$$

We shall show later that

$$EC_{\underset{\sim}{t}} \approx 64(t_2 - t_1)^2\psi^{-4}(t_2 - t_1) \tag{K6b}$$

Then

$$\lambda(\underset{\sim}{t}) = \frac{p(\underset{\sim}{t})}{EC_{\underset{\sim}{t}}} = \frac{1}{32}(t_2 - t_1)^{-1}\psi^2(t_2 - t_1)\exp\left(\frac{-\pi^2}{8\psi^2(t_2 - t_1)}\right).$$

The condition for (K6a) to hold is that $\mathcal{S}$ be bounded away from the diagonal $\{\underset{\sim}{t} : t_1 = t_2\}$, and this hold iff

$$\iint_{0 \leq t_1 < t_2 \leq 1} \lambda(\underset{\sim}{t})\, d\underset{\sim}{t} < \infty.$$

Thus the integral test for (K6a) is

$$\int_{0+} t^{-1}\psi^2(t)\exp\left(\frac{-\pi^2}{8\psi^2(t)}\right)\, dt < \infty. \tag{K6c}$$

To argue (K6b), fix $\underset{\sim}{t} = (t_1, t_2)$ and condition on $\underset{\sim}{t} \in \mathcal{S}$. Let

$$\begin{aligned} b_0 &= (t_2 - t_1)^{\frac{1}{2}}\psi(t_2 - t_1) \\ (y_1, y_2) &= (\min_{t_1 \leq u \leq t_2} B_u, \max_{t_1 \leq u \leq t_2} B_u); \end{aligned}$$

then

$$\frac{1}{2}(y_2 - y_1) = b_0 - L \tag{K6d}$$

for some $L > 0$ which is small relative to b_0. For small u_i (positive or

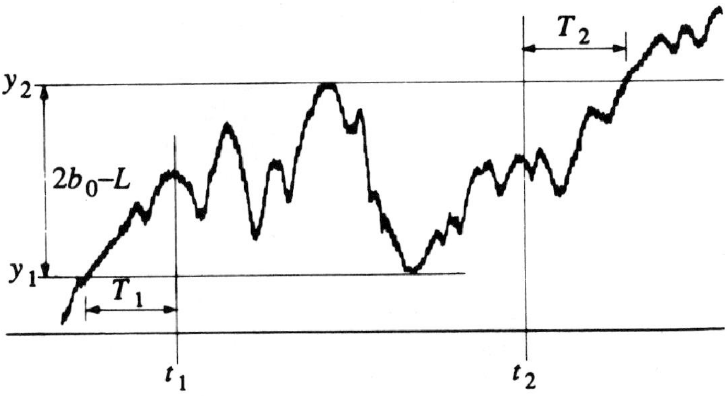

FIGURE K6a.

negative), consider what has to happen to make $t' = (t_1 - u_1, t_2 + u_2) \in \mathcal{S}$. There are two different types of constraint. As in the previous "clump size" argument, we need

$$u_2 \leq T_2 \text{ (say)} \stackrel{\mathcal{D}}{\approx} \text{exponential, mean } \frac{8b_0^2}{\pi^2}, \tag{K6e}$$

where T_2 is the time taken from t_2 until the path escapes from the bounds (y_1, y_2). Similarly,

$$u_1 \leq T_1 \text{ (say)}, \quad \text{where } T_1 \stackrel{\mathcal{D}}{=} T_2. \tag{K6f}$$

The second constraint for $t' = (t_1', t_2') = (t_1 - u_1, t_2 + u_2)$ to be in $\mathcal{S}$ is that $(t_2' - t_1')^{1/2}\psi(t_2' - t_1')$ must be $\geq \frac{1}{2}(y_2 - y_1)$. Using (K6d), this is the requirement that

$$(u_1 + u_2)\frac{d}{dt}(t^{\frac{1}{2}}\psi(t))\Big|_{t_2 - t_1} \geq -L$$

which works out as

$$u_1 + u_2 \geq -U; \quad \text{where } U = 2(t_2 - t_1)\frac{L}{b_0}. \tag{K6g}$$

Thus the clump $\tilde{C}$ containing $\underset{\sim}{t} = (t_1, t_2)$ is approximately the triangular-shaped region of $(t_1 - u_1, t_2 + u_2)$ satisfying (K6e,K6f,K6g), and this has area

$$\tilde{C} \approx \frac{1}{2}(T_1 + T_2 + U)^2. \tag{K6h}$$

Now L (defined at (K6d)), is distributed as $b_0 - R_{t_2-t_1}$ conditioned on $\{R_{t_2-t_1} < b_0\}$, so appealing to (K5d) we see

$$L \overset{\mathcal{D}}{\approx} \text{ exponential, mean } \frac{4b_0^3}{\pi^2(t_2-t_1)};$$

and so

$$U \overset{\mathcal{D}}{\approx} \text{ exponential, mean } \frac{8b_0^2}{\pi^2}.$$

Thus the sum in (K6h) has approximately a Gamma distribution. Recalling from Section A6 that the mean clump size EC is the harmonic mean of $\widetilde{C}$, we calculate from (K6h) that

$$EC \approx \left(\frac{8b_0^2}{\pi^2}\right)^2.$$

This gives the estimate (K6b) asserted earlier.

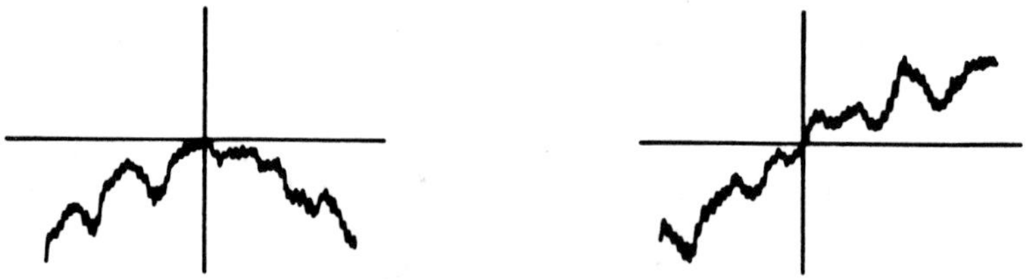

FIGURE K7a.

K7 Example: Local maxima and points of increase. A function $f(t)$ has a local maximum at t_0 if $f(t) \le f(t_0)$ in some neighborhood $(t_0 \pm \epsilon)$ of t_0; and a point of increase at t_0 if $f(t) \le f(t_0)$ on $(t_0-\epsilon, t_0)$ but $f(t) \ge f(t_0)$ on $(t_0, t_0+\epsilon)$. A paradoxical property of Brownian motion is that its sample paths have local maxima but not points of increase: paradoxical because "by symmetry" one would expect the two phenomena to be equally likely. There are slick modern proofs of this result (Section K13), but to my mind they do not make visible the essential difference between the two cases. The heuristic, in the spirit of the original study by Dvoretzky et al. (1961), does.

Fix $\delta > 0$ small. Let $\mathcal{S}_\delta^i$ [resp. $\mathcal{S}_\delta^m$] be the random set of t such that

$$\max_{t-1\le u\le t} B_u \le B_t+\delta \text{ and } \min_{t\le u\le t+1} B_u > B_t-\delta \quad [\text{resp. } \max_{t\le u\le t+1} B_u \le B_t+\delta].$$

For each random set,

$$p = \boldsymbol{P}(t \in \mathcal{S}_\delta) = \boldsymbol{P}^2(\max_{u\le 1} B_u \le \delta) \approx 2\pi^{-1}\delta^2 \quad \text{as } \delta \to 0. \qquad \text{(K7a)}$$

For each random set write $\mathcal{S}_0 = \bigcap_{\delta>0} \mathcal{S}_\delta$. Then $\mathcal{S}_0^i$ [$\mathcal{S}_0^m$] is the set of points of increase [local maxima] of Brownian motion. In the language of our heuristic, the random sets $\mathcal{S}_\delta$ will have some clump rates λ_δ^i, [λ_δ^m]. So the result we are trying to explain (that $\mathcal{S}_0^i$ is empty while $\mathcal{S}_0^m$ is not) is the result

$$\lambda_\delta^i \to 0 \quad \text{but } \lambda_\delta^m \not\to 0 \quad \text{as } \delta \downarrow 0.$$

By (K7a) and the fundamental identity $p = \lambda/EC$, we see that the "paradoxical result" is equivalent to the assertion that the clump sizes C_δ satisfy

$$EC_\delta^m = O(\delta^2) \quad \text{but } EC_\delta^i \neq O(\delta^2). \tag{K7b}$$

We shall now sketch arguments for this assertion.

In each case, fix t_0 and condition on $t_0 \in \mathcal{S}_\delta$. Consider the processes

$$\begin{aligned} Z_u^- &= B_{t_0} + \delta - B_{t_0 - u}; \\ Z_u^+ &= B_{t_0+u} - B_{t_0} + \delta \quad [\text{resp. } Z_u^+ = B_{t_0} + \delta - B_{t_0+u}]. \end{aligned}$$

Conditioning on $t_0 \in \mathcal{S}_\delta$ is precisely the same as conditioning on the processes Z_u^+ and Z_u^- being non-negative on $[0,1]$. By Section K14 these processes behave, for small u, like independent BES(3) processes started with $Z_0 = \delta$. In each case we shall estimate

$$E\widetilde{C} = \int_{|t|\text{small}} \boldsymbol{P}(t_0 + t \in \widetilde{C})\, dt, \tag{K7c}$$

where $\widetilde{C}$ is the clump containing t_0. We consider first the case of near-

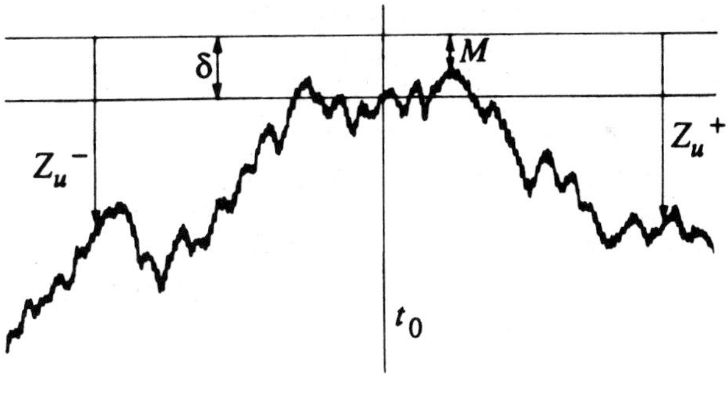

FIGURE K7b.

local-maxima. Let $M = \min_{0 \le u \le 1} Z_u^+$. Then M is approximately uniform on $[0, \delta]$, since $\delta - M$ is distributed as $\max_{u \le 1} B_u$ conditioned on this max being $\le \delta$. Now in order that $t_0 - u$ be in $\mathcal{S}_\delta$ it is necessary that $Z_u^- \le M + \delta$. By (K7c) and symmetry we get a bound

$$E\widetilde{C} \le 2Es(M + \delta) \tag{K7d}$$

where $s(y)$ is the mean sojourn time of Z^- in $[0, y]$. Since Z^- is approximately BES(3), we can quote the mean sojourn time (Section K14) for BES(3) as an approximation: $s(y) \approx y^2 - \delta^2/3$. Evaluating (K7d),

$$E\widetilde{C} \leq 4\delta^2.$$

Since $EC \leq E\widetilde{C}$ (Section A6), we have verified the first part of (K7b).

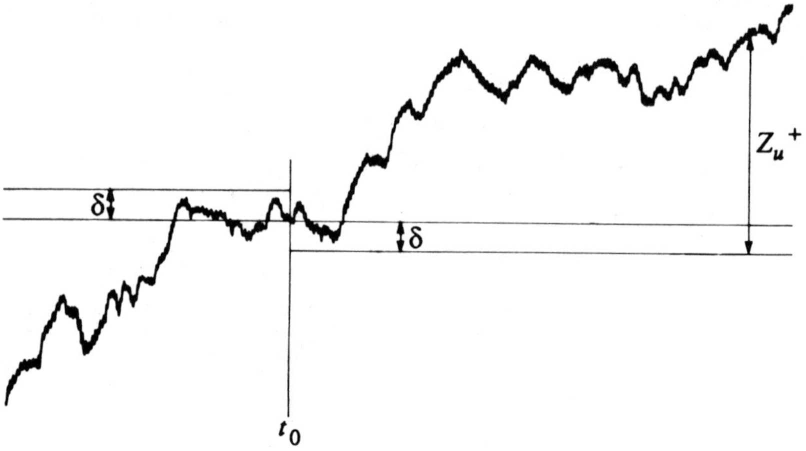

FIGURE K7c.

We now look at the case of near-points-of-increase. Recall we are conditioning on $t_0 \in \mathcal{S}_\delta$. Consider $t > 0$ small: what is a sufficient condition for $t_0 + t$ to be in $\mathcal{S}_\delta$? From the picture, it is sufficient that $Z_t^+ > \delta$, that $\max_{u \leq t} Z_u^+ \leq Z_t^+ + \delta$, and that $\min_{t \leq u < 1+t} Z_u^+ > Z_t - \delta$. By (K7c), $E\widetilde{C} \geq$ the mean duration of time t that the conditions above are satisfied:

$$E\widetilde{C} \geq E\int_0^{o(1)} 1_{(Z_t^+ > \delta)} 1_{(\sup_{u \leq t} Z_u^+ \leq Z_t^+ + \delta)} 1_{(\inf_{u \geq t} Z_u^+ > Z_t^+ - \delta)}\, dt.$$

Approximating Z^+ as BES(3) it is remarkably easy to evaluate this. First condition on Z_t^+, and then consider occupation densities: we get

$$E\widetilde{C} \geq \int_\delta^{o(1)} g^*(y) \boldsymbol{P}_y(T_{y-\delta} = \infty)\, dy$$

where T_b is first hitting time on b, and $g^*(y)$ is mean occupation density at height y where we allow only times t for which $\max_{u \leq t} Z_u^+ \leq y + \delta$. Now a visit to y is "counted" in $g^*(y)$ if the process does not subsequently make a downcrossing from $y + \delta$ to y. Hence $g^*(y) = g(y)P_{y+\delta}(T_y = \infty)$, where $g(y)$ is the unrestricted mean occupation density at y. So

$$E\widetilde{C} \geq \int_\delta^{o(1)} g(y) \boldsymbol{P}_{y+\delta}(T_y = \infty) \boldsymbol{P}_y(T_{y-\delta} = \infty)\, dy.$$

But the terms in the integrand are given by basic 1-dimensional diffusion theory (Section D3):

$$g(y) = 2y; \qquad \boldsymbol{P}_{y+\delta}(T_y = \infty) = \frac{\delta}{y+\delta}; \qquad \boldsymbol{P}_y(T_{y-\delta} = \infty) = \frac{\delta}{y}.$$

Evaluating the integral,

$$E\widetilde{C} \geq 2\delta^2 \log(1/(2\delta)).$$

Optimistically supposing that EC is of the same order as $E\widetilde{C}$, we have verified the second part of (K7b).

Finally, if it is indeed true that $EC \sim a\delta^2 \log(1/\delta)$, we can use the heuristic to give a quantitative form of the non-existence of points of increase. For then the clump rate $\lambda^i_\delta \sim \widehat{a}/\log(1/\delta)$ and so $\boldsymbol{P}(\mathcal{S}^i_\delta \cap [0,T]$ empty) $\approx \exp(-\lambda^i_d T)$. Writing

$$I_T = \inf_{1 \leq t \leq T-1} \sup_{0 \leq u \leq 1} \max(B_{t-u} - B_t, B_t - B_{t+u}),$$

we have $\{I_T > \delta\} = \{\mathcal{S}^i_\delta \cap [0,T] \text{ empty}\}$, and a little algebra gives the limit assertion

$$T^{-1} \log(1/I_T) \xrightarrow{\mathcal{D}} \frac{c}{V} \quad \text{as } T \to \infty, \tag{K7e}$$

where V has exponential(1) distribution and c is a constant.

Remark: Merely assuming that EC and $E\widetilde{C}$ have the same order — as we did above and will do in the next example — is clearly unsatisfactory and potentially erroneous. We haven't done it in any other examples.

K8 Example: Self-intersections of d-dimensional Brownian motion. Let B_t be Brownian motion in dimension $d \geq 4$. Then it is well known that B_t has no self-intersections: here is a (rather rough) heuristic argument. Let $I = \{(s,t) : 0 \leq s < t \leq L, t - s \geq 1\}$, for fixed L. For $\delta > 0$ consider the random sets $\mathcal{S}_\delta = \{(s,t) \in I : |B_t - B_s| \leq \delta\}$ as approximately mosaic processes (which is not very realistic). We want to show that the clump rate $\lambda_\delta(s,t) \to 0$ as $\delta \to 0$.

Since $\boldsymbol{P}(|B_1| \leq x) \sim c_d x^d$ as $x \to 0$, we have by scaling

$$p_\delta(s,t) \equiv \boldsymbol{P}(|B_t - B_s| \leq \delta) \sim c_d |t-s|^{-\frac{1}{2}d} \delta^d. \tag{K8a}$$

Let B^1, B^2 be independent copies of B and for ϵ in $\boldsymbol{R}^d$ define

$$D(\epsilon) = \text{area}(\, (t_1,t_2) : t_i > 0, t_1 + t_2 \leq 1, |B^1_{t_1} - B^2_{t_2} - \epsilon| \leq \delta\,).$$

Let U be uniform on the unit ball. Fix (s,t) and condition on $(s,t) \in \mathcal{S}_\delta$: then conditionally $(B_t - B_s) \overset{\mathcal{D}}{\approx} U\delta$ and the clump distribution $\widetilde{\mathcal{C}}(s,t) \overset{\mathcal{D}}{\approx}$

$D(U\delta)$. In particular

$$EC(s,t) \le E\widetilde{C}(s,t) \approx ED(U\delta) \le ED(0). \qquad \text{(K8b)}$$

Now we can compute

$$\begin{aligned} ED(0) &= \int_0^1 s\boldsymbol{P}(|B_s| \le \delta)\,ds \\ &\sim \delta^4 a_d \quad \text{as } \delta \to 0 \qquad (d \ge 5) \\ &\sim \delta^4 \log(1/\delta) a_d \quad \text{as } \delta \to 0 \qquad (d = 4), \end{aligned}$$

where a_d does not depend on δ. Assuming the inequalities in (K8b) are equalities up to constant multiples, we get clump rates

$$\lambda_\delta(s,t) = \frac{p(s,t)}{EC(s,t)} \sim \begin{cases} (t-s)^{-\frac{1}{2}d}\delta^{d-4}a'_d & (d \ge 5) \\ (t-s)^{-2}/\log(1/\delta)a'_4 & (d = 4) \end{cases}.$$

As required, these rates $\to 0$ as $\delta \to 0$.

COMMENTARY

K9 General references. Chapter 1 of Csorgo and Revesz (1981) contains many results related to our first examples; looking systematically at these results from the viewpoint of our heuristic would be an interesting project.

K10 Modulus of continuity and large increments. Presumably our distributional approximations such as Example K3 can be justified as limit theorems; but I don't know explicit references. The Chung-Erdos-Sirao test is in Chung et al. (1959). Csorgo and Revesz (1981) discuss results of the following type (their Theorem 1.2.1) for the modulus $w(\delta, T)$ of Brownian motion. Let $a_T \nearrow \infty$ and $T/a_T \searrow 0$ as $T \to \infty$. Then

$$\limsup_{T\to\infty} \beta_T w(a_T, T) = 1 \quad \text{a.s.},$$

where $\beta_T^{-2} = 2a_T\{\log(T/a_T) + \log\log T\}$. Similar results are given by Hanson and Russo (1983b; 1983a). These results can be derived from the heuristic, using the ideas of this chapter.

K11 Spikes. I don't know any discussion of "spikes" (Example K4) in the literature. A rigorous argument can be obtained by considering local maxima of fixed width, which form a tractable point process which has been studied by Pitman (unpublished). The integral test is: $w^*(\delta, 1) \leq \delta^{1/2}\psi(\delta)$ for all sufficiently small δ iff

$$\int_{0^+} u^{-2}\psi^3(u)\exp(-2u^2)\,du < \infty.$$

K12 Small increments. In place of our $M(\delta, T)$ at Example K5, it is customary to consider

$$m(\delta, T) = \inf_{0 \leq t \leq T-\delta} \sup_{0 \leq u \leq \delta} |B_{t+u} - B_t|.$$

Csorgo and Revesz (1981) give results of the type (their 1.7.1):

if $a_T \nearrow \infty$ and $T/a_T \searrow \infty$ as $T \to \infty$ then

$$\liminf_{T\to\infty} \gamma_T m(a_T, T) = 1 \quad \text{a.s.},$$

where

$$\gamma_T^2 = 8\pi^{-2}a_T^{-1}(\log(T/a_T) + \log\log T).$$

We have the relation

$$M(\delta, T) \leq m(\delta, T)$$

The quantity M is slightly easier to handle using the heuristic. The basic a.s. limit theorems seem the same for m and M; I don't know if this remains true for the integral test (Example K6).

Ortega and Wschebor (1984) give a range of integral test results for both small and large increment problems. Csaki and Foldes (1984) treat the analogous question for random walk: the heuristic arguments are similar.

K13 Local maxima and points of increase. A continuous function either has a local maximum or minimum on an interval, or else it is monotone on that interval. It is easy to show Brownian motion is not monotone on any interval, and hence to deduce the existence of local maxima. There is a slick argument for non-existence of points of increase which uses the joint continuity properties of local time — see Geman and Horowitz (1980), Karatzas and Shreve (1987) Sec. 6.4.

The original paper showing non-existence of points of increase, Dvoretzky-Erdos-Kakutani (1961), is essentially a formalization of the heuristic argument we sketched; it is generally regarded as a hard to read paper! A simpler direct argument is given in Adelman (1985). None of these methods gives quantitative information: proving our limit assertion (K7e) is an interesting open problem (see also Section K15).

K14 Conditioned Brownian motion is BES(3). The concept of "standard Brownian motion on the space interval $[0, \infty)$ conditioned never to hit 0" may be formalized via a limit procedure in several ways: each gives the same result, the BES(3) process X_t, that is the diffusion with drift $\mu(x) = 1/x$ and variance $\sigma^2(x) \equiv 1$. Recall from Section I22 that BES(3) is also the radial part of 3-dimensional Brownian motion. The standard calculations of Section D3 give some facts used in Example K7:

$$\boldsymbol{P}_y(X_T \text{ hits } x) = \frac{x}{y}, \qquad 0 < x < y \tag{K14a}$$

$$E_x(\text{sojourn time in } [0, y]) = y^2 - \frac{x^2}{3}, \qquad 0 < x < y. \tag{K14b}$$

K15 The second-moment method. As discussed at Section A15, using the bound $EC \leq E\widetilde{C}$ in the heuristic is essentially equivalent to using the second moment method. We estimated $E\widetilde{C}$ in the "point-of-increase" Example K7 and the self-intersection example (Example K8). So in these examples the second-moment method should give rigorous one-sided bounds. That is, one should get an upper bound for I_T in (K7e) of the right order of magnitude, and similarly in Example K8 get an upper bound for

$$\inf_{\substack{0 \leq s < t < T \\ t-s \geq 1}} |B_t - B_s|$$

in $d \geq 4$ dimensions. The former is in Schaper (1988) and the latter in Aizenmann (1985).

K16 Other "a.s." properties. It would be interesting to use the heuristic to study the "slow points" problems in Davis and Perkins (1985).

A modern account of self-intersections is given in Dynkin (1988).

K17 *d* dimensions. Versions of the "modulus of continuity" and "small increments" results for d-dimensional Brownian motion are given by Ugbebor (1980). The a.s. limit results are fairly simple, but the "integral test" results involve some extra complications: it would be interesting to apply the heuristic here.

L Miscellaneous Examples

Most of these examples are similar in spirit to those considered earlier, but did not fit in conveniently. We will give references as we go.

In Chapter B we often used simple random walk as a local approximation to Markov processes on the integers. Here is another simple application in a different setting. The fact we need is that for simple random walk in continuous time with rates a, b, started at 0, the number Z of sojourns at 0 has geometric distribution with parameter $\theta = |b-a|/(b+a)$:

$$\boldsymbol{P}(Z = n) = \theta(1-\theta)^{n-1}; \qquad n \geq 1. \tag{L0a}$$

L1 Example: Meetings of empirical distribution functions. Let F, G be distribution functions with continuous densities f, g. Let J be an interval such that $F(t) = G(t)$ for a unique $t \in J$, and suppose $f(t) \neq g(t)$. Let F_N, G_N be the empirical distribution functions of independent samples of size N from each distribution. The $2N$ sample values divide the line into $2N+1$ intervals; let $Z_N(J)$ be the number of these intervals in J on which $F_N = G_N$. For N large, the heuristic says

$$\boldsymbol{P}(Z_N(J) = n) \approx \theta_J(1-\theta_J)^{n-1}, \quad n \geq 1; \quad \text{where } \theta_J = \frac{|f(t)-g(t)|}{f(t)+g(t)}. \tag{L1a}$$

To argue this, observe that the empirical distribution function evolves locally like the non-homogeneous Poisson process of rate $Nf(t)$. Let T be the first point in J where $F_N = G_N$. For N large, T will be near t and the subsequent difference process $X(u) = F_N(T+u) - G_N(T+u)$ will evolve like the simple random walk with transition rates $Nf(t)$ and $Ng(t)$. So (L1a) follows from the corresponding fact (L0a) for random walk.

By considering different points where $F = G$, one can see that the total number Z_N of intervals where $F_N = G_N$ will be approximately distributed as a sum of independent geometrics. Nair et al. (1986) give details of this and related crossing statistics.

For our next example, take a sample of size n from the uniform distribution on $[0,1]$ and let $U_{n,1}, \ldots, U_{n,n}$ be the order statistics. There is a large literature on asymptotics of various quantities derived from the U's; see

Shorack and Wellner (1986) for an encyclopedic account. Problems related to extremal properties can be handled via the heuristic, and on the heuristic level are very similar to examples we have already seen. We illustrate with one result.

L2 Example: Maximal k-spacing. Given n and k, let

$$M = \max_{1 \le i \le n-k-1} (U_{n,i+k+1} - U_{n,i}).$$

We study M in the case $n \to \infty$, $k = k(n) \to \infty$, $k = o(\log n)$.

Fix n. The $U_{n,i}$ behave approximately like the Poisson process of rate n. In particular, the k-spacings $U_{n,i+k+1} - U_{n,i}$ have approximately the distribution $n^{-1}G_k$, where G_k has Gamma density

$$f_k(x) = e^{-x}\frac{x^k}{k!}.$$

Note the tail estimate

$$\boldsymbol{P}(G_k > x) \sim f_k(x) \quad \text{as } \frac{x}{k} \to \infty. \tag{L2a}$$

This leads to the estimate

$$\begin{aligned} \boldsymbol{P}\left(U_{n,i+k+1} - U_{n,i} > \frac{\log n + c}{n}\right) &\approx f_k(\log n + c) \\ &\approx n^{-1}e^{-c}\frac{(\log n + c)^k}{k!}. \end{aligned} \tag{L2b}$$

Now the process $(U_{n,i+k+1} - U_{n,i})$ as i varies is essentially a moving average process, and as at Example C6 its maximum M will behave as if the terms were i.i.d. So we get

$$-\log \boldsymbol{P}(nM \le \log n + c) \approx n\boldsymbol{P}(U_{n,i+k+1} - U_{n,i} > \frac{\log n + c}{n}).$$

Taking logs again,

$$\log(-\log \boldsymbol{P}(nM \le \log n + c)) \approx -c + k\log(c + \log n) - k(\log k - 1). \tag{L2c}$$

Now, put $c = ak\log(ek^{-1}\log n)$ for fixed a. Then the right side of (L2c) becomes

$$k(1-a)(\log\log n - \log k + 1) + o(k) \to \begin{cases} \infty & \text{for } a < 1 \\ -\infty & \text{for } a > 1 \end{cases}.$$

This gives

$$\frac{nM - \log n}{k\log(ek^{-1}\log n)} \to 1 \text{ in probability.} \tag{L2d}$$

See Deheuvels and Devroye (1984) for extensions to LIL-type results. Note the argument above used only estimates for fixed n. A more sophisticated use of the heuristic is to consider the random subset of $\{1,2,3,\dots\} \times \{0,1\}$ consisting of those (n,x) such that the interval $[x, x+(c+\log n)/n]$ contains less than k of the order statistics $(U_{n,i}; 1 \leq i \leq n)$. This enables the strong laws and LILs to be obtained heuristically.

Here is a simple combinatorial example.

L3 Example: Increasing runs in i.i.d. sequences. Let $(X_i; i \geq 1)$ be i.i.d. real-valued with some continuous distribution. Let L_N be the length of the longest increasing run in $(X_1, \dots, X_N)$:

$$L_N = \max\{\, k : X_n > X_{n-1} > \cdots > X_{n-k+1} \text{ for some } n \leq N\,\}.$$

We can approximate the distribution of L_N as follows. Fix k. Let $\mathcal{S}$ be the random set of n such that $(n-k+1, \dots, n)$ is an increasing run. Then $p = \boldsymbol{P}(n \in \mathcal{S}) = 1/k!$ Given $n \in \mathcal{S}$, the chance that $n+1 \in \mathcal{S}$ equals

$$\begin{aligned} \frac{\boldsymbol{P}((n-k+1,\dots,n+1) \text{ is an increasing run})}{\boldsymbol{P}((n-k+1,\dots,n) \text{ is a run})} &= \frac{1/(k+1)!}{1/k!} \\ &= \frac{1}{k+1.} \end{aligned}$$

In the notation of (A9g) we have $f^+(0) = k/(k+1)$, since the clump containing n ends at n iff $X_{n+1} < X_n$, that is iff $n+1 \in \mathcal{S}$. Applying (A9g),

$$\begin{aligned} \boldsymbol{P}(L_N < k) &= \boldsymbol{P}(\mathcal{S} \cap [1,N] \text{ empty}) \\ &\approx \exp(-\lambda N) \\ &\approx \exp(-pf^+(0)N) \\ &\approx \exp\left(-\frac{N}{(k+1)(k-1)!}\right). \end{aligned} \tag{L3a}$$

Revesz (1983) gives the asymptotics.

The next example is mathematically interesting because it combines a "stochastic geometry coverage" problem treated in Chapter H with a "maximum of process whose variance has a maximum" problem in the style of Chapter D. The model arose as a (much simplified) model of DNA splitting; result (L4c) is discussed in Shepp and Vanderbei (1987).

L4 Example: Growing arcs on a circle. Consider a circle of circumference L. Suppose points are created on the circumference according to a

Poisson process in time and space of rate a; and each point grows into an interval, each end of which grows deterministically at rate $\frac{1}{2}$. Consider

$$\begin{aligned} T &= \text{time until circle is covered by the growing intervals} \\ N_t &= \text{number of components of the uncovered part of the circle at time } t \\ M &= \sup_t N_t. \end{aligned}$$

For the heuristic analysis, let $\mathcal{S}_t$ be the uncovered set at time t. Then

$$\begin{aligned} p(t) &\equiv \boldsymbol{P}(x \text{ uncovered at time } t) && \text{for specified } x \\ &= \exp(-a\int_0^t (t-s)\,ds) && \text{since a point created at time } s \text{ has grown an interval of length } (t-s) \text{ by time } t \\ &= \exp(-\frac{1}{2}at^2). \end{aligned}$$

At time t, the clockwise endpoints of growing intervals form a Poisson process of rate at, and so the clump length C_t for $\mathcal{S}_t$ has exponential(at) distribution, and so the clump rate is

$$\lambda(t) = \frac{p(t)}{EC_t} = at\exp(-\frac{1}{2}at^2). \tag{L4a}$$

Thus N_t satisfies

$$N_t \overset{\mathcal{D}}{\approx} \text{Poisson}(\mu(t)), \qquad \text{for } \mu(t) = L\lambda(t) = Lat\exp(-\frac{1}{2}at^2); \tag{L4b}$$

the implicit conditions being that aL and at^2 are large. In particular,

$$\boldsymbol{P}(T \le t) = \boldsymbol{P}(N_t = 0) \approx \exp(-\mu(t)) \qquad \text{for } t \text{ large} \tag{L4c}$$

(This is actually a special case of (H12a), covering a circle with randomly-sized intervals.)

To analyze M, observe first that $\mu(t)$ has maximum value $\mu^* = L(a/e)^{1/2}$ attained at $t^* = a^{-1/2}$. Consider $b > \mu^*$ such that $\boldsymbol{P}(N_{t^*} > b)$ is small. We want to apply the heuristic to $\widehat{\mathcal{S}}_t = \{\,t : N_t = b\,\}$. So

$$\begin{aligned} p(t,b) &\equiv \boldsymbol{P}(N_t = b) \\ &\equiv (\mu^*)^{-\frac{1}{2}}\phi\left(\frac{b-\mu(t)}{\sqrt{\mu^*}}\right) \qquad \text{for } t \approx t^*, \end{aligned}$$

by the Normal approximation to Poisson (ϕ denotes standard Normal density). Evaluating the integral as at (C21e) gives

$$\int p(t,b)\,dt \approx (b-\mu^*)^{-\frac{1}{2}}(2La^{\frac{3}{2}}e^{-\frac{1}{2}})^{-\frac{1}{2}}\exp\left(\frac{-(b-\mu^*)^2}{\mu^*}\right). \tag{L4d}$$

Now for $t \approx t^*$ the process N_t behaves like the M/M/∞ queue; the arrival rate α being the rate at which new points are created in the uncovered region, so

$$\begin{aligned}\alpha &= a \times \text{length of uncovered region}\\ &\approx aLp(t^*) = aL\exp(-\frac{1}{2}at^{*2}) = aLe^{-\frac{1}{2}}.\end{aligned}$$

Since N_{t^*} has Poisson(μ^*) distribution, the "departure rate per customer" β in the M/M/∞ approximation must be $\beta = \alpha/\mu^*$. So using the local random walk approximation as at Example B4, the mean clump size for $\widehat{\mathcal{S}}_t$ at $t \approx t^*$ is

$$EC \approx (\beta b - \alpha)^{-1} = a^{-\frac{1}{2}}(b - \mu^*)^{-1}.$$

Writing $\lambda_b(t)$ for the clump rate of $\widehat{\mathcal{S}}_t$, the heuristic gives

$$\boldsymbol{P}(M \le b) \approx \exp(-\lambda_b); \tag{L4e}$$

where

$$\begin{aligned}\lambda_b &= \int \lambda_b(t)\,dt = \int \frac{p(t,b)}{EC}\,dt\\ &= (b-\mu^*)^{\frac{1}{2}}(2La^{\frac{1}{2}}e^{-\frac{1}{2}})^{-\frac{1}{2}}\exp\left(\frac{-(b-\mu^*)^2}{\mu^*}\right) \qquad \text{using (L4d).}\end{aligned}$$

L5 Example: The LIL for symmetric stable processes. Write $Z(t)$ for the symmetric stable process of exponent $0 < \alpha < 2$:

$$E\exp(i\theta Z(t)) = \exp(-t|\theta|^\alpha).$$

Chover (1966) gave the "law of the iterated logarithm"

$$\limsup_{t\to\infty} |t^{1/\alpha}Z(t)|^{1/\log\log t} = e^{1/\alpha} \quad \text{a.s.} \tag{L5a}$$

We shall give a heuristic derivation of this and a stronger result, (L5c) below.

Put $X(t) = e^{-t/\alpha}Z(e^t)$. Then

$$\begin{aligned}dX(t) &= e^{-t/\alpha}(e^t)^{1/\alpha}\,dZ(t) - \alpha^{-1}e^{-t/\alpha}Z(e^t)\,dt\\ &= dZ(t) - \alpha^{-1}X(t)\,dt.\end{aligned}$$

Thus X is the stationary autoregressive process of Section C19, whose clump rate for $\{\, t : X_t \ge b\,\}$ was calculated to be

$$\lambda_b \approx Kb^{-\alpha}; \qquad b \text{ large.}$$

Let $b(t) \to \infty$ be a smooth boundary. Then

$$\boldsymbol{P}(X(t) \leq b(t) \text{ ultimately}) = 1 \begin{cases} \text{iff } \int^{\infty} \lambda_{b(t)}\, dt < \infty \\ \text{iff } \int^{\infty} b^{-\alpha}(t)\, dt < \infty \end{cases}.$$

Putting $b_c(t) = t^c$, we get

$$\boldsymbol{P}(X(t) \leq b_c(t) \text{ ultimately}) = 1 \text{ iff } c > \frac{1}{\alpha}. \tag{L5b}$$

But this is equivalent to

$$\limsup_{t\to\infty} \frac{\log X(t)}{\log t} = \frac{1}{\alpha} \quad \text{a.s.}$$

and this in turn is equivalent to

$$\limsup_{t\to\infty} \frac{\log(t^{-1/\alpha} Z(t))}{\log\log t} = \frac{1}{\alpha} \quad \text{a.s.,}$$

which is (L5a).

We could instead consider $b_c(t) = (t \log^c(t))^{1/\alpha}$, and get

$$\boldsymbol{P}(X(t) \leq b_c(t) \text{ ultimately}) = 1 \text{ iff } c > 1.$$

This translates to

$$\limsup_{t\to\infty} \frac{\log(t^{-1/\alpha} Z(t)) - \log_{(2)} t}{\log_{(3)} t} = \frac{1}{\alpha} \quad \text{a.s.,} \tag{L5c}$$

where $\log_{(3)} \equiv \log\log\log$.

L6 Example: Min-max of process. Here is a type of problem where the heuristic provides the natural approach: I do not know any treatment in the literature. Let $X(t_1, t_2)$ be a 2-parameter process and consider

$$M_T = \min_{0 \leq t_1 \leq T} \max_{0 \leq t_2 \leq T} X(t_1, t_2), \quad T \text{ large.}$$

Fix b large, let $\mathcal{S} = \{ (t_1, t_2) : X(t_1, t_2) \geq b \}$ and suppose we can approximate $\mathcal{S}$ as a mosaic process of rate λ and clump distribution $\mathcal{C}$. Then

$$\mathcal{S}_1 = \{ t_1 : \max_{0 \leq t_2 \leq T} X(t_1, t_2) \geq b \}$$

is the projection of $\mathcal{S}$ onto the t_1-axis. So $\mathcal{S}_1$ is approximately a mosaic process with rate λT and clump distribution $\mathcal{C}_1$, where $\mathcal{C}_1$ is the projection of $\mathcal{C}$. And

$$\boldsymbol{P}(M_T \geq b) = \boldsymbol{P}(\mathcal{S}_t \text{ covers } [0, T]).$$

For b in the range of interest, $\mathcal{S}$ will be a sparse mosaic and $\mathcal{S}_1$ will be a high-intensity mosaic. We can use the heuristic to estimate the covering probability above. Let us treat the simplest setting, where $\mathcal{C}_1$ consists of a single interval of (random) length C_1. Then (H12a) gives

$$\boldsymbol{P}(M_T > b) \approx \exp(-\lambda T^2 \exp(-\lambda T E C_1)). \tag{L6a}$$

As a specific example, consider the stationary smooth Gaussian field (Example J7) with

$$EX(0,0)X(t_1,t_2) \sim 1 - \frac{1}{2}(\theta_1 t_1^2 + \theta_2 t_2^2) \quad \text{as } t \to 0.$$

Then (J7d) gives

$$\lambda = (2\pi)^{-1}(\theta_1\theta_2)^{\frac{1}{2}} b\phi(b). \tag{L6b}$$

The arguments in Example J7 lead to

$$\begin{aligned} C_1 &\stackrel{\mathcal{D}}{\approx} b^{-1}\left(\frac{2\xi}{\theta_1}\right)^{\frac{1}{2}}; \quad \xi \stackrel{\mathcal{D}}{=} \text{exponential}(1) \\ EC_1 &\approx b^{-1}\left(\frac{\pi}{2\theta_1}\right)^{\frac{1}{2}}. \end{aligned} \tag{L6c}$$

Substituting (L6b) and (L6c) into (L6a) gives our approximation for the min-max M_T.

L7 2-dimensional random walk In Chapter B we used the heuristic to estimate hitting times for Markov chains, in settings where the target set was small and the chain had a "local transience" property: see Section B2. In particular, this was used for chains which behaved locally like random walks in $d \geq 3$ dimensions (Example B7): but fails completely for processes behaving like simple symmetric walk in 1 dimension (Section B11). What about 2 dimensions? Of course simple random walk on the 2-dimensional lattice is recurrent, but "only just" recurrent. For continuous-time simple symmetric random walk X_t on the 2-dimensional lattice, we have

$$\boldsymbol{P}(X_t = 0) \sim \frac{1}{\pi t} \quad \text{as } t \to \infty$$

and hence

$$E(\#\text{ visits to 0 before time } t) \sim \pi^{-1}\log t; \tag{L7a}$$

this holds in discrete time too. See e.g. Spitzer (1964). It turns out that, in examples where we can apply the heuristic to processes resembling $d \geq 3$ dimensional random walk or Brownian motion, we can also handle the 2-dimensional case by a simple modification. Here is the fundamental example.

L8 Example: Random walk on Z^2 modulo N. This is the analog of Example B7. For N large and states i, j not close, we get

$$E_i T_j \approx 2\pi^{-1} N^2 \log N. \tag{L8a}$$

and the distribution is approximately exponential. To see this, recall from Section B24.3 the "mixing" formalization of the heuristic. In this example, the time τ for the continuous-time walk to approach stationarity is order N^2. The clumps of "nearby" visits to j are taken to be sets of visits within time τ of an initial visit. So by (L7a) the mean clump size is

$$EC \approx \pi^{-1} \log \tau \approx 2\pi^{-1} \log N.$$

Then the familiar estimate (B4b) $ET \approx EC/\mu(j)$, where μ is the stationary (uniform) distribution, gives (L8a). See Cox (1988) for a rigorous argument.

Recall that in dimensions $d \geq 3$ the analogous result (Example B8) was

$$E_i T_j \approx R_d N^d$$

where R_d is the mean total number of visits to 0 of simple symmetric r.w. started at 0 on the infinite lattice. This suggests that results for $d \geq 3$ involving R_d extend to 2 dimensions by putting $R_2 = 2\pi^{-1} \log N$. This works in the "random trapping" example (Example B8): the traps have small density q, and the mean time until trapping in $d \geq 3$ dimensions was

$$ET \approx R_d q^{-1}. \tag{L8b}$$

In 2 dimensions we get

$$ET \approx \pi^{-1} q^{-1} \log(1/q). \tag{L8c}$$

The idea is that, for periodic traps with density q, the problem is identical to the previous example with $qN^2 = 1$; thus in (L8b) we put $R_2 = 2\pi^{-1} \log N = \pi^{-1} \log(1/q)$.

L9 Example: Covering problems for 2-dimensional walks. The study of 2-dimensional walks starts to get hard when we consider covering problems. As at Example F13, consider the time V_N for simple symmetric random walk on Z^d modulo N to visit all states. For $d \geq 3$ we found

$$V_N \sim R_d N^d \log(N^d) \qquad \text{in probability as } N \to \infty$$

and a stronger, convergence in distribution, result. One might expect in $d = 2$ to have

$$V_N \sim v_N = (2\pi^{-1} N^2 \log N) \log N^2 = 4\pi^{-1} N^2 \log^2 N. \tag{L9a}$$

It is easy to see that v_N is an asymptotic upper bound for V_N: use the formula (L8a) for mean hitting times, the exponential approximation for their distribution, and Boole's inequality. It is not clear how to argue a lower bound: for w near v_N the events $\{T_j > w\}$ each have probability around $1/N$ but are locally very dependent.

A related example concerns simple symmetric random walk on the infinite 2-dimensional lattice: what is the time W_N taken to visit every state of a $N \times N$ square, say $S = \{1, 2, \ldots, N-1\} \times \{0, 1, \ldots, N-1\}$? Write $a_N(t)$ for the time spent in S up to time t. It is known that

$$\frac{a_N(t)}{N^2 \log t} \xrightarrow{\mathcal{D}} \xi \quad \text{as } t \to \infty \tag{L9b}$$

where ξ has an exponential distribution. Let V_N be the time spent in S until time W_N; then (L9b) suggests

$$\frac{V_N}{N^2 \log W_N} \overset{\mathcal{D}}{\approx} \xi. \tag{L9c}$$

Now the random walk, observed only on S, behaves like simple symmetric random walk on the interior of S, with some more complicated boundary behavior; this process differs from random walk on $\boldsymbol{Z}^2$ modulo N only in the boundary behavior (both have uniform stationary distribution), so the behavior of the covering time V should be similar. Applying the upper bound v_N of (L9a), we see from (L9c) that $\log W_N$ should be asymptotically bounded by $\xi^{-1} \times 4\pi^{-2} \log^2 N$. In particular,

$$\frac{\log W_N}{\log^2 N} \quad \text{is tight} \quad \text{as } N \to \infty. \tag{L9d}$$

Kesten has unpublished work proving (L9d), the bound for unrestricted random walk covering the $N \times N$ square.

M The Eigenvalue Method

M1 Introduction. Consider a stationary process $(X_t : t > 0)$ and a first hitting time $T = \min\{ t : X_t \in B \}$. Under many circumstances one can show that T must have an exponential tail:

$$\boldsymbol{P}(T > t) \sim A\exp(-\lambda t) \quad \text{as } t \to \infty$$

and give an eigenvalue interpretation to λ. (In discrete time, "exponential" becomes "geometric", of course.) In the simplest example of finite Markov chains this is a consequence of Perron-Frobenius theory reviewed below. See Seneta (1981) and Asmussen (1987) for more details.

Proposition M1.1 *Let $\widehat{\mathbf{P}}$ be a finite substochastic matrix which is irreducible and aperiodic. Then $\widehat{\mathbf{P}}^n(i,j) \sim \theta^n \beta_i \alpha_j$ as $n \to \infty$, where*

> *θ is the eigenvalue of $\widehat{\mathbf{P}}$ for which $|\theta|$ is largest;*
>
> *θ is real; $\theta = 1$ if $\widehat{\mathbf{P}}$ is stochastic, $0 < \theta < 1$ otherwise;*
>
> *α and β are the corresponding eigenvectors $\alpha\widehat{\mathbf{P}} = \theta\alpha$, $\widehat{\mathbf{P}}\beta = \theta\beta$, normalized so that $\sum \alpha_i = 1$ and $\sum \alpha_i\beta_i = 1$.*

Here "substochastic" means $\widehat{\boldsymbol{P}}(i,j) \geq 0$, $\sum_j \widehat{\boldsymbol{P}}(i,j) \leq 1$, and "irreducible, aperiodic" is analogous to the usual Markov (= "stochastic") sense. In the Markov case, $\theta = 1$, $\beta = \underset{\sim}{1}$, $\alpha = \pi$, the stationary distribution, so we get the usual theorem about convergence to the stationary distribution.

Now consider a discrete-time Markov chain with state space J and transition matrix $\boldsymbol{P}$. Let $A \subset J$ and suppose A^c is finite. Let $\widehat{\boldsymbol{P}}$ be $\boldsymbol{P}$ restricted to A^c. Then

$$\widehat{\boldsymbol{P}}^n(i,j) = \boldsymbol{P}_i(X_n = j, T_A > n).$$

So Proposition M1.1 tells us that, provided $\widehat{\boldsymbol{P}}$ is irreducible aperiodic,

$$\boldsymbol{P}_i(X_n = j, T_A > n) \sim \theta^n \beta_i \alpha_j \quad \text{as } n \to \infty, \tag{M1a}$$

where θ, α, β are the eigenvectors and eigenvalues of $\widehat{\boldsymbol{P}}$ as in Proposition M1.1. Let me spell out some consequences of this fact.

> For any initial distribution, the distribution of T_A has a geometric tail: $\boldsymbol{P}(T_A > n) \sim c\theta^n$ as $n \to \infty$. (M1b)

For any initial distribution, $\text{dist}(X_n \mid T_A > n) \xrightarrow{\mathcal{D}} \alpha$ as $n \to \infty$. (M1c)

If X_0 has distribution α then $\text{dist}(X_n \mid T_A > n) = \alpha$ for all n, so T_A has exactly geometric distribution:

$$\boldsymbol{P}_\alpha(T_A = i) = (1-\theta)\theta^{i-1}; \quad E_\alpha T = 1/(1-\theta) \tag{M1d}$$

Because of (M1d), α is called the *quasi-stationary distribution.*

There is an obvious analogue of (M1a) for continuous-time chains. There are analogues for continuous-space processes which we will see later.

It is important to emphasize that the "exponential tail" property for a hitting time T is much more prevalent than the "approximately exponential distribution" property of T. In Section B24 we discussed settings where T should have approximately exponential distribution: in such settings, the eigenvalue method provides an alternative (to our clumping methods) way to estimate the exponential parameter, i.e. to estimate ET. This is what I shall call a "level-1" use of the eigenvalue technique. It is hard to exhibit a convincing example in the context of Markov chains: in examples where one can calculate λ fairly easily it usually turns out that (a) one can do the clumping calculation more easily, and (b) for the same amount of analysis one can find a transform of T, which yields more information.

Instead, we shall mostly discuss "level-2" uses, as follows. The Poisson clumping method in this book treats families (A_i) of rare events. Mostly it is easy to calculate the $\boldsymbol{P}(A_i)$ and the issue is understanding the dependence between the events. Occasionally we need the eigenvalue method to calculate the $\boldsymbol{P}(A_i)$ themselves: this is a "level-2" application. We can then usually rely on

M2 The asymptotic geometric clump principle. Suppose (X_n) is stationary and has no long-range dependence. Suppose we have a notion of "special" strings such that

$$(x_1, x_2, \ldots, x_n) \text{ special implies } (x_1, x_2, \ldots, x_{n-1}) \text{ special;} \tag{M2a}$$

$$\boldsymbol{P}((X_1, X_2, \ldots, X_n) \text{ special}) \sim c\theta^n \quad \text{as } n \to \infty. \tag{M2b}$$

Let L_K be the length of the longest special string $(X_i, X_{i+1}, \ldots, X_{i+L_K-1})$ contained in $(X_1, \ldots, X_K)$. Then

$$\boldsymbol{P}(L_K \le l) \approx \exp(-c(1-\theta)\theta^{l+1}K) \qquad \text{for } K \text{ large.} \tag{M2c}$$

This corresponds to the limit behavior for maxima of i.i.d sequences with geometric tails. More crudely, we have

$$\frac{L_K}{\log K} \to \frac{1}{\log(1/\theta)} \quad \text{a.s.} \quad \text{as } K \to \infty. \tag{M2d}$$

To argue (M2c), fix l and let $\mathcal{S}$ be the random set of times n such that $(X_n, X_{n+1}, \ldots, X_{n+l})$ is special. Applying the clumping heuristic to $\mathcal{S}$, we have $p = \boldsymbol{P}(n \in \mathcal{S}) \approx c\theta^{l+1}$. And $\boldsymbol{P}(n+1 \notin \mathcal{S} \mid n \in \mathcal{S}) \approx 1-\theta$ by (M1b) and (M1c), so (M2c) follows from the ergodic-exit form (A9h) of the heuristic.

Here is the fundamental example.

M3 Example: Runs in subsets of Markov chains. Let X_n be a discrete-time stationary chain with transition matrix $\boldsymbol{P}$ and stationary distribution π. Let B be a finite subset of states. Let P' be $\boldsymbol{P}$ restricted to B. Suppose P' is irreducible aperiodic. Then Proposition M1.1 applies, and P' has leading eigenvalue θ and eigenvectors α, β, say. Let L_K be the longest run of $(X_1, \ldots, X_K)$ in B. Then by (M1a)

$$\boldsymbol{P}(X_1, X_2, \ldots, X_n \text{ all in } B) \sim c\theta^{n-1}, \quad \text{where } c = \sum_{i \in B} \pi_i \beta_i.$$

So our principle (Section M2) yields (replacing c by c/θ)

$$\boldsymbol{P}(L_K \leq l) \approx \exp(-c(1-\theta)\theta^l K) \qquad \text{for } K \text{ large} \qquad \text{(M3a)}$$

or more crudely $L_K / \log K \to 1/\log(1/\theta)$.

This example covers some situations which are at first sight different.

M4 Example: Coincident Markov chains. Let $\boldsymbol{P}^X$, $\boldsymbol{P}^Y$ be transition matrices for independent stationary Markov chains (X_n), (Y_n) on the same state space. Let L_K be the length of the longest coincident run $X_i = Y_i$, $X_{i+1} = Y_{i+1}$, ..., $X_{i+L_K-1} = Y_{i+L_K-1}$ up to time K. This example is contained in the previous example by considering the product chain $Z_n = (X_n, Y_n)$ and taking B to be the diagonal; the matrix P' in Example M3 can be identified with the matrix

$$P'(s,t) = \boldsymbol{P}^X(s,t)\boldsymbol{P}^Y(s,t) \qquad \text{(elementwise multiplication).}$$

So (M3a) holds, for θ and c derived from P' as before.

At Example F9 we saw an extension of this example to block matching. Other extensions, to semi-Markov processes, are given in Fousler and Karlin (1987).

M5 Example: Alternating runs in i.i.d. sequences. Let (X_n) be an i.i.d. sequence with some continuous distribution. Call $x_1, \ldots, x_n$ alternating if $x_1 < x_2 > x_3 < x_4 > \cdots$. Let L_K be the length of the longest alternating string in $(X_1, \ldots, X_K)$. We shall show that

$$\boldsymbol{P}(X_1, X_2, \ldots, X_n \text{ alternating}) \sim 2\left(\frac{2}{\pi}\right)^{n-1} \quad \text{as } n \to \infty; \qquad \text{(M5a)}$$

and then the geometric clump principle (Section M2) gives

$$\boldsymbol{P}(L_K \leq l) \approx \exp\left(-2\left(1-\frac{2}{\pi}\right)\left(\frac{2}{\pi}\right)^l K\right) \qquad \text{for } K \text{ large.} \qquad \text{(M5b)}$$

To show (M5a), note first that the distribution of X is irrelevant, so we may take it to be uniform on $[0,1]$. Write

$$\begin{aligned} Y_n &= \begin{cases} X_n & n \text{ odd} \\ 1-X_n & n \text{ even} \end{cases} \\ Z_n &= Y_n + Y_{n+1} \end{aligned}$$

Then

$$X_1, X_2, \ldots, X_n \text{ alternating iff } Z_1 < 1, Z_2 < 1, \ldots, Z_n < 1. \qquad \text{(M5c)}$$

Consider the process X killed when it first is not alternating. Then Y evolves as the Markov process on $[0,1]$ with transition density $P'(y_1, y_2) = 1_{(y_1+y_2<1)}$. For the continuous-space analogue of Proposition M1.1 we seek an eigenvalue θ and normalized eigenfunctions $\alpha(x)$, $\beta(y)$ such that

$$\begin{aligned} \int_0^1 \alpha(x) P'(x,y)\, dx &= \theta\alpha(y) \\ \int_0^1 P'(x,y)\beta(y)\, dy &= \theta\beta(x). \end{aligned}$$

It is easy to obtain the solution

$$\begin{aligned} \theta &= \frac{2}{\pi} \\ \alpha(y) &= \frac{2}{\pi}\cos(\pi y/2) \\ \beta(x) &= \pi\cos(\pi x/2). \end{aligned}$$

The continuous-space analogue of Proposition M1.1 gives

$$P'^n(x,(0,1)) \sim \beta(x)\theta^n \quad \text{as } n \to \infty.$$

Using (M5c) we have

$$\begin{aligned} \boldsymbol{P}(X_1, X_2, \ldots, X_n \text{ alternating}) &= \int_0^1 P'^{(n-1)}(x,(0,1))\, dx \\ &\sim \theta^{n-1}\int_0^1 \beta(x)\, dx, \end{aligned}$$

and this gives (M5a).

Recall that at Example L3 we discussed increasing runs; they behave differently because clump lengths are not geometric.

Another instance of a continuous-space version of Proposition M1.1 is related to our discussion of "additive Markov processes" at Section C11. Let ξ have density f, $E\xi < 0$, $E\exp(\theta\xi) = 1$. Consider the random walk X with steps ξ, killed on entering $(-\infty, 0]$. This has transition density

$$P'(x,y) = f(y-x); \qquad x,y > 0.$$

Now θ can be regarded as an eigenvalue of P' and, under technical conditions analogous to positive-recurrence, there exist normalized eigenfunctions $\alpha(x)$, $\beta(x)$ such that

$$\begin{aligned} \textstyle\int \alpha(x)P'(x,y)\,dx &= \theta\alpha(y) \\ \textstyle\int P'(x,y)\beta(y)\,dy &= \theta\beta(x) \\ \boldsymbol{P}_x(X_n \in dy) &\sim \beta(x)\alpha(y)\theta^n\,dy \quad \text{as } n \to \infty. \end{aligned} \tag{M5d}$$

Here is an integer-valued example.

M6 Example: Longest busy period in M/G/1 queue. Let X_n be the number of customers when the n'th customer begins service. During a busy period, X evolves as the random walk with step distribution $\xi = A-1$, where A is the number of arrivals during a service. Let θ, β be the eigenvalue and eigenfunction of (M5d). Then

$$\begin{aligned} &\boldsymbol{P}(\text{at least } n \text{ customers served in busy period} X_0 = x) \\ &\qquad = P'^{(n-1)}(x,[1,\infty)) \\ &\qquad \sim \beta(x)\theta^{n-1} \quad \text{as } n \to \infty \end{aligned}$$

So

$$\begin{aligned} &\boldsymbol{P}(\text{at least } n \text{ customers served in busy period}) \\ &\qquad \sim c\theta^{n-1}. \end{aligned}$$

where $c = \sum \pi(x)\beta(x)$ and where π is the stationary distribution. Let L_K be the largest number of customers served during busy period, amongst the first K customers. Then the geometric clump principle (Section M2) gives

$$\boldsymbol{P}(L_K \le l) \approx \exp(-c(1-\theta)\theta^l K) \qquad \text{for large } K. \tag{M6a}$$

We do not want to go into details about continuous time-and-space models, but let us just sketch the simplest examples.

M7 Example: Longest interior sojourn of a diffusion. Let X_t be a 1-dimensional diffusion, restricted to $[a, b]$ by reflecting boundaries, with stationary density π. Let T be the first hitting time on $\{a, b\}$. Then we expect

$$\begin{aligned} \boldsymbol{P}_x(T > t) &\sim \beta(x)\exp(-\theta t) \\ \boldsymbol{P}(X_t \in dy \mid T > t) &\to \alpha(y)\,dy \end{aligned} \qquad \text{(M7a)}$$

where θ, α, β, are the leading eigenvalue and associated eigenfunctions of the generator of the diffusion killed at $\{a, b\}$. In particular, for the stationary process

$$\boldsymbol{P}(T > t) \sim c\exp(-\theta t); \qquad c = \int \beta(x)\pi(x)\,dx.$$

Let L_t be the longest time interval in $[0, t]$ during which the boundary is not hit. The continuous-time analogue of the principle (Section M2) now yields

$$\boldsymbol{P}(L_t < l) \approx \exp(-c\theta e^{-\theta l} t); \qquad t \text{ large.}$$

Naturally L_t can be rescaled to yield the usual extreme-value limit distribution.

As a particular case of (M7a), consider Brownian motion on $[-b, b]$. Then it is easy to calculate

$$\begin{aligned} \theta &= \tfrac{\pi^2}{8b^2} \\ \alpha(x) &= \tfrac{\pi}{4b}\cos\left(\tfrac{\pi x}{2b}\right) \\ \beta(x) &= \tfrac{4}{\pi}\cos\left(\tfrac{\pi x}{2b}\right). \end{aligned} \qquad \text{(M7b)}$$

These were used at Example K5 to study small increments of Brownian motion.

M8 Example: Boundary crossing for diffusions. For a stationary diffusion X on the real line, we expect the first hitting time T on b to have the form

$$\begin{aligned} \boldsymbol{P}(T > t) &\sim c\exp(-\lambda(b)t) && \text{as } t \to \infty \\ \boldsymbol{P}(T < t + \delta \mid T > t) &\sim \lambda(b)\delta && \text{as } t \to \infty, \end{aligned}$$

where $\lambda(b)$ is the leading eigenvalue associated with the diffusion killed at b. If instead we are interested in the first crossing time T of a moving barrier $b(t) \to \infty$ slowly, then similarly we expect

$$\boldsymbol{P}(T < t + \delta \mid T > t) \sim \lambda(b(t))\delta \qquad \text{as } t \to \infty,$$

and hence

$$\boldsymbol{P}(T > t) \sim c' \exp\left(-\int_0^t \lambda(b(s))\,ds\right) \qquad \text{as } t \to \infty.$$

Bass and Erickson (1983) give formalizations of this idea. Of course the arguments at Section D13, where applicable, give stronger and more explicit information about the non-asymptotic distribution of T.

Postscript

The examples were collected haphazardly over the period 1983–1987. Current issues of the probability journals usually have one or two papers dealing with problems related to our heuristic, so it is hard to know when to stop adding examples. There are a few areas which, given unlimited time, I would have like to go into more deeply. My scattered examples on queueing are fairly trite; it would be interesting to study hard examples. The area of physics illustrated by Kramer's problem (Example I14) seems a rich source of potential examples. There is a huge area of "data-snooping statistics" where you have a family of test statistics $T(a)$ whose null distribution is known for fixed a, but where you use the test statistic $T = T(a)$ for some a chosen using the data. Here one can hope to estimate the tail of the null distribution of T, similar to the Kolmogorov-Smirnov type statistics of Chapter J.

In this book I have tried to explain the heuristics and direct the reader to what has been proved, in various special areas. I will be well-satisfied if applied researchers are convinced to add the heuristic as one more little tool in their large toolkit. For theoreticians, I have already made some remarks on the relationship between heuristics and theory: let me end with one more. A mathematical area develops best when it faces hard concrete problems which are not in the "domain of attraction" of existing proof techniques. An area develops worst along the "lines of least resistance" in which existing results are slightly generalized or abstracted. I hope this book will discourage theoreticians from the pursuit of minor variations of the known and the formalization of the heuristically obvious, and encourage instead the pursuit of the unknown and the unobvious.

Bibliography

J. Abrahams (1984a). Distribution of the supremum of the two-parameter Slepian process on the boundary of the unit square. *Stochastic Proc. Appl.*, 18:179–185.

J. Abrahams (1984b). A survey of recent progress on level crossing problems for random processes. In I.F. Blake and H.B. Poor, editors, *Communications and Networks*, Springer-Verlag, New York, 1984.

J. Abrahams and B.L. Montgomery (1985). *The Distribution of the Maximum of Particular Random Fields.* Technical Report, Carnegie-Mellon, 1985.

O. Adelman (1985). Brownian motion never increases: a new proof to a result of Dvoretsky, Erdos and Kakutani. *Israel Journal of Mathematics*, 50:189–192.

R.J. Adler (1981). *The Geometry of Random Fields.* Wiley, New York, 1981.

R.J. Adler (1984). The supremum of a particular Gaussian field. *Annals of Probability*, 12:436–444.

R.J. Adler and L.D. Brown (1986). Tail behavior for suprema of empirical processes. *Annals of Probability*, 14:1–30.

R.J. Adler and G. Samorodnitsky (1987). Tail behavior for the suprema of Gaussian processes with a view towards empirical processes. *Annals of Probability*, 15:1339–1351.

M. Aizenmann (1985). The intersection of Brownian paths as a case study of a renormalization group method for quantum field theory. *Commun. Math. Phys.*, 97:91–110.

D.J. Aldous (1982). Markov chains with almost exponential hitting times. *Stochastic Proc. Appl.*, 13:305–310.

D.J. Aldous (1983a). On the time taken by random walks on finite groups to visit every state. *Z. Wahrsch. Verw. Gebiete*, 62:361–374.

D.J. Aldous (1983b). Random walks on finite groups and rapidly mixing Markov chains. In *Seminaire de Probabilites XVII*, pages 243–297, Springer-Verlag, New York, 1983.

D.J. Aldous (1985). Self-intersections of random walks on discrete groups. *Math. Proc. Camb. Phil. Soc.*, 9:155–177.

D.J. Aldous (1986). Some interesting processes arising as heavy traffic limits in a M/M/∞ storage process. *Stochastic Proc. Appl.*, 22:291–313.

D.J. Aldous (1988a). The harmonic mean formula for probabilities of unions: applications to sparse random graphs. *Discrete Mathematics*. To appear.

D.J. Aldous (1988b). Sparse mosaic limit theory. 1988. In preparation.

D.J. Aldous (1988c). *Stein's Method in a Two-Dimensional Coverage Problem.* Technical Report, U.C. Berkeley, 1988.

D.J. Aldous and P. Diaconis (1986). Shuffling cards and stopping times. *Amer. Math. Monthly*, 8:333–348.

D.J. Aldous and P. Diaconis (1987). Strong uniform times and finite random walks. *Adv. Appl. Math.*, 8:69–87.

V. Anantharam (1988). *The Optimal Buffer Allocation Problem.* Technical Report, Cornell, 1988.

V.V. Anisimov and A.I. Chernyak (1982). Limit theorems for some rare functionals on Markov chains and semi-Markov processes. *Theory Prob. Math. Stat.*, 26:1–6.

M. Aronowich and R.J. Adler (1986). Extrema and level crossings of χ^2 processes. *Adv. Appl. Prob.*, 18:901–920.

R. Arratia, L. Goldstein, and L. Gordon (1987). *Two Moments Suffice for Poisson Approximations: the Chen-Stein Method.* Technical Report, U.S.C., 1987.

R. Arratia, L. Gordon, and M.S. Waterman (1984). An extreme value theory for sequence matching. *Annals of Statistics*, 14:971–993.

R. Arratia, L. Gordon, and M.S. Waterman (1988). *The Erdos-Renyi Law in Distribution, for Coin-Tossing and Sequence-Matching.* Technical Report, U.S.C., 1988.

R. Arratia and M.S. Waterman (1985a). Critical phenomena in sequence matching. *Annals of Probability*, 13:13–23.

R. Arratia and M.S. Waterman (1985b). An Erdos-Renyi law with shifts. *Advances in Math.*, 55:1236–1249.

R. Arratia and M.S. Waterman (1988). *The Erdos-Renyi Strong Law for Pattern-Matching with a Given Proportion of Mismatches.* Technical Report, U.S.C., 1988.

S. Asmussen (1987). *Applied Probability and Queues.* Wiley, New York, 1987.

A.D. Barbour and G.K. Eagleson (1983). Poisson approximation for some statistics based on exchangeable trials. *Adv. Appl. Prob.*, 15:585–600.

A.D. Barbour and L. Holst (1987). *Some Applications of the Stein-Chen Method for Proving Poisson Approximation.* Technical Report, Zurich Univ., 1987.

R.F. Bass and K.B. Erickson (1983). Hitting time of a moving boundary for a diffusion. *Stochastic Proc. Appl.*, 14:315–325.

P. Baxendale (1984). Recurrence of a diffusion process to a shrinking target. *J. London Math. Soc.*, 32:166–176.

R.V. Benveneto (1984). The occurrence of sequence patterns in ergodic Markov chains. *Stochastic Proc. Appl.*, 17:369–373.

S.M. Berman (1982a). Sojourns and extremes of a diffusion process on a fixed interval. *Adv. Appl. Prob.*, 14:811–832.

S.M. Berman (1982b). Sojourns and extremes of stationary processes. *Annals of Probability*, 10:1–46.

S.M. Berman (1983a). High level sojourns of a diffusion process on a long interval. *Z. Wahrsch. Verw. Gebiete*, 62:185–199.

S.M. Berman (1983b). Sojourns of stationary processes in rare sets. *Annals of Probability*, 11:847–866.

S.M. Berman (1984). Sojourns of vector Gausian processes inside and outside spheres. *Z. Wahrsch. Verw. Gebiete*, 66:529–542.

S.M. Berman (1986a). Extreme sojourns for random walks and birth-and-death processes. *Comm. Stat.- Stoch. Mod.*, 2:393–408.

S.M. Berman (1986b). Limit theorems for sojourns of stochastic processes. In *Probability on Banach Spaces V*, Springer-Verlag, 1986.

S.M. Berman (1987). Extreme sojourns of a Gaussian process with a point of maximum variance. *Probab. Th. Rel. Fields*, 74:113–124.

S.M. Berman (1988). Extreme sojourns of diffusion processes. *Annals of Probability*, 16:361–374.

P. Bickel and M. Rosenblatt (1973). Two-dimensional random fields. In P.K. Krishnaiah, editor, *Multivariate Analysis III*, Academic Press, New York, 1973.

J.D. Biggins and C. Cannings (1987). Markov renewal processes, counters and repeated sequences in Markov chains. *Adv. Appl. Prob.*, 19:521–545.

G. Blom and D. Thorburn (1982). How many random digits are required until given sequences are obtained. *J. Appl. Prob.*, 19:518–531.

B. Bollobas (1985). *Random Graphs.* Academic Press, New York, 1985.

M. Bramson and R. Durrett (1988). *Random Walk in Random Environment: a Counter-Example.* Technical Report, Cornell, 1988.

A. Buonocore, A.G. Nobile, and L.M. Ricciardi (1987). A new integral equation for the evaluation of first-passage time probability densities. *Adv. Appl. Prob.*, 19:784–800.

E.M. Cabana and M. Wschebor (1982). The two-parameter Brownian bridge: Kolmogorov inequalities and upper and lower bounds for the distribution of the maximum. *Annals of Probability*, 10:289–302.

J. Chover (1966). A law of the iterated logarithm for stable summands. *Proc. Amer. Math. Soc.*, 17:441–443.

K.L. Chung, P. Erdos, and T. Sirao (1959). On the Lipschitz condition for Brownian motion. *J. Math. Soc. Japan*, 11:263–274.

V. Chvatal and D. Sankoff (1975). Longest common subsequences of two random sequences. *J. Appl. Prob.*, 12:306–315.

P. Clifford, N.J.B. Green, and M.J. Pilling (1987). Statistical models of chemical kinetics in liquids. *J. Roy. Statist. Soc.*, B 49:266–300.

E.G. Coffman, T.T. Kadota, and L.A. Shepp (1985). A stochastic model of fragmentation in dynamic storage allocation. *SIAM J. Computing*, 14:416–425.

R. Cogburn (1985). On the distribution of first passage and return times for small sets. *Annals of Probability*, 13:1219–1223.

J.T. Cox (1988). *Coalescing Random Walks and Voter Model Consensus Times on the Torus in* Z^D. Technical Report, Cornell, 1988.

E. Csaki and A. Foldes (1984). The narrowest tube of a recurrent random walk. *Z. Wahrsch. Verw. Gebiete*, 66:387–405.

M. Csorgo and P. Revesz (1981). *Strong Approximations in Probability and Statistics.* Academic Press, New York, 1981.

R.W.R. Darling and M.S. Waterman (1985). Matching rectangles in d dimensions: algorithms and laws of large numbers. *Advances in Math.*, 55:1–12.

R.W.R. Darling and M.S. Waterman (1986). Extreme value distribution for the largest cube in a random lattice. *SIAM J. Appl. Math.*, 46:118–132.

F.N. David and D.E. Barton (1962). *Combinatorial Chance.* Griffin, London, 1962.

B. Davis and E. Perkins (1985). Brownian slow points: the critical case. *Annals of Probability*, 13:779–803.

R.A. Davis (1982). Maximum and minimum of one-dimensional diffusions. *Stochastic Proc. Appl.*, 13:1–9.

R.A. Davis and S. Resnick (1985). Limit theory for moving averages of random variables with regularly varying tail probabilities. *Annals of Probability*, 13:179–195.

M.V. Day (1983). On the exponential limit law in the small parameter exit problem. *Stochastics*, 8:297–323.

M.V. Day (1987). Recent progress on the small parameter exit problem. *Stochastics*, xxx.

P. Deheuvels and L. Devroye (1984). Strong laws for the maximal k-spacings when $k < c \log n$. *Z. Wahrsch. Verw. Gebiete*, 66:315–334.

G. Deken (1979). Some limit theorems for longest common subsequences. *Discrete Mathematics*, 26:17–31.

P. Diaconis and F. Mosteller (1989). Coincidences. 1989. Book in Preparation.

J. Durbin (1985). The first-passage density of a continuous Gaussian process to a general boundary. *J. Appl. Prob.*, 22:99–122.

R. Durrett (1984). Some general results concerning the critical exponents of percolation processes. *Z. Wahrsch. Verw. Gebiete*, 69:421–437.

R. Durrett (1985). Reversible diffusion processes. In J. Chao and W. Woyczynski, editors, *Probability and Harmonic Analysis*, Dekker, New York, 1985.

R. Durrett (1986). Multidimensional random walks in random environments with subclassical limiting behavior. *Commun. Math. Phys.*, 104:87–102.

A. Dvoretzky, P. Erdos, and S. Kakutani (1961). Nonincrease everywhere of the Brownian motion process. In *Fourth Berkeley Symposium*, pages 103–116, University of California Press, 1961.

E.B. Dynkin (1962). *Markov Processes.* Springer-Verlag, New York, 1962.

E.B. Dynkin (1988). Self-intersection gauge for random walks and for Brownian motion. *Annals of Probability*, 16:1–57.

W. Feller (1968). *An Introduction to Probability Theory and Its Applications.* Volume 1, Wiley, New York, third edition, 1968.

L. Flatto, A. Odlyzko, and D.B. Wales (1985). Random shuffles and group representations. *Annals of Probability*, 13:154–178.

D.E. Fousler and S. Karlin (1987). Maximal success durations for a semimarkov process. *Stochastic Proc. Appl.*, 24:203–224.

D. Freedman (1971). *Brownian Motion and Diffusion.* Holden-Day, San Francisco, 1971.

P. Gaenssler (1983). *Empirical Processes. Lecture Notes 3*, Institute of Mathematical Statistics, Hayward, CA, 1983.

J. Galombos (1978). *The Asymptotic Theory of Extreme Order Statistics.* Wiley, New York, 1978.

J. Gani (1988). Extremes of random processes in applied probability. *Adv. Appl. Prob.*, 20:1–13.

C.W. Gardiner (1983). *Handbook of Stochastic Methods.* Springer-Verlag, New York, 1983.

D.H. Gates and M. Westcott (1985). Accurate and asymptotic results for distributions of scan statistics. *J. Appl. Prob.*, 22:531–542.

D. Geman and J. Horowitz (1980). Occupation densities. *Annals of Probability*, 8:1–67.

H.U. Gerber and S.R. Li (1981). The occurrence of sequence patterns in repeated experiments and hitting times in a Markov chain. *Stochastic Proc. Appl.*, 11:101–108.

I.B. Gertsbakh (1984). Asymptotic methods in reliability theory: a review. *Adv. Appl. Prob.*, 16:147–175.

L. Gordon, M.F. Schilling, and M.S. Waterman (1986). An extreme value theory for long head runs. *Probab. Th. Rel. Fields*, 72:279–287.

P. Groeneboom (1988). Brownian motion with a parabolic drift and Airy functions. *Probab. Th. Rel. Fields.*

L.J. Guibas and A.M. Odlyzko (1980). Long repetitive patterns in random sequences. *Z. Wahrsch. Verw. Gebiete*, 53:241–262.

B. Hajek (1982). Hitting-time and occupation-time bounds implied by drift analysis with applications. *Adv. Appl. Prob.*, 14:502–525.

P. Hall (1985a). Distribution of size, structure and number of vacant regions in a high-intensity mosaic. *Z. Wahrsch. Verw. Gebiete*, 70:237–261.

P. Hall (1985b). On the coverage of k-dimensional space by k-dimensional spheres. *Annals of Probability*, 13:991–1002.

P. Hall (1988). *Introduction to the Theory of Coverage Processes.* Wiley, New York, 1988.

D.L. Hanson and R.P. Russo (1983a). Some more results on increments of the Wiener process. *Annals of Probability*, 11:1009–1015.

D.L. Hanson and R.P. Russo (1983b). Some results on increments of the Wiener process with applications to lag sums of i.i.d. random variables. *Annals of Probability*, 11:609–623.

J.M. Harrison and R.J. Williams (1987). Multidimensional reflecting Brownian motions having exponential stationary distributions. *Annals of Probability*, 15:115–137.

M.L. Hogan and D. Siegmund (1986). Large deviations for the maxima of some random fields. *Adv. Appl. Math.*, 7:2–22.

W.T.F. Den Hollander (1984). Random walks on lattices with randomly distributed traps. *J. Stat. Phys.*, 37:331–367.

L. Holst (1986). On birthday, collector's, occupancy and other classical urn problems. *International Statistical Review*, 54:15–27.

D.L. Huber (1983). Trapping of excitation in the average T-matrix approximation. *J. Stat. Phys.*, 30:345–354.

J.J. Hunter (1983). *Mathematical Techniques of Applied Probability.* Academic Press, New York, 1983.

D.L. Iglehart (1972). Extreme values in the GI/G/1 queue. *Ann. Math. Stat.*, 43:627–635.

J.P. Imhof (1986). On the time spent above a level by Brownian motion with negative drift. *Adv. Appl. Prob.*, 18:1017–1018.

R. Isaac (1987). A limit theorem for probabilities related to the random bombardment of a square. *Acta. Math. Hungar.* to appear.

D. Isaacson and R. Madsen (1976). *Markov Chains.* Wiley, New York, 1976.

V.A. Ivanov, G.I. Ivchenko, and Y.I Medvedev (1984). Discrete problems in probability theory. *J. Soviet Math.*, 31:2759–2795.

S. Janson (1986). Random coverings in several dimensions. *Acta Math.*, 156:83–118.

C. Jennen (1985). Second-order approximations to the density, mean and variance of Brownian first-exit times. *Annals of Probability*, 13:126–144.

N.L. Johnson and S. Kotz (1977). *Urn Models and Their Applications.* Wiley, New York, 1977.

J.P. Kahane (1985). *Some Random Series of Functions.* Cambridge University Press, second edition, 1985.

I. Karatzas and S.E. Shreve (1987). *Brownian Motion and Stochastic Calculus.* Springer-Verlag, New York, 1987.

S. Karlin and F. Ost (1987). Counts of long aligned word matches among random letter sequences. *Adv. Appl. Prob.*, 19:293–351.

S. Karlin and F. Ost (1988). Maximal length of common words among random letter sequences. *Annals of Probability*, 16:535–563.

S. Karlin and H.M. Taylor (1975). *A First Course in Stochastic Processes.* Academic Press, New York, 1975.

S. Karlin and H.M. Taylor (1982). *A Second Course in Stochastic Processes.* Academic Press, New York, 1982.

N. Karmarkar, R.M. Karp, G.S. Lueker, and A.D. Odlyzko (1986). Probabilistic analysis of optimum partitioning. *J. Appl. Prob.*, 23:626–645.

R.M. Karp and J.M. Steele (1985). Probabilistic analysis of heuristics. In E.L. Lawler et al, editor, *The Traveling Salesman Problem*, Wiley, New York, 1985.

J. Keilson (1979). *Markov Chain Models — Rarity and Exponentiality.* Springer-Verlag, New York, 1979.

F.P. Kelly (1979). *Reversibility and Stochastic Networks.* Wiley, New York, 1979.

J.G. Kemeny and J.L Snell (1959). *Finite Markov Chains.* Van Nostrand, Princeton NJ, 1959.

J. Kemperman (1961). *The First Passage Problem for Stationary Markov Chains.* Institute of Mathematical Statistics, 1961.

H. Kesten (1987). Percolation theory and first-passage percolation. *Annals of Probability*, 15:1231–1271.

J.F.C Kingman (1976). Subadditive processes. In *Ecole D'Ete V*, Springer-Verlag, New York, 1976.

C. Knessl, B.J. Matkowsky, Z. Schuss, and C. Tier (1985). An asymptotic theory of large deviations for Markov jump processes. *SIAM J. Appl. Math.*, 45:1006–1028.

C. Knessl, B.J. Matkowsky, Z. Schuss, and C. Tier (1986a). A finite-capacity single-server queue with customer loss. *Comm. Stat.- Stoch. Mod.*, 2:97–121.

C. Knessl, B.J. Matkowsky, Z. Schuss, and C. Tier (1986b). Asymptotic analysis of a state-dependent M/G/1 queueing system. *SIAM J. Appl. Math.*, 46:483–505.

V.F. Kolchin, B.A. Sevast'yanov, and V.P. Chistyakov (1978). *Random Allocations.* Winston, Washington, D.C., 1978.

D.V. Korolyuk and D.S. Sil'vestrov (1984). Entry times into asymptotically receding domains for ergodic Markov chains. *Theory Prob. Appl.*, 19:432–442.

M.R. Leadbetter (1983). Extremes and local dependence in stationary sequences. *Z. Wahrsch. Verw. Gebiete*, 65:291–306.

M.R. Leadbetter, G. Lindgren, and H. Rootzen (1983). *Extremes and Related Properties of Random Sequences and Processes.* Springer-Verlag, New York, 1983.

R. Leadbetter and H. Rootzen (1988). Extremal theory for stochastic processes. *Annals of Probability*, 16:431–478.

H.R. Lerche (1986). *Boundary Crossing of Brownian Motion. Lecture Notes in Statistics 40*, Springer-Verlag, New York, 1986.

S.R. Li (1980). A martingale approach to the study of occurrence of sequence patterns in repeated experiments. *Annals of Probability*, 8:1171–1176.

G. Lindgren (1984a). Extremal ranks and transformations of variables for extremes of functions of multivariate Gaussian processes. *Stochastic Proc. Appl.*, 17:285–312.

G. Lindgren (1984b). Use and structure of Slepian model processes for prediction and detection in crossing and extreme value theory. In J.T. De Oliveira, editor, *Statistical Extremes and Applications*, Reidel, 1984.

B.F. Logan and L.A. Shapp (1977). A variational principle for random Young's tableaux. *Advances in Math.*, 28:206–222.

M. Maejima (1982). Some limit theorems for sojourn times of strongly dependent Gaussian processes. *Z. Wahrsch. Verw. Gebiete*, 60:359–380.

B.J. Matkowsky, Z. Schuss, and E. Ben-Jacobs (1982). A singular perturbation approach to Kramer's diffusion problem. *SIAM J. Appl. Math.*, 42:835–849.

B.J. Matkowsky, Z. Schuss, and C. Tier (1984). Uniform expansion of the transition rate in Kramer's problem. *J. Stat. Phys.*, 35:443–456.

P.C. Matthews (1988a). Covering problems for Brownian motion on spheres. *Annals of Probability*, 16:189–199.

P.C. Matthews (1988b). Covering problems for Markov chains. *Annals of Probability*. To appear.

D. Mollison (1978). Markovian contact processes. *Adv. Appl. Prob.*, 10:85–108.

T.F. Mori (1988a). More on the waiting time until each of some given patterns occurs as a run. 1988.

T.F. Mori (1988b). On the waiting time until each of some given patterns occurs as a run. 1988.

J.A. Morrison (1986). Asymptotic analysis of the waiting-time distribution for a large closed processor-sharing system. *SIAM J. Appl. Math.*, 46:140–170.

A. Naess (1984). The effect of the Markov chain condition on the prediction of extreme values. *J. Sound Vibration*, 94:87–103.

V.N. Nair, L.A. Shepp, and M.J. Klass (1986). On the numbers of crossings of empirical distribution functions. *Annals of Probability*, 14:877–890.

J.I. Naus (1979). An indexed bibliography of clusters, clumps and coincidences. *International Statistical Review*, 47:47–78.

J.I. Naus (1982). Approximations for distributions of scan statistics. *JASA.*, 77:177–183.

T. Nemetz and N. Kusolitsch (1982). On the longest run of coincidences. *Z. Wahrsch. Verw. Gebiete*, 61:59–73.

T. Norberg (1987). Semicontinuous processes in multidimensional extreme value theory. *Stochastic Proc. Appl.*, 25:27–55.

G.L. O'Brien (1987). Extreme values for stationary and Markov sequences. *Annals of Probability*, 15:281–291.

B. Oksendal (1985). *Stochastic Differential Equations.* Springer-Verlag, New York, 1985.

J.T. De Oliveira (1984). *Statistical Extremes and Applications.* Reidel, Boston, 1984.

E. Orsingher (1987). On the maximum of random fields represented by stochastic integrals over circles. *J. Appl. Prob.*, 24:574–585.

E. Orsingher and F. Battaglia (1982). Probability distributions and level crossings of shot noise models. *Stochastics*, 8:45–61.

J. Ortega and M. Wschebor (1984). On the increments of the Wiener process. *Z. Wahrsch. Verw. Gebiete*, 65:329–339.

N.U. Prabhu (1980). *Stochastic Storage Systems.* Springer-Verlag, New York, 1980.

R. Pyke and D.C. Wilbour (1988). *New Approaches for Goodness of Fit Tests for Multidimensional Data.* Technical Report, Univ. Washington, 1988.

C. Qualls and H. Watanabe (1973). Asymptotic properties of Gaussian random fields. *Trans. Amer. Math. Soc.*, 177:155–171.

S. Resnick (1987). *Extreme Values, Regular Variation and Point Processes.* Springer-Verlag, New York, 1987.

P. Revesz (1983). Three problems on the length of increasing runs. *Stochastic Proc. Appl.*, 15:169–179.

S.A. Roach (1968). *The Theory of Random Clumping.* Methuen, London, 1968.

L.C.G. Rogers and D. Williams (1987). *Diffusions, Markov Processes and Martingales.* Volume 2, Wiley, New York, 1987.

H. Rootzen (1978). Extremes of moving averages of stable processes. *Annals of Probability*, 6:847–869.

H. Rootzen (1986). Extreme value theory for moving average processes. *Annals of Probability*, 14:612–652.

H. Rootzen (1988). Maxima and exceedences of stationary Markov chains. *Adv. Appl. Prob.* To appear.

S.M. Ross (1983). *Stochastic Processes.* Wiley, New York, 1983.

I. Rychlik (1987). A note on Durbin's formula for the first-passage density. *Stat. Prob. Letters*, 5:425–428.

E. Samuel-Cahn (1983). Simple approximations to the expected waiting time for a cluster of any given size, for point processes. *Adv. Appl. Prob.*, 15:21–38.

C. Schaper (1988). PhD thesis, U.C. Berkeley, 1988.

S. Schumacher (1985). Diffusions with random coefficients. In R. Durrett, editor, *Particle Systems, Random Media and Large Deviations*, American Math. Soc., 1985.

Z. Schuss (1980). *Theory and Applications of Stochastic Differential Equations.* Wiley, New York, 1980.

P.K. Sen and M.J. Wichura (1984a). Estimation of first crossing time distribution for Brownian motion processes relative to upper class boundaries. *Sankhya*, 46:24–34.

P.K. Sen and M.J. Wichura (1984b). Estimation of first crossing time distribution for n-parameter Brownian motion processes relative to upper class boundaries. *J. Multivariate Anal.*, 15:201–221.

E. Seneta (1981). *Non-Negative Matrices and Markov Chains.* Springer-Verlag, New York, 1981.

L.A. Shepp (1972a). Covering the circle with random arcs. *Israel Journal of Mathematics*, 11:328–345.

L.A. Shepp (1972b). Covering the line with random intervals. *Z. Wahrsch. Verw. Gebiete*, 23:163–170.

L.A. Shepp and S.P. Lloyd (1966). Ordered cycle lengths in a random permutation. *Trans. Amer. Math. Soc.*, 121:340–357.

L.A. Shepp and R. Vanderbei (1987). A probabilistic model for the time to unravel a strand of DNA. 1987. In preparation.

G.R. Shorack and J.A. Wellner (1986). *Empirical Processes with Applications to Statistics.* Wiley, New York, 1986.

D. Siegmund (1985). *Sequential Analysis: Tests and Confidence Intervals.* Springer-Verlag, New York, 1985.

D. Siegmund (1986). Boundary crossing probabilities and statistical analysis. *Annals of Statistics*, 14:361–404.

D. Siegmund (1988). Approximate tail probabilities for the maxima of some random fields. *Annals of Probability*, 16:487–501.

B. Silverman and T.C. Brown (1978). Short distances, flat triangles and Poisson limits. *J. Appl. Prob.*, 15:815–825.

H. Solomon (1978). *Geometric Probability.* SIAM, Philadelphia, 1978.

S. Song and M. Yor (1987). Inegalites pour les processus self-similaires arretes a un temps quelquonque. In *Seminaire Des Probabilites XXI*, Springer-Verlag, 1987.

F. Spitzer (1964). *Principles oj Random Walk.* Van Nostrand, Princeton, NJ, 1964.

C. Stein (1986). *Approximate Computation of Expectations. Lecture Notes* 7, Institute of Math. Statistics, Hayward, CA, 1986.

D. Stoyan, W.S. Kendall, and J. Mecke (1987). *Stochastic Geometry and its Applications.* Wiley, New York, 1987.

D.W. Stroock and S.R.S. Varadhan (1979). *Multidimensional Diffusion Processes.* Springer-Verlag, New York, 1979.

M.S. Taqqu (1979). Convergence of integrated processes of arbitrary Hermite rank. *Z. Wahrsch. Verw. Gebiete*, 50:53–83.

A. Tsurui and S. Osaki (1976). On a first-passage problem for a cumulative process with exponential decay. *Stochastic Proc. Appl.*, 4:79–88.

O.O. Ugbebor (1980). Uniform variation results for Brownian motion. *Z. Wahrsch. Verw. Gebiete*, 51:39–48.

E. Vanmarcke (1982). *Random Fields.* MIT Press, Cambridge, MA, 1982.

S.R.S Varadhan (1984). *Large Deviations and Applications.* SIAM, Philadelphia, 1984.

A.M. Vershik and S.V. Kerov (1977). Asymptotics of the Plancherel measure of the symmetric groups and the limiting form of Young tableaux. *Soviet Math.*, 18:527–531.

A.M Vershik and A.A. Schmidt (1977). Limit measures arising in the theory of symmetric groups I. *Theory Prob. Appl.*, 22:70–85.

J.B. Walsh (1981). A stochastic model of neural response. *Adv. Appl. Prob.*, 13:231–281.

M. Weber (1980). Analyse asymptotique des processus Gaussiens stationnaires. *Ann. Inst. Henri Poincare*, 16:117–176.

K. Yamada (1985). Some exponential type bounds for hitting time distributions of storage processes. *Stochastic Proc. Appl.*, 19:101–109.

A.M. Zubkov and V.G. Mikhailov (1974). Limit distribution of random variables associated with long duplications in a sequence of independent trials. *Theory Prob. Appl.*, 19:172–179.

Index

aggregation 173
autoregressive process 48, 59, 70

Berman's method 101
Bessel process 187, 236
binary strings 139
birth-and-death process 25
birthday problem 108ff
  blocks 124, 130
  Markov chains 118ff
  $M/M/\infty$ 128
  random walk 119, 130
block-matching: see matching
boundary-crossing 83ff
boundary layer expansion 188
Brownian bridge
  increments 213, 217
  maxima 82, 86, 198ff
  upturns 214, 217
Brownian motion 220ff
  boundary crossing 97ff, 100
  large increments 212, 221, 234
  law of iterated logarithm 85
  local maxima 230, 235
  lower boundary 182
  modulus of continuity 220, 234, 236
  points of increase 230, 235
  quadratic boundary 88
  self-intersections 233
  small increments 225ff, 235, 236
  sojourn distribution 72, 104
  sphere 183, 188
  spikes 224, 234
  tangent approximation 99
Brownian sheet 191, 202ff
  power formula 212

card shuffling 27, 41, 120
χ^2 process 68, 186, 188
Chung-Erdos-Sirao test 221
clique 132, 142
clumps 2ff
  exponential distribution 22
  geometric distribution 22, **121ff**, 247
clusters in random scatter 149ff
  1-dimensional 164
coincidences 110
  see also birthday problem
combinatorics
  exponential extrema 131ff
  processes 118ff
  simple 106ff
compound Poisson approximation 4ff, 20, 43, 73, 103
conditioning on semi-local maxima 10, 66, 90ff, 141
coupon-collector's problem: see **covering**
coverage in stochastic geometry
  discs 153, 165, 166
  multiple 159, 165
  non-uniform 160, 165
  1-dimensional 162, 166

sphere surface 161
squares 157
covering
Brownian motion on sphere 183
coupon-collector's problem 113ff
cube 133
i.i.d. blocks 127
M/M/∞ 128
random walk 125ff, 130
2-dimensional walks 244

diffusion 73ff
boundary crossing 83ff, 97ff
conditioned 105
first passage time 76ff, 101
hitting small ball 171, 187
interior sojourn 250
multidimensional 167ff
near misses 171
random environment 94
reversible 169
under a potential 78, 93, 173ff, 187
drift-jump process 57, 71
Durbin's formula 104

eigenvalue method 246ff
empirical distribution 191, 237, 238
ergodic-exit method 12, 32, 51, 55, 58, 71, 101, 128, 163, 239
exit problems 187
extreme value theory 44ff, 69
heuristic 48, 51
smooth processes 54

fundamental identity 6

Gaussian fields 190ff
isotropic 205, 219
product rule 196, 217
self-normalized 212
Gaussian process 63ff, 81, 86ff

harmonic mean formula 8, 20, 143, 146, 150ff, 206, 219, 226, 230, 232
bounds from 207
holes in random scatter 149

I5 problem 60

jitter 42

Kolmogorov-Smirnov test (multidimensional analogs) 191, 198ff
Kramers' equation 177

law of iterated logarithm
symmetric stable 241
2 parameter 215, 217
locally Brownian 73, 167, 185

marked clump method 22, 222ff
Markov chain 17, 23ff
additive 52, 58, 70
birthday problem 118ff
hitting sizable subset 32
hitting small subset 30, 42
hitting time 24ff, 38
matching problem 120ff
recurrent potential estimate 41
runs in subset 248
matching 111
i.i.d. blocks 122, 130
Markov chain 120ff, 130
min-max of 2-parameter process 242
monotonicity convention 64
mosaics 2ff, 165, 166
moving average 48, 70, 238

natural outer bound 131

order statistics 238
Ornstein-Uhlenbeck process 80
 boundary-crossing 85, 89ff, 98
 multidimensional 170
 multiparameter 196, 205ff
 smoothed 95
 unstable 93, 181

partitioning graphs 143, 147
partitioning numbers 134, 147
patterns 26, 31, 41
percolation 136, 138, 147
point process 47, 69
POIS 20
Poisson line process 151
Poissonization 107, 116, 155
potential well 93, 176ff, 187

quasi-Markov estimate 21, 105, 164
quasi-stationary distribution 247
queues
 G/G/1 52
 M/G/1 busy period 250
 M/M/1 1ff
 M/M/1 state-dependent 79
 M/M/1 tandem 28, 35
 M/M/∞ combinatorics 128
 M/M/∞ storage 91
 alternating server 37

random fields 190ff
random graph 132, 142ff, 147
random number test 110, 116
random permutation 111, 140, 147
random scatter 149ff, 165
 empty discs 149
 empty squares 154
 empty rectangles 156
 non-uniform 160
random walk
 birthday problem 119
 covering 125ff
 dispersal 127
 hitting times 28, 38
 simple 237
 2-dimensional 243ff
regular graph 38, 115
reliability model 33
renewal-sojourn method 11, 24, 41, 73, 93, 162, 171ff
reversibility 42, 169
Rice's formula 54, 64, 70
 Brownian analog 185
 d-dimensional 188
 d-parameter 217
runs
 alternating, in i.i.d. sequence 248
 increasing, in i.i.d. sequence 239
 in subsets of Markov chains 248
 i.i.d. 31

second moment method 18, 131, 218, 236
shot noise 195, 217
Slepian's inequality 206
Slepian model 101
sojourn distribution: see compound Poisson
spacings 238
stable process 59, 241
Stein's method 165
stochastic geometry 149ff
storage example 52, 91
subsequences 141

timesharing computer model 33, 41
tree-indexed processes 145ff, 149

uniform dist. Poisson process 56

Applied Mathematical Sciences

cont. from page ii

55. *Yosida:* Operational Calculus: A Theory of Hyperfunctions.
56. *Chang/Howes:* Nonlinear Singular Perturbation Phenomena: Theory and Applications.
57. *Reinhardt:* Analysis of Approximation Methods for Differential and Integral Equations.
58. *Dwoyer/Hussaini/Voigt (eds.):* Theoretical Approaches to Turbulence.
59. *Sanders/Verhulst:* Averaging Methods in Nonlinear Dynamical Systems.
60. *Ghil/Childress:* Topics in Geophysical Dynamics: Atmospheric Dynamics, Dynamo Theory and Climate Dynamics.
61. *Sattinger/Weaver:* Lie Groups and Algebras with Applications to Physics, Geometry, and Mechanics.
62. *LaSalle:* The Stability and Control of Discrete Processes.
63. *Grasman:* Asymptotic Methods of Relaxation Oscillations and Applications.
64. *Hsu:* Cell-to-Cell Mapping: A Method of Global Analysis for Nonlinear Systems.
65. *Rand/Armbruster:* Perturbation Methods, Bifurcation Theory and Computer Algebra.
66. *Hlaváček/Haslinger/Nečas/Lovíšek:* Solution of Variational Inequalities in Mechanics.
67. *Cercignani:* The Boltzmann Equation and Its Applications.
68. *Temam:* Infinite Dimensional Dynamical System in Mechanics and Physics.
69. *Golubitsky/Stewart/Schaeffer:* Singularities and Groups in Bifurcation Theory, Vol. II.
70. *Constantin/Foias/Nicolaenko/Temam:* Integral Manifolds and Inertial Manifolds for Dissipative Partial Differential Equations.
71. *Catlin:* Estimation, Control, and the Discrete Kalman Filter.
72. *Lochak/Meunier:* Multiphase Averaging for Classical Systems.
73. *Wiggins:* Global Bifurcations and Chaos.
74. *Mawhin/Willem:* Critical Point Theory and Hamiltonian Systems.
75. *Abraham/Marsden/Ratiu:* Manifolds, Tensor Analysis, and Applications, 2nd ed.
76. *Lagerstrom:* Matched Asymptotic Expansions: Ideas and Techniques.
77. *Aldous:* Probability Approximations via the Poisson Clumping Heuristic.
78. *Dacorogna:* Direct Methods in the Calculus of Variations.